OVERSIZE STO SIZE

**ACPL ITEM DISCARDED**

```
665.3 W89
WORLD CONFERENCE ON LAURIC
OILS (1994 :
PROCEEDINGS OF THE WORLD
  CONFERENCE ON LAURIC OILS
```

**ALLEN COUNTY PUBLIC LIBRARY**
FORT WAYNE, INDIANA 46802

You may return this book to any location of
the Allen County Public Library.

DEMCO

# Proceedings of the World Conference on Lauric Oils: Sources, Processing, and Applications

# Proceedings of the World Conference on Lauric Oils:
# Sources, Processing, and Applications

Editor

**Thomas H. Applewhite**

AOCS PRESS

Champaign, Illinois

**AOCS Mission Statement**

To be a forum for the exchange of ideas, information, and experience among those with a professional interest in the science and technology of fats, oils, and related substances in ways that promote personal excellence and provide high standards of quality.

**AOCS Books and Special Publications Committee**

E. Perkins, chairperson, University of Illinois, Urbana, Illinois
T. Applewhite, Austin, Texas
J. Bauer, Texas A&M University, College Station, Texas
T. Foglia, USDA–ERRC, Philadelphia, Pennsylvania
W. Hausmann, Lou Ana Foods, Inc., Opelousas, Louisiana
Y.-S. Huang, Ross Laboratories, Columbus, Ohio
L. Johnson, Iowa State University, Ames, Iowa
J. Lynn, Lever Brothers Co., Edgewater, New Jersey
G. Maerker, Oreland, Pennsylvania
G. Nelson, Western Regional Research Center, San Francisco, California
F. Orthoefer, Riceland Foods Inc., Stuttgart, Arkansas
J. Rattray, University of Guelph, Guelph, Ontario
A. Sinclair, Deakin University, Geelong, Victoria, Australia
T. Smouse, Archer Daniels Midland Co., Decatur, Illinois
G. Szajer, Akzo Chemicals, Dobbs Ferry, New York
L. Witting, State College, Pennsylvania


Copyright © 1994 by AOCS Press. All rights reserved. No part of this book may be reproduced or transmitted in any form or by any means without written permission of the publisher.

The paper used in this book is acid-free and falls within the guidelines established to ensure permanence and durability.


**Library of Congress Cataloging-in-Publication Data**

World Conference on Lauric Oils (1994: Manila, Philippines)
    Proceedings of the World Conference on Lauric Oils: sources,
processing, and applications/Thomas H. Applewhite.
      p. cm.
    Includes bibliographical references and index.
    ISBN 0-935315-56-X
    1. Coconut oil—Congresses.   2. Palm oil—Congresses.
I. Applewhite, Thomas H.   II. Title.
TP684.C7W67   1994
665'.35—dc20                            94-37324
                                        CIP


Printed in the United States of America with vegetable oil–based inks.

# Foreword

For the first time since the international, once-every-four-years, lauric oils conference was established, the Philippines has hosted the 1994 "World Conference and Exhibition on Lauric Oils: Sources, Processing, and Applications." Actually a privilege long overdue, the selection of Manila as the venue of this year's congress recognized that country's status as the largest supplier of coconut-based lauric oil in the world.

With such a conducive and appropriate setting for the meeting, we had a fruitful series of lectures and discussions and obtained a deeper perspective on the sourcing, processing, and applications of lauric oils.

For technical, manufacturing, and marketing people from vegetable oil processing and oleochemical companies, the conference and exhibition provided a rare opportunity to learn the latest developments in this important industry, as well as to discuss business prospects with suppliers.

*E. Charles Leonard*
*General Chairperson*
*World Conference and Exhibition*
*on Lauric Oils*

# Preface

These Proceedings mark another first for the American Oil Chemists' Society and for AOCS Press. This conference devoted to the lauric oils demonstrates the versatility and international flavor of the AOCS World Conference format. Here are nearly all of the papers presented at this meeting, along with the remarks of the Philippine President Fidel U. Ramos. The topics are far-ranging and should provide readers with an insight into the lauric oil complex and its impact on the lives and economic well-being of nearly one-third of the Philippine population. Also presented were topics on new products from laurics, new sources of laurics, and new ideas about health and nutrition as related to laurics. I hope that each reader finds something of interest here. A few presentations were not offered for publication; those interested in those particular topics should contact the authors directly for more information.

*T.H. Applewhite*
*Bailey's Harbor, Wisconsin*

# Contents

**Opening Session**
Conference Chairperson: E. Charles Leonard

Climbing High with Coconuts
*Philippine President Fidel V. Ramos*   2

The Role of the ASEAN Region in the World Lauric Oil Industry
*C. F. Habito and D.C.E. Erfe*   4

**Session 1: Marketing and Economics of Lauric Oils**
Chairperson: Shaw Skillings

The Marketing and Economics of Coconut Oil
*J.L. Arranza*   7

The Marketing and Economics of Palm Kernel Oil
*Y. Basiron and M.N.H. Amiruddin*   15

The Socioeconomic Aspects of Lauric Oils Production: World Bank Involvement in Oil Palm and Coconut
*D.J. Meadows*   22

Quality Aspects of Shipping and Handling Lauric Oils and Oleochemicals
*A.F. Mogerley and L.J. Rogers*   25

Nonfood Uses of Lauric Oils
*R.J. McCoy*   31

**Session 2: The Sources of Lauric Oils**
Chairpersons: Alfredo D. Yniguez, Jr., and E. Charles Leonard

Identifying New Sources of Coconut Oil
*W.G. Padolina*   39

Potential Sources of Lauric Oils for the Oleochemical Industry
*N. Rajanaidu and B.S. Jalani*   47

The Development and Commercialization of High-Lauric Rapeseed Oil
*A. Baum*   51

**Session 3: The Processing of Lauric Oils**
Chairperson: Lutz Haertel

Fractionation of Lauric-Based Fatty Acids Achieving Consistent Product Quality
*K.P. Ho and M.B. Subramaniam*   57

The Fractionation of Lauric Oil Components for Cocoa Butter Substitutes
*S. Wong*   60

The Production of Fatty Alcohols and Their Amino Derivatives from Coco Fatty Acid Methyl Esters
*M. Matsuda, M. Horio, K. Tsukada, K. Sotoya, H. Abe, and R. Tsushima*   64

Catalytic Hydrogenation of Lauric Oils and Fatty Acids
*R.S. Murthy*   72

**Session 4: Applications of Lauric Oils**
Chairpersons: Augustine S.H. Ong, L.H. Princen, A.J. Kaufman, and Robert Modler

Chemical and Physical Properties of Palm Kernel Oil
*C.L. Chong and W.L. Siew*   79

Characteristics and Properties of Malaysian Palm Kernel-Based Specialty Fats
*T.S. Tang and F.C.H. Oh*   84

Formulation of Lauric Oil-Containing Food Products and Their Performance
*E.M. Goh*   98

Nutritional Aspects of Lauric Oils
*M.I. Gurr*   104

Health Effects of Lauric Oils Compared to Unsaturated Vegetable Oils
*E.A. Emken*   110

Health Aspects of Coconut Oil
*C.S. Dayrit*   119

Uses and Applications of Alkyl Polyglycoside in Personal Care Products
*J. Fallon*   127

Use of Lauric Oil Nitrogen Derivatives in Laundry Products
*F.E. Friedli, M.M. Watts, A. Domsch, P. Frank, and R.D. Pifer*   133

The Increasing Importance of Methyl Ester Sulfonates as Surfactants
*N.M. Rockwell and Y.K. Rao*   138

A New Generation of Imidazoline-Derived Amphoteric Surfactants
*R. Vukov, D. Tracy, M. Dahanayake, P.J. Derian, J.M. Ricca, and F. Marcenac*   147

Nondetergent Applications of $C_8$–$C_{14}$ Amines and Derivatives
*H.F.G. Patient*   155

Special Surfactants for Personal Care Products
*K. Sotoya, Y. Yokota, and A. Fujiu*   160

**Poster Presentations**
Chairperson: R.F. Wilson

A Two-Stage Countercurrent Bleaching Process for Edible Oils and Fats
*P. Transfeld and M. Schneider*   168

Dietary Fat Composition Alters Whole Body Utilization of $C_{16:0}$ and $C_{10:0}$
*M.T. Clandinin, L.C.H. Wang, R.V. Rajotte, M.A. French, Y.K. Goh, and E.S. Kield*   171

# Discussion

In a departure from the normal format, Fidel V. Ramos, the President of the Philippines, invited the conference delegates to the Presidential Palace where he provided his opening remarks. Following this presentation, the President invited one delegate from each participating country and the organizing committee to meet over coffee for informal, frank discussion of all aspects surrounding the topics of the conference.

The 45-minute discussion covered topics ranging from the problems of the productivity of small landholders to the development of new coconut products not based on the oil. There was considerable comment on the cooperation between industry and government, and the President assured the delegates that his administration was devoted to such approaches. The need for rejuvenation of the coconut complex centered on the problems of replacing 100 million senile trees, plus other approaches for improving the income of the farmers. It was stressed that over 20 million people (*ca.* one-third of the population) depend on the coconut and its products for their livelihood in the Philippines.

In a effort to improve the situation, cooperatives are being developed, agronomic practices improved, and much research is being devoted with government support to these problems. There will be a new research organization on coconut patterned after the Malaysian organization on palm oil. More international cooperation will be sought, and removal of trade barriers pursued. The President was very optimistic about the progress to date, was grateful to the American Oil Chemists' Society for bringing industry, academia, and government officials together in this World Conference for the in-depth discussions on lauric oils, and proposed that we nurture these newly found communication links in the future.

# Climbing High with Coconuts

**Philippine President Fidel V. Ramos**

*Editor's Note:* Due to pressing affairs of state, His Excellency Fidel V. Ramos, President of The Philippines, who originally was scheduled to offer the keynote address at this conference, was unable to be present during the morning session on Monday, February 21, 1994. However, President Ramos graciously invited all of the conference participants to the Presidential Palace, Malacañang, that afternoon where he offered the following address. In addition, the President invited representatives from the 31 countries that provided delegates for the conference to join him afterwards for refreshments and a 1-hour informal discussion of affairs related to the conference. He also presented each of these delegates with a jewelry box made from coconut products and autographed copies of two of his books. Immediately following the President's speech, Dr. E. Charles Leonard, Chairman of the Conference and James C. Lyon, Executive director of the American Oil Chemists' Society, presented President Ramos, a graduate of the University of Illinois, with a booster's cap, T-shirt with the U of I logo, and a beautiful, framed, montage of the University campus.

---

On behalf of our people and government, I welcome you all to Malacañang, Manila and the Philippines. I thank you for choosing our national capital as the site for this most important World Conference on Lauric Oils.

Besides affording me the chance to show you this palace of our people. This gives us the opportunity to exchange ideas with you on this first day of your wide-ranging conference. Your agenda is central to the future of lauric oil in the world economy.

It is your task to lay the foundations for stability and growth of this global industry we share in common. In your conference, you must address the problems we face separately and together, particularly on how to exploit the opportunities now opening up to our industry, as the world comes out of recession into a more liberalized trading and investment environment.

In the case of the Philippines, our central contribution to the lauric oil industry is coconut oil, of which we are the biggest producer and exporter in the world. The coconut industry is a major pillar of our national economy. It is the largest net dollar earner for our country. Eleven of our 15 regions are coconut-producing areas. And some 20 million of our people—nearly a third of our population—are directly or indirectly dependent on the coconut industry for a living.

Thus we attach great importance to our current program to revitalize and modernize the coconut industry and increase its productivity and production. It is in light of the preeminence of the coconut to our people's livelihood and to the national economy that I have assigned a top team of cabinet officials and coconut experts to dialogue, exchange views, and learn from our foreign friends during your brief stay in the Philippines. I am certain that all this will result in meaningful and profitable contributions for each of you and the lauric oil industry as a whole.

I am told that there are concerns within your community about the cutting of coconut trees and the declining coconut production in the Philippines. Allow me to correct this misimpression. What is really happening is that we are cutting down senile trees that are no longer useful and productive. But as we cut down, we also replant the areas with new trees. And, we are fertilizing existing fruit-bearing trees.

All of these efforts form part of our revitalization program for the coconut industry. These measures are designed not only to maintain our coconut production, but also to enhance and increase productivity. Our target is to increase production by 3–3.5 million tons a year through the year 2000, using financial support from the World Bank and other financial and technological sources.

Like other countries engaged in the production of lauric oils, we in the Philippines continue in search of new ideas—especially those that can improve efficiency, productivity, utilization, and quality. We assure both users and suppliers of lauric oil that free enterprise and intensive research are at the forefront of our coconut industry, thus opening a brighter promise for the future. We will pursue our program to establish a central research agency for the coconut industry, similar to the PORIM in Malaysia for palm oil. We are gearing the industry towards downstream products, which can serve as the main engines to propel the coconut sector more progressively into the twenty-first century.

In this modernization and revitalization effort, we welcome the participation of foreign investors. Today, we can truly claim that we have one of the best and most hospitable environments for foreign investment in the world. And this is borne out not only by new investors coming in, but also by old-time investors who are staying put and expanding because they are profiting from doing business in our country.

We treasure the importance of coordinating and cooperating with our partners in ASEAN in this effort. The ASEAN region is the biggest supplier of lauric oil in the world market. The Philippines is well represented in ASEAN organizations concerned with the industry. Mr. Rodolfo Jimenez is currently the chairman of the ASEAN Oleochemical Manufacturers Group (AOMG), and Mr. Jesus Arranza, Chairman of the United Coconut Associations of the Philippines (UCAP), chairs the ASEAN Vegetable Oils Group (AVOG).

We are also active in the development of the Asia–Pacific coconut community. I wish to add that the formation of these ASEAN organizations is a plan not to cartelize, but to make our region a more reliable supplier of lauric oil.

### The Confidence of Our Asia–Pacific Neighbors

The Philippines is on the road back to sustained progress. We recently cited in our 1993 Report on Economic Performance the indicators that tell us this is so: in terms of growing GNP trend, and reduced inflation and interest rates. A stable Philippine peso, more modern telecommunications, increased foreign and domestic investment, a 25% growth in tourism, and the best-performing stock exchange in Asia. Of these indicators, there is one of which I am particularly proud. And that is the renewed confidence of our Asia–Pacific neighbors are now showing in the Philippines.

Japan remains our largest investor, but the United States, China, South Korea, Indonesia, Malaysia, Taiwan, Thailand, Australia, Hong Kong, and Singapore also have made new commitments to expand, invest in, and conduct a greater volume of business in our agricultural and industrial growth centers. Foreign investment and other cooperative arrangements encompassing Mindanao, East Malaysia, Eastern Indonesia, and Brunei Darussalam have made possible an East ASEAN growth area (EAGA), now rapidly becoming a reality. Lest we forget—the EAGA is full of all kinds of palm trees, especially coconuts.

### The Coconut Industry and "Philippines 2000!!!"

One vision, one idea governs our efforts in the coconut industry and in other sectors of our economy. We seek to link up speedily with the world economy, because in that course lies the real hope for our development. As we strive to compete economically, so must we cooperate and collaborate with others. There is room for all to develop.

Our people's shared vision, called "Philippines 2000!!!," is anchored on the medium-term development plan of my administration for the modernization of our country. The enhancement of our coconut industry constitutes a major goal of "Philippines 2000!!!" Because it involves so many people and covers a large part of our country, our coconut industry has become an important vehicle for ensuring the empowerment of our people and the democratic sharing of the benefits of progress. No one will be left behind as this industry grows.

Ladies and gentlemen, I again bid you the warmest of welcomes to Manila. May you have a most meaningful and productive conference, and may this visit be only one of your many visits to the Philippines in the future. Thank you and good day.

# The Role of the ASEAN Region in the World Lauric Oil Industry

Cielito F. Habito and Doreen Carla E. Erfe

National Economic Development Authority, NEDA Bldg., Amber Ave., Pasig, Metro Manila, The Philippines.

## Abstract

The Philippines continues to be the largest supplier in the world market for coconut oil. However, its dominance is precarious. Coconut oil is but a small portion of the world lauric oil market. The Philippine coconut industry is adjusting to increased competition from palm kernel oil and other large producers of coconut oil. The Philippines has a monumental task before it in maintaining its leadership in coconut production and trade. The task involves the allocation of public expenditures to improve the productivity of the coconut tree farms and to increase investments in road infrastructure and other postproduction facilities, which will certainly lead to lower marketing costs, greater income for producers, and more competitive prices for exporters.

---

The ASEAN region is a major source of oils and fats, particularly the lauric oils. The region includes the Philippines and Indonesia which produce about one-half of the world's coconut and coconut oil. Malaysia, on the other hand, accounts for about one-half of the world's palm oil production and 70% of the palm oil trade. The region, therefore, plays a dominant role in the development of the lauric oil industry.

The Philippines prides itself as a major player in the world market for coconut products. While it is second only to Indonesia in terms of coconut production, it ranks first in terms of copra production and export in the world market. It accounts for approximately 39% of the world's coconut oil supply and about 60% of the world's coconut oil and copra exports.

The bulk of the Philippines' coconut production is exported. In 1992, 72% or 1.6 million metric tons (MMT) (in copra terms) was exported. Copra and coconut oil comprised 92% of the coconut products exported for that year. The industry accounts for about 44% of the country's total agricultural export earnings, and 7% of total export earnings. Aside from being a major income producer, the industry likewise provides direct or indirect employment for a substantial portion (21 million or about 30%) of the country's population. Given this, the coconut industry is often referred to as one of the pillars of the Philippine economy. Through its numerous forward and backward linkages, the sector plays a key role in the pursuit of the country's globalization and agro-industrialization goals.

Inasmuch as the performance of the oils and fats industry in the region is heavily dependent on global trends, constant monitoring of and necessary adjustments to the developments in the world trade scenario involving these products is of paramount importance. Stronger economic and technical cooperation is necessary for the region to maintain its competitiveness.

## The World Market for Lauric Oils and the Philippine Coconut Industry

The United States and the European Community (EC) continue to be the major markets for lauric oils. Lauric oils coming from coconuts and palm kernels are used mainly for industrial purposes. Refined lauric oils are used in processed foods, such as confectioneries, coffee whiteners, shortenings, margarines, and filled milk. On the other hand, crude lauric oils are used in the production of methyl esters, which are further processed into fatty alcohols and acids for end-products, such as detergents, cosmetics, and soaps and other toiletries. In recent years, there has been an increasing demand for crude oil due to the fast-paced developments in the oleochemical industry.

The oils and fats industries, particularly the coconut-based industries in the ASEAN region, rely heavily on global trade. These industries, therefore, are highly influenced by developments in international trade. Copra and coconut oil, for instance, compete with other commodities on the world market for oils and fats. The main substitutes include soybean and its oil, palm kernel oil, rapeseed and rapeseed oil, sunflower, and groundnut. Over the years, copra and coconut oil have come to account for only 5% of the total oils and fats traded on the world market. Soybean, palm, and rapeseed oils, on the other hand, have larger shares of the trade with 27, 24, and 21%, respectively.

Inasmuch as the coconut oil share of the world market is small and can be replaced, its price is heavily influenced by the prices of the other major oils. This results in a highly fluctuating supply of oils in the world market. However, world demand is inelastic since industrial requirements for fats and oils are more or less stable.

The Philippines, as well as other producers of coconut oil in the region, often have to contend with declining market shares, especially when the demand for other vegetable oils increases. The Philippines is usually at a disadvantage, since copra and coconut oil comprise the bulk of its exports.

Substitutes of coconut oil, which are mainly produced in the United States and the EC, are known to receive production and export subsidies. These subsidies cushion the producers from world-price fluctuations. The trade of coconut products is also affected by tariff and quantitative restrictions which are imposed by its major markets in the United States and EC. Protectionist moves of other oil and fat producers, specifically the soybean producers, also have a tremendous impact on coconut exports. It may be recalled that a well-funded campaign alleging coconut oil to have cholesterol-enhancing effects has affected the coconut-exporting countries, including the Philippines, severely.

## Prospects for the Lauric Oil Industry in the ASEAN Region

Despite setbacks, the general scenario for the production of lauric oils in the ASEAN region is relatively favorable for the years to come. Consumption of lauric oils, which has been stable over the past years, has been rising in recent years due mainly to the increasing demands of the oleochemical industry. Trade barriers imposed on these products, especially for coconut-based products, should be eliminated eventually. The recently concluded Uruguay Round (UR) of multilateral negotiations under the aegis of the General Agreement on Tariffs and Trade (GATT) is expected to bring about further liberalization and expansion of world trade, including the lauric oils market. Also, the agreements made during the UR is expected to make the GATT system more responsive to an evolving international economic environment. For the ASEAN region, the UR is of strategic importance since it dealt with a number of vital issues having far-reaching implications for the region's competitiveness and development process. The UR is expected to pave the way to an even playing field among exporting countries.

The UR will have a significant impact on the coconut-exporting countries of the region. The realization of the EC commitment to cut subsidies for rapeseed and other vegetable oils will result in a reduction of the output for these products, and thus allow the ASEAN region, particularly the Philippines, to increase its share in the lauric oil market in Europe. Furthermore, should the United States and EC ultimately reduce and eventually eliminate farm subsidies, overproduction will be minimized. The sale of surplus oils and fats on the global market at lower than market prices will, therefore, be reduced. Consequently, this will boost the competitiveness of lauric oils, specifically coconut oil. To elaborate this point further, the implementation of a U.S. proposal to remove its import duties and export subsidies on oilseeds and their products along with the other member countries of GATT will also increase and diversify the lauric oil market. The optimistic situation for lauric oil trade calls for greater economic, as well as technical cooperation, among the ASEAN member countries. The creation of the ASEAN Free Trade Area (AFTA), which provides the region with the opportunity to deal with the world on an equal footing, is a step toward greater economic integration of the region. The rewards under AFTA are plentiful: by liberalizing trade through the adoption of the Common Effective Preferential Tariff (CEPT) scheme, the member countries will have access to larger markets, not only within the region but also internationally. The cooperation/competition arrangement of having common tariffs on specific goods means greater production volumes, higher quality goods, lower prices, increased employment, and a host of other economic and social benefits.

Technical cooperation, particularly in the fields of research, development, and technological transfer, is also critical for the region to maintain its competitiveness especially in terms of production and trade of agricultural and industrial products. In the coconut sector, the establishment of an enabling mechanism that facilitates the exchange of expertise in research on priority problems affecting productivity, processing, and manufacturing of coconut products in the region, is deemed an important and necessary venture.

The Philippines has a monumental task to maintain its leadership role in coconut production and trade. The tasks include the implementation of a number of policy measures and programs to save what has been called the "sunset industry" of the country. In order to increase productivity in the sector, there is a need for continued improvement of product quality, farm management, processing and postharvest technology. The institutionalization of a fertilization, replanting, and rehabilitation program will likewise increase productivity and ensure the dominance of Philippine coconut in the world market. There is also a need to promote economies of scale in production through estate-type plantations managed by corporations, cooperatives, or individuals. A viable credit program for producers which provides funds for intercropping and livelihood projects, fertilization, postharvest facilities, and for the acquisition of appropriate technologies for the production of coconut-based products will have to be instituted. The previously mentioned requirements need to be complemented with greater public expenditures on research and development along with investments on infrastructure and other support facilities. If everything discussed previously were implemented, narrower marketing margins; higher incomes for producers, processors, and manufacturers; and more competitive prices for exporters could be realized.

## Coconut: A Sunset Industry?

Many discouraging events have led some analysts to conclude that it is time to consider shifting away from the production of coconuts, an industry which from time immemorial has been a source of livelihood for a substantial portion of the Filipino population. Among these events are the observed decline in the share of coconut oil in the world fats and oils market (and the rapid increase in the share of a close substitute, palm kernel oil); the supposed superiority of palm oil to coconut in terms of oil productivity per hectare; a campaign by the American Soybean Association against tropical oils, reducing the demand for coconut oil in the United States; tighter quality standards on copra and copra meal imports into Europe; and attempts to raise duties on coconut oil in European markets.

While it is true that the coconut oil share of the total oils and fats trade continues to decline, this need not be taken with alarm for at least three reasons. First, the declining market share does not necessarily reflect a declining market for coconut oil in absolute terms. It more likely reflects an absolute increase in the markets for competing oils like soybean oil and palm oil, primarily due to their higher production levels as indicated in FAO figures over the past few years.

The main reason coconut oil's share has declined is that the Philippines, responsible for over 70% of world coconut oil exports, has not increased production as fast as palm oil and soybean oil producers have in past years. It would hardly be comforting for the Philippines to have coconut oil maintain or even increase its market share, if this were achieved by increased exports from Sri Lanka or the South Pacific islands. Hence, the declining market share need not be worrisome, because coconut oil maintains its niche in the oils and fats market, and all of the coconut oil placed on the world market will continue to be bought. Chemists attest that coconut oil will never lose its specialized applications due to its unique chemical properties.

Second, the decline in the market share (and perhaps even in absolute volume) of coconut-oil exports from the Philippines can be traced to a significant extent to changes in the composition of the coconut-product exports. For example, increasing amounts of green coconuts have been exported (especially to Taiwan) in recent years. Desiccated coconut and nontraditional coconut-product exports, like coconut milk, are also growing in importance. Hence, while there may be less exportation of coconut oil, fewer coconuts are not necessarily being exported. More coconuts may be exported in a higher value form. These statistics are worth examining closely before one makes a judgment about the long-term prospects of the coconut-export industry.

Third, there remains a large domestic market for Philippine coconut products that is essentially untapped. A dramatic indication of this is the observation that Indonesian coconut production is about as large as Philippine production, and yet Indonesia consumes virtually all of its output domestically. Even if one considers that Indonesia's population is about four times larger than that of the Philippines, the fact that the Philippines exports the majority of its production while Indonesia hardly exports any indicates that Filipinos are consuming far fewer coconut products than the Indonesians. Therefore, the potential for domestic market expansion is high.

Another source of concern for coconut production is the much higher oil productivity of the oil palm compared to the coconut. Thus, the coconut is seen as an "inefficient" oil crop. On this basis, some are advocating a shift to oil palm production. While it is true that the oil palm is able to yield several times more oil per hectare than coconut given our typical yields, this comes at the cost of thicker foliage growth, rendering intercropping infeasible. On the other hand, studies have shown that farmers on intercropped coconut land actually obtain the larger portion of their incomes from the intercrops, rather than the coconut itself.

If the primary concern is the level of income of coconut farmers, and not the level of coconut production per se, then the future of coconut farmers need not be bleak. As long as the coconuts have value, it will remain worthwhile to grow coconuts with the intercrops, because once planted and established the coconuts cost virtually nothing to maintain. Through time, the age-old nickname "lazy man's crop" continues to apply to coconuts. What has changed is that the coconut farmer can no longer afford to be lazy to survive; the areas under the coconut trees need to be put to good use as well.

# The Marketing and Economics of Coconut Oil

Jesus L. Arranza[a]

United Coconut Associations of the Philippines, PCRDF Bldg., Ortigas Complex, Pasig., Metro Manila, The Philippines

## Abstract

The world supply of lauric oils should increase in spite of weather-based fluctuations of the coconut supply. Presently, coconut oil accounts for a major portion of lauric oil production. The steady gains in palm oil production, however, indicate a steady increase in the palm kernel oil share of the lauric supply.

Since usage of coconut/palm kernel oils is far more specialized, a premium price over most oils and fats should be maintainable for most periods. The premiums, however, must prove reasonable so that demand can be sustained.

Coconut oil demand covers the edible and nonedible or industrial applications. The edible sector has been the subject of health concerns, and these have affected in some way the consumer perception of food products derived from coconut oil. Consequently, future availability for the nonedible sector could outpace demand.

As for attempts to develop substitutes (i.e., *cuphea* and high-laurate rapeseed), a vigorous program to commercialize them may be minimized as long as lauric users are assured a steady supply of coconut oil at reasonable prices.

---

Coconut oil is still the world's major source of lauric acid. In the late 1980s, an antitropical oil campaign was launched in the United States. The campaign simplistically classified coconut oil as a highly saturated fat and therefore unhealthy. It spurred a shift by many processors away from coconut oil in their food formulations to vegetable oils indigenous to the United States.

The United Coconut Associations of the Philippines (UCAP) initiated a defense. It entered into a 5-year research agreement with scientists from the New England Deaconess Hospital, a Harvard Medical School affiliate, to investigate the health and nutritional advantages of coconut oil, along with a parallel research group in the Philippines composed of renowned specialists in the fields of medicine and nutrition. The United Coconut Associations of the Philippines also set up of the United States Council for Coconut Research/Information (USCCRI), based in Washington, D.C., to correct public misimpressions about coconut oil and to promote the health advantages of the product. Close monitoring was undertaken of proposed labeling bills in the U.S. Congress, Senate, FDA, and USDA that could affect the marketability of coconut oil in the United States.

[a]Formerly Jesus L. Chua.

The actions UCAP undertook were successful. Attempts to pass laws reducing the marketability of coconut oil were unsuccessful. The Nutrition Labeling and Education Act of 1990 did not select coconut oil for pejorative labeling. The research done in the United States on the dietary use of coconut oil confirmed that coconut oil is not a cholesterol enhancer. A modest dissemination of the medical and nutritional advantages of coconut oil to the American public has been started. Many U.S. firms have returned to coconut oil as a source material. As a result, they are getting the same product qualities, the unique mouthfeel and taste, that only coconut oil can impart to their products.

## Properties of Coconut Oil

Coconut oil is a naturally saturated vegetable oil and as such, it is desired for specific purposes.

1. It is very resistant to oxidative rancidity, resulting in foods prepared with coconut oil being stable in terms of flavor. Thus, it is preferred as a replacement for butter oil or filled milk.
2. Coconut oil has a high melting point, therefore it is used in solid coatings for confectionery and baked foods.
3. Because of its high medium-chain triglyceride content, coconut oil is used in tailor-made, medical foods for people who have difficulty in utilizing ordinary fats as sources of energy. It is also used in infant formulas.
4. It has advantages in bath soaps where a pure whiter, glossy product is desired. The use of coconut oil in soap inhibits yellowing as the soap ages.
5. Coconut oil yields the highest percentage of glycerol of all commonly used oils. Thus, it is useful for making soap with no additional treatment. Most other oils, except tallow and palm, produce soaps that are quite soft.

## New Uses for Coconut Oil

Researchers from the USDA have been conducting experiments using lauric acid and chitosan, a material found in the shells of crabs and shrimp, to produce a protective film to improve the shelf life of fruits and vegetables by slowing the transfer of water and gas into and out of produce. Preliminary results show that when chitosan was used in combination with lauric acid, fruits and vegetables maintained their water content 5 days longer than the untreated produce.

The Philippine Department of Science and Technology has developed biodegradable textile softeners from coconut oil in lieu of using imported softeners. The softening oils not only can condition local clothing materials (ramie), but all types of fibers as well.

Efforts of UCAP also are expected to yield new nutritional health foods from coconut oil. These efforts are based on research findings which indicate some health and nutritional advantages of coconut oil because of its high content of medium-chain triglycerides.

The local coconut industry has joined the worldwide trend toward the use of environmentally friendly fuel sources such as vegetable oil-based methyl esters (rapeseed and soybean oils in the case of Western Europe and the United States). A joint UCAP/PCRDF (Philippine Coconut Research and Development Foundation) study is testing the feasibility of using coconut-based methyl esters as a substitute for diesel fuel. An AUV (Asian Utility Vehicle) is being run entirely with this fuel; the vehicle has travelled more than 70,000 kilometers in Metro Manila and neighboring provinces, and still is running with no difficulties.

## Coconut Oil Supplies

### Production

In 1991, the land area worldwide devoted to coconut production was 10.9 million hectares and produced 46.6 billion nuts, the equivalent of 8.7 million metric tons (MMT) of copra. The average harvest per hectare was 4,264 nuts, or equivalent to 800 kg of copra or 504 kg of oil per hectare. However, in 1991 world copra production was only 4.9 MMT, which was 56.3% of global coconut production.

Indonesia has the largest planting area at 3.5 million hectares, or 32.1% of the world total (Table 1). The Philippines and India follow at 3.1 million and 1.5 million, or 28.4 and 13.8%, respectively. The Philippines ranks second in coconut production at 11.2 billion nuts or 24.0% of world total; Indonesia leads with a production volume of 11.69 billion coconuts (25.1% of the global production), and India follows with 9.7 billion (a 20.8% share).

World production of copra has shown a moderate increase during the past three decades. The volume has increased 69% from 3.24 MMT in 1960 to 5.48 MMT in 1990. This is mainly attributed to the 131% expansion in Indonesia. Growth in Philippine copra production was less pronounced at 76%. Global output of copra in 1992 has been estimated at 4.71 MMT, with the Philippines accounting for 39%, Indonesia 27%, and India 8%. These are the three main copra producers and account for three-fourths of the world total.

### Exports

Over the years, there has been a shift in the global oil and fat trade from raw materials (oilseeds) to vegetable oils. This has allowed for higher quality oil products on the world market, since the raw materials are processed in the countries where the oilseeds are harvested. Coconut-producing countries have followed this trend. They are now exporting less copra and more coconut oil (Fig. 1). In 1992, global copra exports stood at 263,000 metric tons compared to 1.72 MMT in 1960. Meanwhile, coconut oil exports worldwide have increased from only 273,000 MT in 1960 to 1.549 MMT in 1992.

Over the last 30 years, the Philippines and Indonesia have continuously increased their shares of the world coconut oil supply (Fig. 2). In 1960, the Philippines accounted for only 10% of the 1.95 MMT worldwide coconut oil output, while Indonesia produced 12% of the total. In 1970, the Philippines had expanded its share to 24% of the 2.03 MMT global coconut oil output, while Indonesia held 16%. By 1990, the Philippine's share grew to 42% and Indonesia produced 25% of the world's total. Meanwhile, U.S. and Western Europe coconut oil production has gone down. In 1960, both territories accounted for a combined 38% share of global total. This has dwindled to 19% in 1970, 6% in 1980, and only 2% in 1990.

Global coconut oil exports in 1990 were recorded at 1.7 MMT, which represented 50% of world's coconut oil production. This is in contrast to only 273,000 MT exported in 1960, equal to only 14% of global output (Fig. 3). This agrees with the increased crushing activities in coconut-producing countries and the diminished world trade in copra. For example, the growing importance of Indonesia as a coconut oil exporter cannot be ignored. Its shipments of coconut oil have steadily increased from a negligible 1% share of global production in 1970, to 3% in 1980, 11% in 1990 (Fig. 4), and 22% in 1992.

The phasing out of coconut oil mills in Western Europe and the United States as a result of global trends toward exporting processed products instead of raw materials, has triggered a massive inflow of coconut oil into countries which had produced coconut oil from imported copra. In 1992, Western Europe imported a total of 578,000 MT of coconut oil, up from 411,000 MT in 1980; 162,000 MT in 1970; and, 113,000 MT in 1960. U.S. imports have also manifested an increase from 71,000 MT in 1970, to

TABLE 1
Global Spectrum Coconut Area and Productivity (1991)

| Country | Area (million hectares) | Percent of World Acreage | Average Productivity (kg copra/hectare) |
|---|---|---|---|
| Indonesia | 3.5 | 32.1 | 666 |
| Philippines | 3.1 | 28.4 | 666 |
| India | 1.5 | 13.8 | 956 |
| Others | 2.8 | 25.7 | |
| Mexico | | | 2,022 |
| Ivory Coast | | | 2,667 |
| Total | 10.9 | 100.0 | |

Source: Asian and Pacific Coconut Community (APCC).

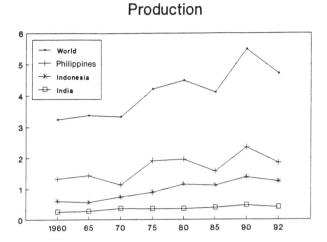
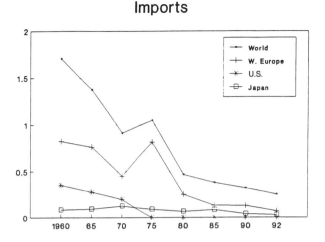
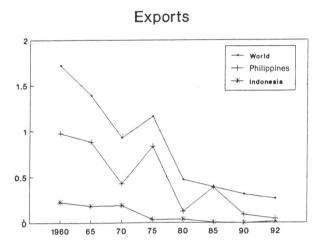
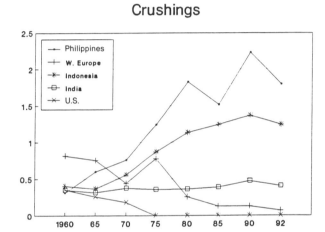

**Fig. 1.** World copra movements (MMT).

399,000 MT in 1980, 452,000 MT in 1990, and 502,000 MT in 1992.

## Demand for Coconut Oil

In most periods in the past, coconut oil has provided the world with reasonably priced lauric oil for edible and nonedible applications. This is in spite of Philippine inflation that has triggered an increase in production cost, in harvesting coconuts, and drying coconut meat for copra (Table 2). Theoretically, the price of copra should increase because of inflation. However, inflation did not trigger a rise in domestic copra prices, so the cost of the raw materials to produce coconut oil has remained unaffected by the decreasing buying power of the Philippine peso. In the final analysis, coconut oil has remained inexpensive despite inflation at home. In real terms, it has become cheaper over the years.

There has been a decline in the consumption of edible coconut oil products in the United States, as indicated by data from the United States Department of Agriculture (USDA) and *Oil World*. Between 1985 and 1987, the percentage of edible utilization to total coconut oil consumption was between 32 and 33%. From 1988 to 1991, the share of edible utilization fluctuated between 18 and 28%. However, in 1991 there was a slight recovery to 20% (Table 3).

The antitropical oils campaign in the United States was responsible for the decline in edible utilization from 1986–1990. The slight recovery in 1991 was the result of UCAP actions undertaken from 1989–1992. Meanwhile, nonedible, industrial applications, generally have sustained demand at well over 300,000 MT/yr. What is ironic, is that research has shown that tropical oils, including coconut oil, are not as unhealthy as claimed by some business interests seeking to promote some vegetable oils at the expense of others.

Current demand for lauric oils is estimated at 4 MMT/yr, and it should reach 5 MMT in the coming years. The steady

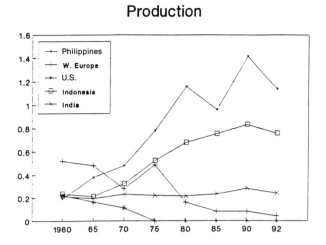
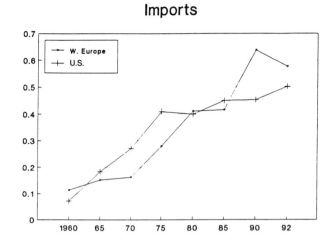
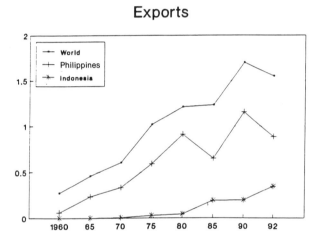
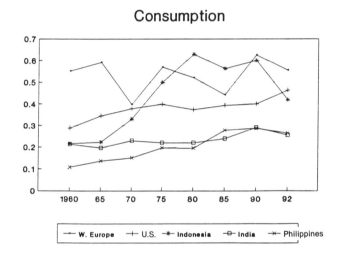

**Fig. 2.** World coconut oil movements (MMT).

growth in palm kernel oil should assure manufacturers that worldwide requirements for lauric oils will be met. In the Philippines there are long-range plans to sustain or even increase productivity, so that the country's position as a major supplier of coconut oil can be maintained. Biotechnology may soon evolve new sources of lauric oils, particularly from rapeseed and *Cuphea*.

## Price of Lauric Oils

Coconut oil exists in highly competitive markets. Its competitiveness has been eroded due to its shrinking share in the world fats and oils market. It is perceived that soybean oil and palm oil have a major influence in setting all oil prices. In 1992, the worldwide export of vegetable oils (including the oil equivalent of the oilseeds) and animal fats was 35.7 MMT. Of this, soybean oil accounted for 9.6 MMT (27.0%) and palm oil at 8.6 MMT (24.0%). Coconut oil exports were recorded at only 1.7 MMT (4.8%). Palm kernel oil (PKO), the coconut oil partner in the lauric pair, was at 0.8 MMT (2.3%).

Extremely high coconut oil prices seem to be mere memories. The last on record was in 1984 at over U.S. $0.60/lb. Since then, prices have been lower, with the peak reaching U.S. $0.33 in early 1992. The primary reason for the lower prices is the increasing volume of palm kernel oil available to supplement lauric oil sources.

Before 1982, when combined world production of coconut and palm kernel oils were between 2 and 3.5 MMT, palm kernel accounted for no more than 20% of the total. However since 1983, in spite of an increase in total production to over 4 MMT, palm kernel oil's share rose to over 20% of the total, and even reached 35% in 1992.

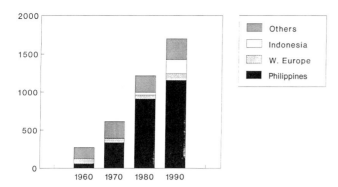

Fig. 3. World coconut oil exports by major exporting countries (thousand MT).

Coconut oil prices fluctuate widely (Fig. 5). The fluctuations are due to changes in the annual coconut harvest, creating variations in the supply and demand balance. Low coconut supplies caused the price increases in 1974, 1979, 1984, and early 1992 (Fig. 5). An abundance of coconuts caused coconut oil prices to drop in 1972, 1976, 1982, and 1986. Some sources claim that cyclical changes in availability prevailed in the 1970s and the first half of the 1980s. But the cycle seemed to have been broken in the late 1980s and 1990s.

Coconut oil is generally considered an expensive oil, and most of the time, it is sold at a premium over soybean and palm oils. This price premium was sustained from 1976 through mid-1985 because of lower-than-expected

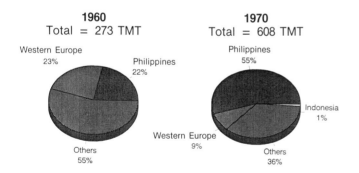

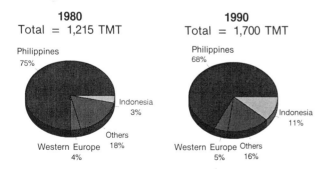

Fig. 4. World coconut oil exports by major exporting countries (% of total, TMT = thousand metric tons).

**TABLE 2**
**Philippine Real and Nominal Copra Prices (shrinking copra prices)**

| Year | Copra Prices (P/ck) Nominal | Copra Prices (P/ck) Real | Consumer Price, Peso | Purchasing Power |
|---|---|---|---|---|
| 1960 | — | — | — | — |
| 1961 | — | — | — | — |
| 1962 | 45.16 | 227.64 | — | — |
| 1963 | 54.20 | 209.21 | 23.8 | 4.20 |
| 1964 | 56.74 | 219.02 | 25.9 | 3.86 |
| 1965 | 66.12 | 246.63 | 26.8 | 3.73 |
| 1966 | 57.92 | 206.20 | 28.1 | 3.56 |
| 1967 | 61.90 | 207.98 | 29.7 | 3.36 |
| 1968 | 74.10 | 244.53 | 30.3 | 3.30 |
| 1969 | 66.69 | 217.41 | 30.7 | 3.26 |
| 1970 | 93.05 | 268.99 | 35.3 | 2.83 |
| 1971 | 87.91 | 204.83 | 43.0 | 2.33 |
| 1972 | 66.76 | 143.53 | 46.5 | 2.15 |
| 1973 | 182.68 | 336.13 | 54.2 | 1.84 |
| 1974 | 363.45 | 501.56 | 72.7 | 1.38 |
| 1975 | 146.80 | 189.49 | 77.6 | 1.29 |
| 1976 | 168.34 | 198.64 | 84.8 | 1.18 |
| 1977 | 255.87 | 273.78 | 93.2 | 1.07 |
| 1978 | 304.13 | 304.13 | 100.0 | 1.00 |
| 1979 | 405.60 | 348.82 | 116.5 | 0.86 |
| 1980 | 215.96 | 157.65 | 137.0 | 0.73 |
| 1981 | 178.41 | 115.97 | 154.0 | 0.65 |
| 1982 | 182.69 | 104.13 | 176.9 | 0.57 |
| 1983 | 350.79 | 182.41 | 190.5 | 0.52 |
| 1984 | 918.05 | 321.33 | 286.4 | 0.35 |
| 1985 | 455.17 | 127.45 | 352.6 | 0.28 |
| 1986 | 288.12 | 80.67 | 355.3 | 0.28 |
| 1987 | 539.23 | 145.59 | 368.7 | 0.27 |
| 1988 | 730.14 | 182.54 | 401.0 | 0.25 |
| 1989 | 692.86 | 159.36 | 443.5 | 0.23 |
| 1990 | 464.72 | 88.30 | 513.6 | 0.19 |
| 1991 | 679.60 | 108.74 | 609.7 | 0.16 |

**TABLE 3**
**U.S. Coconut Oil Consumption (thousand MT)**

| Year | Edible | Nonedible | Total | % Edible |
|---|---|---|---|---|
| 1985 | 125 | 267 | 392 | 32 |
| 1986 | 151 | 313 | 463 | 33 |
| 1987 | 145 | 302 | 447 | 32 |
| 1988 | 106 | 340 | 446 | 24 |
| 1989 | 96 | 253 | 349 | 28 |
| 1990 | 73 | 327 | 400 | 18 |
| 1991 | 77 | 317 | 394 | 20 |

*Sources:* USDA Oct. 1992; *Oil World Annuals* 1991 and 1993.

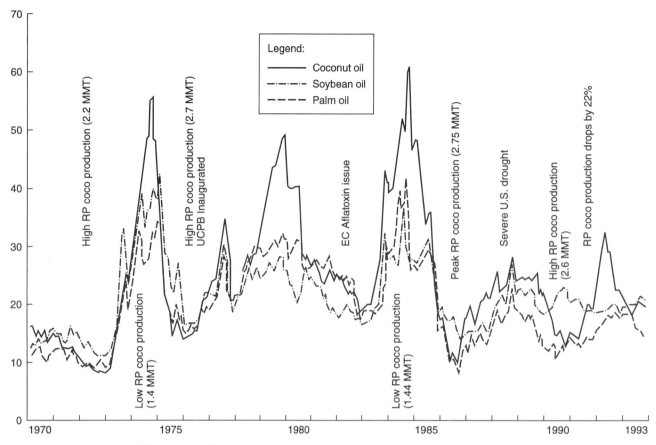

Fig. 5. Comparative monthly price trends in U.S. cents per pound (1973–1993).

coconut production and because policies in the Philippines initiated cooperative marketing among exporters to enhance the returns of coconut farming. However, for most of 1986, coconut oil suffered a discount vis-á-vis soybean oil partly because of overstocking in both the local and international markets caused by the large coconut harvest in the Philippines. Another reason was the antitropical oils campaign.

## Concerns for Coconut Oil

### Growing Palm Kernel Oil Volume

Coconut oil's gradual displacement by PKO, a derivative of the kernel of the palm fruit, in the lauric sector is real. Considering the rapid increase in palm oil output over the last two decades, it is possible that PKO could increase production to quantities greater than present-day coconut oil levels. Figures already indicate that the PKO market share of total lauric oil exports has been increasing. In 1990, total exports of coconut oil and PKO were 2.833 MMT (including the oil equivalent of oilseeds), the latter's share was 947,000 MT or 33% of the total. Compare the 1990 values with those from 1970 and 1980. In 1980, out of a total export production of 2.03 MMT, PKO captured a 25% share or 502,000 MT. The values for 1970 show PKO at 431,000 MT, a 20% share of the 2.19 MMT total. Although the quantity of coconut oil exported in 1990 represented an increase of 38% from 1960, palm kernel oil has experienced a much greater growth of 120% since 1960.

### Rapeseed, a New Lauric Source

Biotechnology is highly practiced in developed countries and can detrimentally effect third world countries. Coconut oil is about to experience competition from rapeseed-derived lauric oil. The inclusion of lauric fatty acid into rapeseed was made possible by biotechnology. Commercialization of this rapeseed variety may begin in 1995. Should this happen, lauric oil users would have more vegetable-oil options, and the coconut oil share of lauric oil exports would drop substantially. Lauric rapeseed would certainly have an advantage over other lauric oils, since the former is not a perennial crop. Thus, lauric rapeseed output may be increased in a relatively short time, depending on the requirements of the market. Also, since rapeseed is grown mostly in the European Community and Canada, and is beginning to be grown on U.S. farms, the decision to support an indigenous lauric oil in these areas is far more convenient than importing lauric oils from the tropics. This could have devastating effect on the economy of the Philippines.

Competing vegetable oils and fats produced in developed countries are heavily subsidized by their governments. In the EC, subsidies for planting rapeseed have given European planters the impetus to sustain production of rapeseed breeds tailored for edible use because of the low content of erucic acid. Such subsidies could also support the commercialization of bioengineered rapeseed for a lauric oil source.

## The Cuphea Threat

Aside from high-lauric rapeseed, efforts are ongoing to develop another source of lauric oil called *Cuphea,* a native to Mexico and the southern United States. In 1984, Henkel reported that Roebbelen of the University of Gottingen, West Germany, had done extensive research to select and breed species that can be grown as commercial crops. This work is being continued by Wirsinger in the United States. The first trials to grow *Cuphea* on a larger scale were reportedly successful.

From what we gather, more work is still required to develop this oilseed into an industrial crop. Should *Cuphea* be commercially successful, the U.S. market for coconut oil may shrink, particularly now that much effort is underway to implement the North American Free Trade Agreement (NAFTA).

## Prospects for Coconut Oil

### Product Diversification

Coconut oil is a major income source of Philippine coconut exports. In 1992, with total earnings from coconut exports of U.S. $706 million, coconut oil contributed U.S. $487 million, or 69%. But there are other products of economic significance. Copra meal, a by-product of oil milling, contributed U.S. $49 million; desiccated coconut, U.S. $87 million; and coco chemicals, U.S. $33 million. Products from the coco shell (charcoal and activated carbon) contributed U.S. $25 million.

There is much promise in the oleochemical sector. Back in the mid-1960s, there were concerns that the industrial usage of coconut oil could be displaced by lower priced petroleum-based inputs. But now, growing environmental concern has placed preference for biodegradable materials, so that the industry now anticipates a growing demand for coconut-based detergents and fatty chemicals. Already, European and U.S. companies have constructed more oleochemical plants in Malaysia and Indonesia; and in the Philippines, an additional plant will be built to join a rapidly expanding oleochemical sector. It may not be long before Asia is a major supplier of oleochemicals and a major consumer of lauric oils.

A coconut farmer's income is heavily dependent on coconut oil prices and since the vegetable oil market is an extremely volatile one, farm income from copra is unpredictable. If other products from the coconut tree are developed in larger quantities than they are presently, farm income could increase significantly and have fewer fluctuations; the increased returns to the coconut farms would be enough to encourage coconut farmers to maintain currently productive coconut trees and to replace senile trees.

### Oleochemicals

Executive Order 259 issued in the early 1980s decreed that local detergent manufacturers must gradually increase the coconut fatty alcohol sulfate (CFAS) content of their surfactant products to 20% by 1989, 40% by 1990, and 60% by 1991. Since early 1993, two multinational firms operating in the Philippines, the Philippine Refining Company (PRC), a Unilever subsidiary; and Procter and Gamble (P&G) Philippines, Inc., have expanded their requirements for coconut-based products with the introduction of "environmentally friendly" detergents containing more CFAS. The surfactant formulations contain at least 60% CFAS, which is 100% biodegradable, unlike its predecessor—the hard alkyl benzene sulfonate (HABS)—which biodegrades much more slowly.

### Farm Productivity

The PCA (Philippine Coconut Authority) has a program to improve coconut productivity via fertilization and to develop seedlings of high-yield varieties. These programs are funded by World Bank loans. The replanting program entails the replacement of senile coconut trees with promising local cultivars and/or hybrids. A total of 137 nurseries were established in 1991, four of which were used to germinate seedlings. The target is 3,500 hectares for plantings, with 900 hectares for hybrids and 2,600 hectares for local varieties.

The rehabilitation component of PCA involves the fertilization of low-yield coconut trees. The target of 94,159 hectares should benefit approximately 101,000 farmers. To date, about 2.8 million coconut trees were fertilized, and 25,000 coconut farmers have benefited.

In the short term, the world supply of coconut oil should remain adequate despite fluctuations brought about by the inherent nature of the coconut crop. However, the Philippine government has redefined its objectives to ensure that coconut farmers receive adequate returns from their farms, improve productivity, and encourage incentives to make coconut oil more competitive with other oils and fats.

### Global Trade Liberalization

The global trends toward trade liberalization will have a profound influence on the marketing of lauric oils. More specifically, there are indications that the reduction and eventual removal of import duties and, hopefully, the gradual phasing out of expensive subsidies in the United States and Western Europe will occur improving access to lauric oils from the Philippines, Malaysia, and Indonesia. Lauric oil consumers would have lower raw material costs because of reduced import-tax burdens. A "level playing field" would result from the ensuing minimal farm subsidies, and prices of lauric oils would be more competitive as the prices of other oils and fats would equilibrate with the real demand.

The formation of the AFTA (ASEAN Free Trade Area) is a dilemma for the coconut industry. It compels the Philippines to adhere to tariff reductions on coconut oil and other animal and vegetable oils. This could prove disadvantageous to the local coconut-based products, because it may allow the entry of less-expensive vegetable oils which may jeopardize the local market for cooking oil, soap, and other products which presently are produced mostly from local coconut oil. On the other hand, adherence to free trade practices among ASEAN member countries may bring about a more diversified market for Philippine coconut oil, which presently is highly dependent on the U.S. and Western European markets.

The "Zero for Zero" initiative proposed by the United States in GATT removes export subsides on U.S. soybean and other oil and fat exports. But this also opens the Philippines to possible flooding by foreign-sourced vegetable oils, which may upset the local coconut oil market. However, since the Philippine coconut industry is export oriented, some sectors believe that it eventually may equalize, since coconut oil exports from the Philippines are not subsidized.

If U.S. and EC subsidies on production and exports for major oilseeds were minimized, if not totally removed, coconut-supplying countries may be able to compete effectively in a real free market, thereby securing better prices for coconut oil and improving farm income. This certainly would encourage higher productivity and ensure an adequate supply of lauric oils to the world markets.

The main producers of lauric oils in the ASEAN, Malaysia, Indonesia, and the Philippines, have recently adopted an ASEAN Common Contract on FOB terms for vegetable oils from the region, such as palm oil, palm kernel oil, and coconut oil. This common contract, however, should not alarm U.S. and Western European buyers of coconut oil, since buyers/sellers are not compelled to depart from using the NIOP and FOSFA contract forms. The ASEAN Vegetable Oils Club (AVOC) is the vehicle for the implementation of the common contract. The organization is not meant to form a cartel for the trading of tropical oils, but to make the industry more efficient and for ASEAN to have its own identity.

# The Marketing and Economics of Palm Kernel Oil

**Yusof Basiron and Mohd Nasir Hj. Amiruddin**

Palm Oil Research Institute of Malaysia (PORIM), P.O. Box 10620, 50720 Kuala Lumpur, Malaysia

## Abstract

Palm kernel oil is a co-product of palm oil, with a production volume 10–13% that of palm oil. Being a lauric oil, it competes with coconut oil in world markets where the two important applications are in oleochemical and soap production and in specialty fats for confectioneries. In recent years, production of both palm and palm kernel oils have expanded rapidly. In 1993, Malaysian palm kernel oil production expanded by about 20% to reach close to 1 million tons. Palm kernel oil is marketed via a chain of numerous crushers, refiners, local and overseas dealers and brokers, and finally the oleochemical or confectionery end users. Time differences between producer and consumer countries and factors such as shipment volume and freight cost-control influence the marketing and economics of the palm oil trade. Competition with coconut oil and the diversity of market outlets, including the existence of large oleochemical plants in the producer countries, are other factors that characterize the palm kernel oil market. An understanding of the technical properties and marketing procedures will enable users to benefit by using palm kernel oil as an alternative to coconut oil.

---

In the lauric oil industry, attention has been focused on the rapid expansion of palm kernel oil supply during the last decade. World production of the palm kernel oil increased 9.0% per annum from 0.767 million metric tons (MMT) in 1984 to 1.533 MMT in 1992. This is contrasted with a lower 3.9% per annum growth in coconut oil production from 2.058 MMT to 2.790 MMT during the same period. Consequently, the share of palm kernel oil in the lauric market has increased from 27% in 1984 to 35% in 1992. The abundance of palm kernel oil has stimulated users to focus on using it as an alternative to coconut oil in soap, oleochemical, and specialty fat production. In the past, coconut oil was the main raw material for these industries.

Palm kernel oil (PKO) is a co-product to palm oil available at a ratio of 10–13 MT of PKO for every 100 MT of palm oil produced. Palm oil producers regard the kernel that they produce as a by-product which does not enter their calculations for determining production cost for palm oil; therefore, its cost of production is not easily determined. The palm kernel crushers have to distribute the crushing cost of the palm kernel to PKO and the palm kernel meal in order to arrive at an estimated cost of production of PKO. Despite the difficulty in establishing the production cost for PKO, its value in terms of market price is easily determined. Since PKO has very similar chemical and physical characteristics to coconut oil, the prices of the two oils are always very close, implying that the market recognizes PKO as being similar to coconut oil.

The development of PKO market has been influenced by its position as a co-product in palm oil production. It has an added advantage when competing with coconut oil, both from considerations of price and rapidly increasing supply. Processing PKO closely follows the trends in the development of palm oil. In the past, crude PKO was the main form exported. With the establishment of large physical refining capacities in the major palm oil-producing countries, more PKO is refined before being exported, just like palm oil.

The supply pattern of PKO was subsequently affected by the rapid expansion of oleochemical industries in South-East Asia. Many of the multinational oleochemical companies have established joint venture oleochemical plants in the lauric oil-producing countries. As a result, less PKO is exported in either the crude or refined form.

## Production

World production of palm kernels in 1991/1992 is estimated at 3.626 MT. The major producers of kernels are Malaysia (53%), Indonesia (18%), and Nigeria (13%). Production of PKO (40–50 kernel%) is expected to continue to

**TABLE 1**
**World Production of Palm Kernel Oil (1000 MT)**

| Country | 1984 | 1987 | 1992 | % | 1993[a] | % |
|---|---|---|---|---|---|---|
| EC-12 | 46.9 | 36 | 3 | (0.2) | 1 | (0.05) |
| Cameroon | 13.1 | 20 | 24 | (1.6) | 23 | (1.3) |
| Ivory Coast | 13.0 | 23.2 | 30 | (1.9) | 31 | (1.8) |
| Nigeria | 60 | 104 | 171 | (11.2) | 178 | (10.0) |
| Zaire | 20.3 | 18.9 | 23 | (1.5) | 23 | (1.3) |
| Colombia | 11 | 14.5 | 29 | (1.9) | 31 | (1.8) |
| Indonesia | 90.5 | 145.8 | 277 | (18.1) | 352 | (19.9) |
| Malaysia | 430 | 583.0 | 812 | (52.9) | 956 | (54.1) |
| Philippines | 4.5 | 4.4 | 6 | (0.4) | 7 | (0.4) |
| Thailand | 6.3 | 12.3 | 25 | (1.6) | 29 | (1.6) |
| Others | 71.0 | 44.3 | 133 | (8.7) | 136 | (7.8) |
| Total | 766.6 | 1006.4 | 1533 | | 1767 | |

[a]Forecast.
*Sources: Oil World Annual 1989; Oil World Annual 1992; Oil World Weekly,* November 26, 1993.

increase in the near future with the projected expansion in palm oil production (Table 1). It is forecast that the production of PKO by the year 2000 will be 2.175 MMT (Table 2).

The high productivity of the oil palm is demonstrated by the high yield of PKO in addition to palm oil. A hectare of oil palm yields an average 0.424 MT of PKO, and about 0.45 MT of palm kernel meal. The yield of PKO alone is comparable to the yield of any other oil-bearing seed and with another 3–5 MT of palm oil as the co-product, the productivity of the oil palm is exceedingly high (Table 3).

In Malaysia, production of PKO is carried out by palm kernel crushers, who buy the kernels from the palm oil mills. There are 61 palm kernel crushers with a total capacity of 9,952 MT/d; most of them have only a small kernel-crushing capacity. There is an overcapacity problem, and some 25 of the crushers with a total capacity of 2,095 MT/d are not in operation.

There are two methods of PKO extraction being employed in Malaysia. These are by mechanical means using a screw press and by solvent extraction. The mechanical extraction process is the most popular. Although the extraction rate of the oil is slightly lower compared to the solvent method, the oil remaining in the meal is highly valued by the animal-feed compounder and hence, there is a compensating effect through a higher price obtained for the meal.

**TABLE 2**
**World Production of Oils and Fats (1000 MT)**

| Source | 1984 | 1987 | 1992 | 2000[a] |
|---|---|---|---|---|
| Annuals | 33,737 | 39,565 | 45,780 | 57,294 |
| Soybean oil | 13,276 | 15,563 | 16,890 | 23,317 |
| Rapeseed oil | 5,226 | 7,373 | 9,240 | 10,829 |
| Groundnut oil | 3,262 | 3,432 | 3,940 | 4,629 |
| Sunflower oil | 5,877 | 6,796 | 8,000 | 9,962 |
| Corn oil | 1,029 | 1,277 | 1,590 | 1,780 |
| Castor oil | 389 | 367 | 460 | 456 |
| Linseed oil | 781 | 817 | 660 | 843 |
| Sesame oil | 551 | 659 | 660 | 702 |
| Cottonseed oil | 3,346 | 3,308 | 4,340 | 4,776 |
| Perennials | 10,757 | 14,038 | 18,713 | 24,808 |
| Palm oil | 6,280 | 8,032 | 12,200 | 17,498 |
| Palm kernel oil | 767 | 1,066 | 1,533 | 2,175 |
| Coconut oil | 2,058 | 3,144 | 2,790 | 3,306 |
| Olive oil | 1,652 | 1,796 | 2,190 | 1,829 |
| Animals | 19,090 | 19,315 | 19,120 | 23,164 |
| Tallow and Grease | 6,372 | 6,435 | 6,840 | 7,735 |
| Lard | 4,878 | 5,237 | 5,350 | 6,857 |
| Butter as fat | 6,290 | 6,201 | 5,910 | 7,012 |
| Fish oil | 1,550 | 1,442 | 1,020 | 1,560 |
| Total | 63,584 | 72,918 | 83,613 | 105,266 |

[a]Forecast.
Sources: Oil World Weekly (Various Issues) 1992/1993; Oil World Annual 1991; Oil World 1958-2007.

**TABLE 3**
**Average Productivity of Various Major Oil Crops**

| Crop | 1992/1993 Oilseed Yield (kg/ha) | Oil Conversion Factor (%) | Oil Equivalent Per ha (kg) |
|---|---|---|---|
| Soybean | 2042.9 | 18–19 | 377.9 |
| Cottonseed | 1016.0 | 18–20 | 193.0 |
| Groundnut | 796.2 | 45–50 | 378.2 |
| Sunflower seed | 1233.1 | 40–45 | 524.0 |
| Rapeseed | 1285.9 | 40–45 | 546.5 |
| Sesameseed | 370.5 | 45–50 | 175.9 |
| Palm oil | 16,000 | 20 | 3200 |
| Palm kernel oil | 891.9 | 45–50 | 423.4 |
| Copra | 520.7 | 65–68 | 346.3 |

Sources: Oil World Weekly, January 1993; Fuels and Chemicals From Oilseeds.

The fractionation of PKO to produce palm kernel stearin and palm kernel olein is also carried out by some of the refiners and specialty fat producers. Palm kernel stearin obtained from the fractionation process can be hydrogenated to improve its melting characteristics for special product requirements. The residual palm kernel olein is usually less valuable, and it can be used with or without hydrogenation in certain confectionery products.

## Development of the Palm Kernel Oil Trade

Malaysia and Indonesia continue to be the two most important exporters of PKO, with 61.4% and 28.7% of the export market share, respectively (Table 4). With them accounting for over 90% of the PKO export trade, and the other countries each having less than a 3% share, the sources of PKO are practically limited to the two countries.

There are three major applications for PKO. The development of its market has been influenced by the changes occurring in these end uses. Traditionally, the first applica-

**TABLE 4**
**World Exports of Palm Kernel Oil (1000 MT)**

| Country | 1984 | 1987 | 1992 | % | 1993[a] | % |
|---|---|---|---|---|---|---|
| EC-12 | 55 | 35.5 | 24 | (3.1) | 22 | (2.5) |
| Ivory Coast | 9.8 | 17.2 | 23 | (2.9) | 14 | (1.6) |
| Indonesia | 14.7 | 87.3 | 223 | (28.4) | 255 | (28.7) |
| Malaysia | 391 | 505.5 | 452 | (57.6) | 545 | (61.4) |
| Philippines | 0.1 | 0.1 | 4 | (0.5) | 6 | (0.7) |
| Others | 63.7 | 33.5 | 59 | (7.5) | 46 | (5.1) |
| Total | 534.3 | 679.1 | 785 | | 888 | |

[a]Forecast.
Sources: Oil World Annual 1989; Oil World Annual 1992; Oil World Weekly, November 26, 1993.

tion of PKO is in making soap. Soap formulations usually include about 20% lauric oil, and PKO or coconut oil are normally used. As the technical merits of PKO become better known, more countries are using it in place of coconut oil in soap formulation. In addition, PKO tends to sell at a lower price than coconut oil, and this attracts the soap manufacturers to consider using PKO.

The second application is in the production of specialty fats. Because of the special melting properties of lauric fats, they are useful in the formulation of cocoa butter substitutes for confectionery use. However, the consumption of PKO in the specialty fat industry is limited by the size of the specialty fat market. In 1984 the specialty market was estimated to be 40,000 MT annually. This would have increased to about 100,000 MT by 1992. This market is shared by PKO, coconut oil, and cocoa butter equivalent fats produced from palm oil midfraction or hydrogenated soybean oil. Recently, lower cocoa butter prices have reduced the growth of the specialty fat market.

The third major outlet for PKO is in the production of fatty acids and other derivatives. These are targeted for the oleochemical industry. A whole range of fatty acids, glycerine, esters, alcohols, and amines are produced, and they are the inputs to the detergent, cosmetic, and many other industries (1). Oleochemical plants are well established in Europe, the United States, and Japan. These plants are integrated with the production of consumer end-products such as shampoo, toiletries, and specialty chemicals for many industries. Although the markets for oleochemical products is concentrated in the developed countries, the demand for oleochemicals in the less-developed countries is growing, especially in the 1990s.

With the increasing availability of PKO, it was noted that the basic oleochemicals could be produced less expensively in the lauric oil-producing countries. New oleochemical plants established in these countries employ the latest technology; having a large choice of locally available raw materials, they operate more efficiently and are more competitive. As a result, Malaysia, Indonesia, and the Philippines, the world's major producers of palm kernel and coconut oils, have set up many new oleochemical plants, mostly on a joint venture basis with established oleochemical companies. A major percentage of the output from these plants is shipped back to the parent companies in Europe, the United States, and Japan. This is because most of the end-product manufacturing activities using the oleochemicals are still concentrated in the developed countries.

The trade pattern for PKO follows the distribution of industries involved in the three major applications. Because of the concentration of consumers in Europe, the United States, and Japan, these countries are the major markets for PKO. Not all the potential markets for PKO have exploited the availability of the commodity and its various subproducts. Some of the new importing countries are still unaware of the interchangeability between PKO and coconut oil. Import tariffs for PKO are often higher than those of coconut oil with the result that for these countries the usage of PKO is unnecessarily curtailed.

## Price Competitiveness and Marketing Channels

In the 1970s, the price per metric ton of PKO was at a premium of about U.S. $18 to that of coconut oil. This was reversed during the 1980s mainly due to the increasing availability of PKO and the fluctuating supply and occasional shortages of coconut oil. Over the years, it has been observed that the prices of the two oils are closely correlated (Fig. 1). Even the prices of the coconut and palm kernel meals are also moving in sync. (Fig. 2).

Recently, with the slight discount of PKO price in comparison to that of coconut, there is an added attraction for users to consider importing PKO. In Malaysia, PKO is made available in a wide range of crude, semiprocessed, and fully processed material (Table 5). In addition, there are various fatty acids and other oleochemicals which can be the intermediate raw materials for many industries. Even the less costly palm kernel fatty acid distillates are available for those looking for less expensive materials for making soap.

The export availability of various palm kernel products is changing rapidly. In the 1970s, users of PKO would import the kernel for crushing to produce the oil and meal. The construction of crushing facilities in the producing countries changed the system to that of the exportation of crude PKO and meal instead of the kernel. The pattern was again affected by the increasing availability of refined

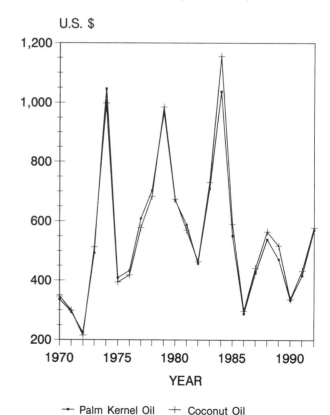

Fig. 1. Prices of palm kernel oil and coconut oil (U.S. $/MT) 1970–1992.

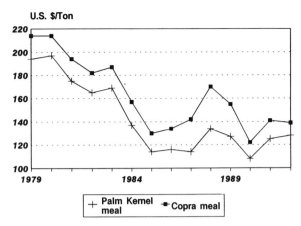

Fig. 2. Prices of palm kernel and copra meal.

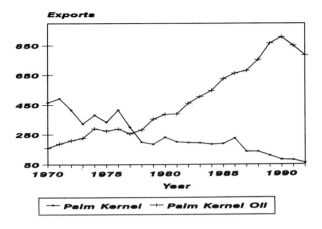

Fig. 3. World export of palm kernel and palm kernel oil 1970–1992.

PKO for export instead of crude PKO. Finally, export of PKO is declining as the result of increasing amounts being used locally by the joint venture oleochemical plants. The changing situations are shown in Figs. 3 and 4.

Exporters of PKO organized themselves under various crushing associations to establish trading rules and specifications (Table 6). There are at least 47 PKO exporters in Malaysia, and 27 of them are members of the Malaysian Edible Oil Manufacturers' Association (MEOMA). MEOMA has established standard specifications for local trade in palm kernel products. In the case of international trading, the FOSFA contract is widely used, while the NIOP contract is used mainly in the United States.

Usually, the crushers sell their output to local refiners. The crushers also may choose to sell crude PKO to local dealers for direct export. Refiners usually make use of local or foreign dealers to receive buying enquiries. Because of the large concentration of buyers in Europe and the United States, the European and American brokers and dealers perform a very important function, looking after the needs of their customers. They are more aware of the stock levels of their customers and alert them to the possibility of making additional purchases at the appropriate time. Pooling orders enables the dealers to ship in bulk volume and to book less costly shipping methods to transport the oil from South-East Asia to Europe or the United States. Price monitoring is another important function carried out by the dealers to advise their customers of the latest market situation. Because of the large time differential between South-East Asia and Europe and the United States, dealers usually manage to obtain price and supply information by being prepared to communicate with suppliers at odd hours of the night.

Producers of PKO products also conduct direct marketing by contacting their own customers. This is particularly true for specialty fat products which must be marketed with technical input to explain the properties of the fats and how they can suit the customers' requirements. Some new markets being developed include Western Asia and China.

## Future Developments

There has been concern about the declining exportability of PKO and coconut oil resulting from the large oleochemical-manufacturing capacities being established in the producer countries. The oleochemical-plant capacity in Malaysia will have reached 730,000 MT by the end of 1993, up from 150,000 MT in 1984. This could use all of the local production of PKO with nothing left to export. The long term reduction in direct export of PKO was pre-

TABLE 5
Malaysian Exports of Palm Kernel Oil by Product Category for the Years 1988, 1990, and 1991 (MT)

| Product | 1988 | 1990 | 1991 |
|---|---|---|---|
| Crude palm kernel oil | 311,340 | 301,267 | 256,477 |
| RBD palm kernel oil | 183,636 | 288,983 | 288,299 |
| RBD palm kernel olein | 10,081 | 34,358 | 39,190 |
| RBD palm kernel stearin | 6,307 | 18,645 | 35,840 |
| Other PPKO | 15,759 | 46,474 | 51,479 |
| Total | 527,123 | 689,727 | 671,285 |

Source: Palm Oil Update—PORLA.

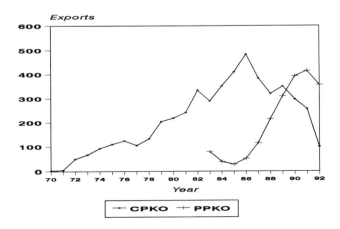

Fig. 4. Malaysian exports of palm kernel oil 1970–1992.

**TABLE 6**
**MEAMA Standard Specifications for Palm Kernel Products**

| Product | Specification | |
|---|---|---|
| 1. Crude palm kernel oil | FFA (as lauric) | 5.0% maximum |
| | Moisture and insoluble | 0.5% maximum |
| | Iodine value (Wijs) | 19 maximum at time of shipment |
| 2. Crude palm kernel olein | FFA (as lauric) | 5.0% maximum |
| | Moisture and insoluble | 0.5% maximum |
| | Iodine value (Wijs) | 25 maximum |
| | Cloud point | 8°C maximum |
| 3. Crude palm kernel stearin | FFA (as lauric) | 5.0% maximum |
| | Moisture and insoluble | 0.5% maximum |
| | Iodine value (Wijs) | 9 maximum |
| 4. RBD palm kernel olein | FFA (as lauric) | 0.1% maximum |
| | Moisture and insoluble | 0.1% maximum |
| | Iodine value (Wijs) | 16.2–19.2 |
| | Lovibond Color (5 1/2") | Red 2 maximum |
| | Taste | Bland |
| 5. RBD palm kernel olein | FFA (as lauric) | 0.1% maximum |
| | Moisture and insoluble | 0.1% maximum |
| | Iodine value (Wijs) | 25 maximum |
| | Lovibond color (5 1/2") | Red 1.5 maximum |
| | Cloud point | 8°C maximum |
| 6. RBD palm kernel stearin | FFA (as lauric) | 0.1% maximum |
| | Moisture and insoluble | 0.1% maximum |
| | Iodine value (Wijs) | 9 maximum |
| | Lovibond color (5 1/2") | Red 1.5 maximum |
| 7. Distilled palm kernel fatty acids | Acid number | 238–253 |
| | Sap number | 240–255 |
| | Titer | 20–28 |
| | Iodine value (Wijs) | 16–20 |
| | Color (Gardner) | 3 G maximum |
| 8. Palm kernel | Fair average quality | |
| | Profat | 22% minimum |
| | Moisture | 12% maximum |
| 9. Palm kernel extraction pellets | Fair average quality | |
| | Profat | 16% minimum |
| | Moisture | 14% maximum |
| 10. Lauric acid 95% | Acid number | 278–294 |
| | Sap number | 279–285 |
| | Titer | 41–44 |
| | Iodine value | 0.5 maximum |
| | Color (5 1/2" Lovibond) | 0.5 Red 5 Yellow |
| 11. Myristic acid 95% | Acid number | 243–249 |
| | Sap number | 244–250 |
| | Titer | 52–54 |
| | Iodine value | 0.5 maximum |
| | Color (5 1/2" Lovibond) | 0.5 Red 5 Yellow |

dicted by most of the multinational oleochemical companies, and a simple solution was for them to establish joint venture oleochemical plants in the PKO-producing countries, which helps ensure a stable supply. Perhaps it is a natural development for the basic oleochemical products to be produced in the palm kernel-producing countries, as this leads to greater efficiency and improved competitiveness.

In the past, the market has been in the developed countries (Table 7). With the growing supply, PKO will have to be marketed more widely in the developing countries. In view of their increasing affluence, some of the developing countries will emulate the usage pattern of the developed countries for oleochemical products. It is thus projected that the developing countries will be the growth markets

TABLE 7
Imports of Palm Kernel Oil (1000 MT)

| Country | 1970[a] | 1980 | 1990 | 1991 | 1992 |
|---|---|---|---|---|---|
| EEC | 105.0 | 215.0 | 425.3 | 405.8 | 354.3 |
| United States | 37.0 | 82.8 | 154.3 | 145.7 | 167.5 |
| Japan | — | 4.0 | 36.7 | 29.3 | 42.9 |
| South Africa | — | 8.4 | 29.0 | 24.4 | 24.0 |
| Brazil | 1.0 | — | 1.5 | 25.2 | 16.0 |
| Singapore | — | 12.4 | 14.6 | 29.2 | 24.3 |
| Turkey | — | — | 19.7 | 17.2 | 19.0 |
| Canada | 5.0 | 8.9 | 16.1 | 17.4 | 16.2 |
| Egypt | — | 1.6 | 16.9 | 6.0 | 7.7 |
| Pakistan | — | — | 2.5 | 2.6 | 2.3 |
| Others | 18.0 | 66.5 | 151.2 | 139.1 | 142.2 |
| World | 166.0 | 399.6 | 867.8 | 841.9 | 816.4 |

[a]Source: Oil World; Oil World 25 years.

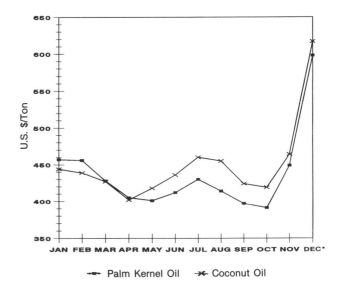

Fig. 5. 1993 Prices of palm kernel oil and coconut oil.

for oleochemical products in the near future, and they are the next important destinations for the export of PKO and its products.

An important consideration is for users to realize the techno-economic value of PKO and its various derivatives. PKO can be economically and technically superior to other competing oils in certain applications, provided the right product type is selected and the advantages of the efficient trading channels are fully utilized. For example, PKO can produce a larger range of cocoa butter fat substitutes (2) because it can be hardened by hydrogenation to different melting points, due to the higher unsaturation level of PKO compared to coconut oil. It must also be realized that as long as PKO is regarded as a co-product, its value is going to be arbitrarily determined, and during the period of increasing supply, it is likely that it will be marketed at a lower price than coconut oil even though it has its own desirable technical properties.

Another development is the launching of crude PKO futures contracts at the Kuala Lumpur Commodity Exchange (KLCE) on October 29, 1992. It thus provides the first lauric oil futures market in the world. Sophisticated facilities supported by state of the art in computer technology are provided for hedging and price discovery at the exchange. The prices derived at the KLCE are widely used as a basis for pricing cash market transactions.

The contract size has been changed from 15 MT to 25 MT/lot effective on August 16, 1993. PKO delivered into port tank installations shall not exceed 3.5% free fatty acids and from port tank installations shall not exceed 3.75%. Moisture and impurities shall not exceed 0.5%. Iodine value shall be between 16.50–18.75. Color (5 1/4" lovibond cell) range shall be 4–8 Red and 60 Yellow, maximum. The delivery point shall be at the port tank installation at Port Klang. About 7215 lots of crude PKO were traded at the KLCE during 1993.

Those trading in lauric oils are welcome to use the PKO futures in Malaysia. As an incentive, an exchange levy of RM2.50/lot (about U.S.$1.00) is waived for the first 40 contracts/day/broker once the broker has made 40 contracts in a day.

## Prospects for 1994

Prices for PKO have recovered from a low of U.S.$391/MT in October 1993, to reach U.S.$598/MT towards the end of December, 1993 (Table 8). The big increase in PKO production projected for 1993 was initially having a bearish effect on lauric oil prices (Fig. 5). Although PKO supply increased over 17% during 1993, the price prospects for 1994 had to depend on coconut oil production. A shortfall in coconut oil production was anticipated for 1994, owing to lower yields from drought-

TABLE 8
1993 Prices of Palm Kernel Oil and Coconut Oil (U.S.$/MT)

| Month | Palm Kernel Oil Malaysia CIF Rott | Coconut Oil Philippine/Indonesia CIF Rott |
|---|---|---|
| January | 457 | 444 |
| February | 456 | 439 |
| March | 428 | 427 |
| April | 405 | 402 |
| May | 401 | 418 |
| June | 412 | 436 |
| July | 430 | 460 |
| August | 414 | 455 |
| September | 397 | 424 |
| October | 391 | 419 |
| November | 449 | 464 |
| December[a] | 598 | 617 |

[a]Source: Oil World Weekly. Average of 3 weeks.

affected trees in the Philippines. With only a moderate increase predicted in PKO production, a shortfall in lauric oil supply is projected for 1994, and this has stimulated price increases even in late 1993. The price outlook for PKO for 1994 would be "firm" as expected from the supply shortage to be experienced by the lauric price leader—coconut oil.

## Conclusions

PKO markets have developed as a result of the rapid expansion of its production as a co-product of the palm oil industry. The market development is also influenced by consumption pattern spearheaded by the presence of established consumers in Europe, the United States, and Japan. Further market expansion is projected to occur in the developing countries in the coming decade.

Palm kernel oil is well accepted for use in soap, specialty fat products, and in oleochemical manufacturing. Marketing channels for securing the supply of PKO are based on the normal network of crushers, refiners, dealers and brokers, and the final customers. The problems of different time zones and the long distances between producing and consuming countries are overcome by an efficient dealer and broker community serving the PKO trade. This is further assisted by the presence of joint venture oleochemical plants in the PKO-producing countries. They have helped to enhance the efficient marketing of palm kernel products and improve the opportunity for producing countries to enjoy greater value added through the export of semi-processed products compared to the export of only raw commodities as in the past.

## References

1. Appalasami, S., and R.J. de Vries, *Palm Oil Developments*, 14 (1990).
2. Timms, R.E., *Specialty Fats from Palm and Palm Kernel Oils*, The Proceedings of the International Conference on Oils, Fats and Waxes, Auckland, 1983.

# The Socioeconomic Aspects of Lauric Oils Production: World Bank Involvement in Oil Palm and Coconut

David J. Meadows

World Bank, 700 18th Street, N.W., Washington, D.C. 20433, USA.

## Abstract

The paper discusses the involvement of the World Bank in the development of oil palm and coconut over the past three decades and contrasts the facts with the common perceptions of its role in promoting these crops. The proportion of World Bank financing in the projects it has supported is also discussed. The different project approaches that have been adopted are considered, ranging from large-scale industrial estates, through nucleus estate/smallholder schemes, to the support of in situ small holders.

The social benefits derived from the investments in the lauric crops financed by the World Bank are presented in terms of employment generation on estates, livelihood enhancement on small holdings and downstream income generation. The benefits to the environment resulting from the projects are enumerated.

## The World Bank's Role

As a World Bank Tree Crop Specialist, I am not qualified to discuss pure socioeconomics. Therefore I will interpret this subject very broadly. I will discuss the World Bank's role in financing the development of coconut and oil palm over the past three decades; then describe the different models of developing those crops which the World Bank has supported; and only then examine the socio-economic implications of lauric oil production, particularly that arising from World Bank financing.

The World Bank, both directly and through the International Development Association (IDA), which provides credits to the poorest countries on concessional terms, has been seen as a major player in the development and rehabilitation of lauric oil crops. Financing for coconut and oil palms has been made available as loans to many countries in Asia, Oceania, humid tropical Africa, and to a lesser extent, in Central and South America. Indeed, as many people are aware, the World Bank's support for those crops has from time to time attracted criticism from a well-known nonlauric oil lobby in the United States and from environmental advocates concerned with the perceived conversion of tropical rain forests to plantations.

However, it is a common mistake, perhaps even within the World Bank, to overestimate its influence on the vigorous development of coconut and oil palm. A recent World Bank study has shown that during the past 30 years, World Bank-financed projects were responsible for only 8.6% of the 290,000 hectares (ha) of coconut estimated to have been planted or replanted during that period. In the case of oil palm, the World Bank's projects involved 10.5% of the 235,000 ha planted and replanted during the three decades. In effect, World Bank lending for these crops helped finance a market-determined level of total investment in them. In doing so, it offered the poor countries and the poorer people in the borrowing countries the opportunity to participate in the growth that occurred. There is no evidence that the World Bank projects' share of the areas planted to any plantation crop has influenced the actual growth rate of that crop; indeed, that has been the role of the market.

In lending for development projects, the World Bank establishes limits on the proportion of external financing to total project costs. These vary between countries, being higher in the case of the poorest. These limits apply even when, as is frequently the case, other external cofinanciers participate. The recipients' governments must meet the remainder of the costs and, of course, are responsible for repayment of the loan or IDA credit. In the case of smallholder projects, a part of the cost is contributed by the smallholders in the form of sweat equity. Taking into account the contributions of cofinanciers, who participated in many projects, the World Bank financial contribution averaged about 60% of the total cost of the projects it supported. World Bank loans and IDA credits therefore directly financed only about 5% of the coconut development and rehabilitation and 6.3% of the oil palm investment that occurred over the past three decades.

There is little doubt that the World Bank's technical involvement, economic appraisal, and attraction of cofinancing has had more influence on the development of the lauric oil tree crops than the purely financial figures suggest. It also should be remembered that the rigorous appraisal process, which the World Bank applies to its projects, has promoted the use of improved planting material and better agricultural practices, resulting in higher yields and more efficient production.

## Lauric Tree-Crop Development Models

The World Bank has financed several approaches to the development of the lauric oil tree crops. The one most commonly perceived by the public is the establishment of big plantation estates. These have been most appropriate in the case of oil palm, where the large investment for effi-

---

The views expressed are those of the author.

cient factory facilities, and the logistics and discipline required for timely harvesting and processing to obtain a quality product, make economy of scale very important. With current copra production techniques, there is no similar justification for coconut estates, although other factors may still make the estate model appropriate. These would include large scale development at remote sites, where the superior managerial and logistical facilities available through the estate approach can be decisive.

Smallholder projects have always been an important element of the World Bank tree-crop portfolio, and can usually be divided into two types. In some, new areas of oil palm, or more commonly coconut, were developed. Frequently, these were in the form of land settlements, and included schemes in which landless rural labor or even, in some cases, urban poor were moved to new sites to grow these crops. In other projects, resident smallholders were provided with project facilities to plant and grow the new crops themselves.

In general, experience has shown that the latter model, provided that the extension services and delivery of inputs are well organized, produces the best results. Quite frequently, inexperienced settlers feel little ownership for holdings which are established for them, and in which they have directly contributed little of their own effort. In general where settlement on new sites is involved, the lesson learned has been the sooner the future farmer becomes directly involved on a specific holding, the better the outcome is likely to be. It is worth mentioning at this point that tree crops, particularly those with year-round harvest (and therefore income), such as coconut and oil palm, provide a key element in successful settlement. Few farmers are going to abandon a long-term source of income.

The other type of smallholder project frequently financed by the World Bank, has been the rehabilitation of existing smallholdings. In the case of oil palm, this has of course involved replanting, which is necessary every 25 years or so, as the palms grow too high to harvest. Unfortunately with coconut, there has been less pressure for replanting, although it is urgently required in many countries. One example of a World Bank-financed coconut smallholder rehabilitation project is the Small Coconut Farms Development Project, which is being implemented by the Philippines Coconut Authority. This is a large and ambitious project intended to launch a 20-year program to rehabilitate the country's coconut industry. Recognizing that the coconut industry, which supports at least one-quarter of the population, was in serious decline, the Government of the Philippines negotiated a loan of $121.8 million from the World Bank in order to finance the replanting with new, high-yield hybrid coconuts and rehabilitation of mature, but not senile, coconut palms by the application of fertilizer. The latter is very important because, until sufficient hybrid coconut seed is available, stabilization of the currently declining coconut production can only be achieved through the large-scale application of fertilizer. In spite of setbacks, the project has already benefited 126,000 small farms, improving the welfare of over one-half million poor people.

Because I have mentioned hybrids, it is appropriate to refer to a hybrid between the estate and smallholder models. The nucleus estate, a form of development which has frequently been financed by the World Bank in many countries, is established to provide the processing and marketing facilities and overall management while satellite smallholdings are either developed, or existing smallholders are encouraged to plant the crop and participate in the enterprise. Typically, the estate provides planting material, inputs, and management expertise and organizes the transport of the product. Payment to the smallholder for the crop is net income after deduction of processing and other costs, including credit recovery.

This latter aspect makes Nucleus Estate development particularly attractive to financing agencies, as cost recovery from conventional smallholder projects can be a major difficulty. In the lauric tree crops, this is especially true of coconut, where market outlets may exist for copra, nuts, or even toddy, allowing the farmers to circumvent credit repayment. Oil palm poses fewer problems in this respect, as the farmer has to send his crop to the processing plant, which is usually part of the project.

## Social Benefits

What are the social benefits of these investments? It is frequently asserted that only assistance to smallholders has a benign effect, but the provision of secure employment for a reasonable wage, together with the provision of housing, medical facilities, and schools, which are typical of estate development, are just as valuable to a poor family as the more hazardous prospect of independence on one's own holding. Thus the estates which the World Bank has financed have provided employment for management, technical, and administrative staff; mechanics; drivers; and field and factory workers. There is nothing intrinsically wrong in working for a salary or wage. Indeed, most people are happy to be doing just that. But there is a perception that plantation estates somehow have an unpleasant colonial connotation which is, by definition, detrimental to the employees.

Smallholders participate in the projects because they expect their livelihood to improve. They invest their labor and frequently acquire credit in order to plant crops which will yield marketable products for about 20 or 50 years, or to rehabilitate an elderly or inadequately attended crop. They do so in anticipation of better income, and the World Bank only finances such projects after rigorous analysis to ensure that adequate economic returns to the farmers and to the project as a whole can be expected.

Both forms of development clearly have downstream effects. The salaries and wages earned on estates and the higher incomes of the smallholders enter the local economies benefiting both merchants and the providers of services, including utilities. The products require transport and, in the case of copra, further processing. All of which creates employment and income. The extent to which fluctuations in that income (occasioned most frequently at this time by fluctuations in commodity prices) can be clearly seen in the Philippines. At times of low income, the rural economy is severely affected. The sale of products ranging from beer to life-saving drugs changes significantly with the fortunes of the coconut industry. There is no doubt that successful lauric tree-crop projects have a beneficial effect on the livelihood of the participants and many others far beyond the farm or estate gate.

Current lauric tree-crop projects either involve the rehabilitation or replanting of existing stands of oil palms or coconuts, or finance new planting in areas of logged-over scrub or fire climax grasslands. The effect is to reestablish a tree canopy which, together with the planting of leguminous ground cover crops, results in significant improvement in the local environment. A forest system is recreated, resulting in a better microclimate, abatement of fires, and reduction of soil erosion.

Important as these improvements are, they are overshadowed by the benefits of stabilizing the farming population. As previously mentioned, the availability of a long-term year-round income supersedes shifting cultivation and the continued destruction of forests which it causes. No one is going to walk away from a source of continuing cash to incur the physical labor of clearing tropical rain forest to cultivate annual crops, which cannot be grown for more than 2 or 3 years before the top soil is exhausted.

# Quality Aspects of Shipping and Handling Lauric Oils and Oleochemicals

## Albert F. Mogerley and Liam J. Rogers

Hudson Tank Terminals Corporation, P.O. Box 2549, Newark, NJ 07114, USA

## Abstract

Quality aspects of shipping and handling lauric oils and oleochemicals are discussed. Both the food and nonfood industries have been undergoing serious reviews regarding transportation modes. This includes the interface with bulk liquid, deepwater terminals and other storage facilities. This paper is intended to provide current information and insights to the reader. It will briefly review current practices and assurances, then move to considerations for the future. The industry is on the threshold of changes which may be forced upon it through international convention or harmonization. National governments, responsible for food industry standards and practices, are deliberating over new international standards of quality assurance in handling, processing, and manufacturing procedures. Details of those efforts will be discussed in this paper.

---

Food and nonfood industries using lauric oils and oleochemicals as raw materials or intermediates have been subjected to comprehensive reviews of quality assurance programs for all modes of transportation. This includes storage and handling at deepwater bulk liquid terminals and other warehousing facilities. The initiation of these reviews during recent years has been brought about by industry groups voluntarily, as a result of competitive necessity, and by governments, as a result of international convention, harmonization, or the need to respond to abuses through a regulatory route.

This paper separates the quality aspects of shipping and handling lauric oils and oleochemicals into two categories: product quality, for example, analytical specifications; and service quality, such as performance systems. The following statements are applicable to all food and nonfood substances produced from vegetable and animal fats and oils.

The first category, product quality, identifies fundamental specifications, either inherent or achieved through processing, in which the properties of the substance are stated and/or specified. Product quality is the conformance of products to specifications which have been determined and agreed upon between a seller and a buyer. Proven quality measurement and control systems are used to develop the operational procedures involved. Product quality, therefore, becomes a measurable reality. Any deviation from normal procedures during the processing activity results in the nonconformance to product specifications.

Another influence on product specifications and/or formulations result from proposals made to a governmental body by special interest groups claiming to be more representative of a community than the government or private sector. The potential ramifications of a perceived problem create duress before a thorough and objective examination of the issues can occur. Increased public knowledge of the problem contributes to both political and emotional pressures. There is a tendency to overflow national borders and affect long-term relationships between consumers and suppliers whose previous purpose was to focus on product quality. This kind of special interest interference can be eliminated by industrial and governmental integrity. If interference occurs, the issue remains in flux and will take an inordinate amount of time to resolve.

Present analytical techniques and technology make it possible to measure foreign substances present in concentrations as low as 1 part per trillion (ppt). The definition of product quality, therefore, becomes dependent on what is acceptable to the buyer and maintainable by the producer or supplier. It is essential that a producer be cognizant of a customer's needs. Similarly, it is necessary that a customer involve producers/suppliers as soon as a change in specifications is needed or desired. Quality specifications can then be established by contract or other acceptable means. Whether or not that quality is met is not the primary subject of this paper. This now brings us to the second category, service quality, which includes a number of additional views.

Many firms perform business functions subsidiary to production and distribution for the owners of bulk liquid products. Notably included are firms that provide ocean transportation and those that provide deepwater terminal facilities for lauric oils and oleochemicals. These firms may own and operate sophisticated tankers to transport parcels across the oceans, and/or they may own and operate bulk liquid terminal facilities to store and redistribute the products. Others may own and operate overland transportation equipment to satisfy various transportation requirements. Firms providing these services to the industry are typically known as service businesses, and as such, perform at a level of service quality.

Owners of the lauric oils and oleochemicals transported and handled, literally may not be able to see the quality of service required or expected. However, performance of quality service can be gauged by a number of standards or systems pertinent to the concepts of quality assurance. Quality service is not easy to achieve and, at times, may be very difficult to manage. Because it is important to the future of the service business and a long-term commitment to the purchaser of the service, the concept of a service–supplier partnership can be undertaken. Then, it is possible that the two firms will commit to a long-term relationship to address all quality assurance matters. The two organizations recognize that a program of disciplined actions must be designed and implemented to bring about continuous improvement in all areas of business.

At any point during the shipping and handling of a material, a "snapshot" can be taken of product quality. The product can be tested for industry specifications and a comparison can be made to the shipped quality or a merchantable quality. However, in order to observe service quality, one either has to wait for the outcome of the service or attempt to examine the system upon which the performance is based. Waiting for the outcome of a service is generally not an acceptable form of quality control. This leaves examination of systems and/or procedures as the most viable way of defining service quality. To illustrate, this would be similar to defining the quality of a bag of potato chips based on whether the oil used in frying them met analytical quality specifications. Although the oil is an integral part of the quality of the chips, it is not the complete picture. Similarly, while the systems in place to ensure service quality are of vital importance, without successful implementation they are meaningless. Performance is seen as more important than theory. This dilemma exemplifies why it is much more difficult to measure service quality.

For many years, owners and operators of service organizations that furnish essential transportation, storage, and distribution services for owners of bulk liquid substances have engaged in a dialogue about the ability of a service organization to enter a partnership arrangement and to continually review service performance. The perceived risk of product degradation, rather than the real risk, became increasingly important as more and more of these substances were handled and transported in bulk liquid form. Many service organizations, especially those looking for a competitive edge, clearly understood that performance would determine if the relationship between the customer and the supplier would continue. Service organizations, traditionally smaller companies than those they serve, have the ability to implement quality considerations and the principles of quality management on a timely basis. The basic partnership may be expanded so that a service organization may also be involved in a buyer's "preferred supplier" program.

The most important concerns of the product owner are that product quality, service quality, and integrity are maintained, and that on-time delivery occurs. Indeed, the customer is often a division within the product owner's firm.

There are many attempts being made, both direct and indirect, to assess, analyze, and standardize service quality. Many of these endeavors originate within world bodies, the European Communities (EEC), the United States, other national standard bodies, and industry associations. These initiatives are known by various acronyms and other titles:

ISO 9000 Quality Standard Series; ISO 5555 Sampling Methods; IMO Annex II (International Maritime Organization); Codex Alimentarius Commission; EEC Directive 93/43/EC; SFTA (Sanitary Food Transportation Act of 1990); OPA 90 (Oil Pollution Act of 1990); FOSFA International—Prior Cargo Listings; and National Institute of Oilseed Products—Prior Cargo Listings. A synopsis of the listed statutes, directives, regulations, and standards is provided.

## ISO 9000 Quality Standard Series

The International Organization for Standardization (ISO) published the ISO 9000 series in 1987. The standards are concise and straightforward. During the past 5 years, in an effort to improve overall European product quality, the EC has recommended that businesses conform to the appropriate ISO 9000 Standard. Also, the EC continues to consider new or revised legislation related to these standards.

The standards were originally intended for use by suppliers and customers, with the expectation that they would agree to use the hierarchical standard that best met their needs. ISO 9000 is introductory in nature and provides guidance for selecting 9001, 9002, or 9003. ISO 9004 provides guidance on developing quality systems and discusses product liability and safety issues. Many firms that already have comprehensive quality assurance programs are quietly revising quality system documentation to assure that all ISO 9000 elements would be visible in an audit by a third party.

All U.S. Food and Drug Administration (FDA) Current Good Manufacturing Practices (CGMP) include many elements from the ISO 9000 Standards. The U.S. Government has not established a uniform approach for oversight or accreditation of private businesses that engage in quality assessment. Under the European system, certificates of conformity with the standard may be issued by government agencies or by private registrars that have been certified to perform assessments. Discussions are taking place between the U.S. Department of Commerce and the EC concerning the criteria for mutual recognition of conformance assessments.

The European experience suggests that countries may adopt consensus standards of international organizations for regulatory purposes more readily than using one nation's system as a model. The FDA's CGMP and other rules have served as international prototypes, however, future leadership in such product quality regulatory areas may be best demonstrated by participating in standards development.

At present, considerable confusion exists as to whether ISO 9000 is a voluntary standard or a regulatory requirement. In the United States, many FDA-unregulated corporate quality assurance managers agree that market forces, particularly the desire to be competitive in Europe, are the primary motivation for corporate decisions to initiate an ISO 9000-based quality program. The European Commission (EC) adopted the ISO 9000 series as a voluntary standard, European Norm (EN) 29000, and has encouraged internal markets to apply the standard as a means of improving the quality and competitiveness of European products. As they become effective, specific EEC directives for certain product categories will incorporate the EN 29000 Standard into the EEC regulatory scheme.

There is growing worldwide recognition that total quality management (TQM) provides a superior mechanism for market place competitiveness, as opposed to finished product testing or inspection, and this creates the need for parties to communicate reliable quality information. The international trend is also towards horizontal rather than vertical standards. Vertical standards are specific in nature or product oriented while horizontal standards are written for broader applications, such as classes of products, processes, or systems.

This flexible approach, using guidance to encourage "state of the art" processes and controls, is consistent with current TQM philosophy, which reportedly will be incorporated in the next revision of the ISO 9000 Standards. As technology advances, industry has the primary responsibility of continuing to improve process-control measures. This is an important consideration in the United States because longstanding U.S. national policy supports the development of standards and the administration of standards-related activities by the private sector. This policy can be compared with that of many European governments which are more actively involved in aiding and protecting the private sector through quasi-governmental bodies and subsidiaries.

Once recognized, standards should be used preferentially in procurement and regulatory activities in the interest of promoting trade as specified in the provisions of the Standards Code in the General Agreement on Tariffs and Trade (GATT). Knowledge of international definitions, standards, codes, and widely accepted guidelines enhances cooperation, communication, and participation in standards development.

## ISO 5555:1991(E) Methods for Sampling of Animal and Vegetable Fats and Oils

This ISO standard describes methods of sampling crude or processed animal and vegetable fats and oils of any origin and whether the sample is liquid or solid. It also describes the apparatus used for this process. ISO 5555 is included in the Compendium of Test Methods for the European Oleochemicals and Allied Products Group (APAG). This ISO standard has been adopted by many government and private sector organizations throughout the world, and referenced in commodity buy/sell contracts as the method to be employed by samplers, superintendents, and surveyors. The standard includes a "Temperature Limits" Table that shows minimum and maximum temperature ranges which should be maintained for various substances at the time of vessel loading, discharge, and when samples are customarily drawn.

European- and UK-based marine surveying organizations are proposing uniform community-wide standards, for the EEC. If adopted by the EEC Council, a directive incorporating ISO 5555 will impact bulk liquid-sampling methods throughout the community, as well as, the world. All government and private sector organizations should participate in the development of uniform sampling standards for fats, oils, and oleochemicals.

## International Maritime Organization (IMO)

The International Maritime Organization (IMO) is an agency of the United Nations. It is based in London; consists of 144 member countries, and develops international conventions, codes, and recommendations governing maritime transportation.

The International Maritime Organization has adopted approximately 30 international conventions. One of the conventions which is now law in most nations is the International Convention for Prevention of Pollution from Ships, 1973, as modified by the 1978 protocol relating thereto (MARPOL 73/78). This protocol established regulations for ocean going tankers and onshore reception facilities.

Regulations under MARPOL 73/78 include annexes concerning oil—Annex I; noxious liquid substances (NLS)—Annex II; and ship-generated garbage, including medical waste—Annex V. These regulations provide for the disposal of residues and mixtures remaining in ships' cargo tanks and ship-generated garbage into onshore reception facilities to reduce the amount of cargo residue and refuse discharged into the sea.

MARPOL 73/78 Annex II classifies liquid products transported in bulk by ships into four categories, from Category A (most hazardous) to Category D (least hazardous) according to their potential for harming the marine environment. Category D NLS includes animal, vegetable, and marine fats and oils and most oleochemical substances above $C_{12}$.

During early 1993, The Netherlands submitted comments to an IMO subcommittee on bulk chemicals (BCH) that vegetable oil residue and residue from other "lipophilic substances" discharged from ships may be harmful to marine bird life. Though the scientific evidence supporting The Netherlands' contention is debatable and inconclusive, the subcommittee considered the matter during their meeting in London, September, 1993.

The Marine Environment Protection Committee (MEPC) recognized that such substances are currently categorized as category D "Noxious Liquid Substances," and that no quantity limitations are prescribed for the discharge of cargo residues in this category. The Committee felt, however, that the discharge of residues of such substances from ships should be limited in order to minimize the adverse effects of such discharge on seabirds. Noting that the bulk chemicals subcommittee had been instructed to undertake a complete review of all the regulations for the control of pollution by noxious liquid substances carried in bulk, as contained in Annex II of MARPOL 73/78, the MEPC decided to wait for the outcome of that review before making a final decision regarding a possible amendment of the discharge standards of animal, vegetable, and marine fats and oils.

In view of the potential adverse effects of such substances on seabirds, as indicated previously, the committee agreed that all efforts to reduce the discharge of animal, vegetable, and marine fats and oils from ships should be made immediately. Recommendations were made to ship operators to comply with cargo-residue discharge provisions, for example, when unloading animal and fish oils and fats and vegetable oils, the existing cargo-unloading capability should be used in such a way as to minimize the cargo residue left in the tank. The stripping facilities should always be used to their full capacity and in accordance with the ship's P & A Manual as applicable. When there is a choice, the tank with the best available stripping system should be used; and the discharge of cargo residues should be made below the waterline and in a depth of water of not less than 25 meters.

The main purpose of describing this is to state that if The Netherlands' proposal is adopted, and more stringent restrictions are placed on shipowners transporting lauric oils and oleochemicals, those shipowners would have no option but to pass increased costs on to charterers and product owners if these cargo residues cannot be discharged at sea. It is reasonable to assume that a reduced number of vessels will be available to transport oils, fats, and oleochemicals. It is also expected that there will be a commercial impact if these substances are reclassified into another MARPOL Annex II category. All quality procedures relative to the carriage of oils, fats and oleochemicals will require extensive review at that time.

## Codex Alimentarius Commission

The Codex Committee on Fats and Oils approved a "Code of Practice for Storage and Transport of Edible Fats and Oils in Bulk" during 1987. The principal purpose of the Code is to reduce or eliminate sources of degradation to edible oils and fats. The Code, originally prepared in 1983–85 by the Palm Oil Research Institute of Malaysia (PORIM), set forth a number of standards and guidelines to minimize changes in quality occurring during storage and transport of the oils and fats.

The Code includes instructions and suggestions relative to materials of construction, product-heating recommendations, operating procedures, and other advisory material. Draft revisions to the Code were recently reviewed by the Codex Committee on Fats and Oils during the course of their September, 1993 meetings in London.

The new Codex Alimentarius Committee on Food Import and Export Inspection and Certification Systems, which held its first meeting in Australia, has avoided the issue of import/export certification. The committee's final report notes that "several delegations ... state that it was not appropriate for the Codex Alimentarius Commission to become a certification body, as the Commission had other well-defined and internationally recognized responsibilities which were carried out within the framework of limited resources." The group did not rule out future consideration of the report at its next meeting at the end of June. Catherine Carnebale, of FDA's Center for Food Safety and Applied Nutrition, *Food Chemical News*, predicted increased pressure on Codex to become a certification body.

## Council Directive 93/43/EEC

Article 1 of this EEC Directive lays down the general rules of hygiene for foodstuffs and the procedures for verification of compliance with these rules.

Article 2 states that "food hygiene" shall mean all measures necessary to ensure the safety and wholesomeness of foodstuffs. The measures shall cover all stages after primary production during preparation, processing, manufacturing, packaging, storing, transportation, distribution, handling, and offering for sale or supply of foodstuffs to the consumer.

Section 2 of Article 3 of the Directive asserts that food-business operators shall identify any step in their activities which is critical to ensuring food safety and ensure that adequate safety procedures are identified, implemented, maintained, and reviewed on a basis of named principles, used to develop a system of Hazard Analysis and Critical Control Points (HACCP).

Article 10 states that if a hygiene problem likely to pose a serious risk to human health arises or spreads in the territory of a third country, the Commission, either on its own initiative or at the request of a member state, shall take the following measures without delay, depending on the seriousness of the situation: suspend imports from all or part of the third country concerned and, where necessary, from the transit third country; and/or lay down special conditions for foodstuffs from all or part of the third country concerned.

Chapter IV of the Annex, regarding Transport, makes two important points: conveyances and/or containers used for transporting foodstuffs must be kept clean and maintained in good repair and condition, in order to protect foodstuffs from contamination; and receptacles in vehicles and/or containers must not be used for transporting anything other than foodstuffs where this may result in the contamination of foodstuffs. Bulk foodstuffs in liquid, granular, or powder form must be transported in receptacles and/or containers/tankers reserved for the transport of foodstuffs.

## Sanitary Food Transportation Act of 1990— United States

The Sanitary Food Transportation Act (SFTA) [P.L.101-500] is an act to prohibit certain food transportation practices and to provide for regulation by the Secretary of Transportation that will safeguard food and certain other products from contamination during motor or rail transportation, and for other purposes.

An Advanced Notice of Proposed Rulemaking (ANPRM), published February 20, 1991 in the Federal

Register (56 FR-6934), set forth proposed regulations on which the agency sought comment. The proposal was issued by the Research and Special Programs Administration (RSPA), U.S. Department of Transportation (DOT).

A Notice of Proposed Rulemaking (NPRM), published May 21, 1993 in the Federal Register (58 FR-29698), set forth proposed regulations addressing the safe transportation of food products in highway and rail transportation. The intended effect of this rulemaking is to increase the safety level associated with the transportation of FDA-regulated food products.

The Department of Transportation is still reviewing the comments presented to the agency from approximately 100 responders as of this date. It is believed DOT will not have the final regulations drafted for several months.

## Oil Pollution Act of 1990—United States

As a result of major petroleum spills into the marine environment, most notably the Exxon Valdez spill into Prince William Sound during March 1989, the U.S. Congress passed the Oil Pollution Act of 1990 (OPA-90). The Oil Pollution Act of 1990 (P.L.101-380) provided for the delegation of authority to the Department of Transportation (DOT) and the Environmental Protection Agency (EPA) to establish procedures, methods, and other requirements to prevent the discharge of oil and other hazardous substances from onshore facilities into the waters of the United States. In addition, the United States Coast Guard (USCG) drafted regulations pertaining to vessel construction standards, the removal of spills by vessels and onshore facilities, and other related rules.

As a result of OPA-90, a proliferation of rules affecting the transport and storage of oil has occurred. By definition, this is "oil of any kind," and causes animal and vegetable fats and oils, including their by-products, to be regulated in methods similar to petroleum oils and hazardous substances. The regulations require onshore facilities and vessels to prepare extensive "Response Plans" to deal with spills. The regulations may also require response plans to deal with spills from railroad tank cars and tank trucks.

While it may be difficult to literally interpret the direct impact that OPA-90 regulations have on the quality aspects of shipping and handling fats and oils, the reality is that these and all other regulations dealing with transportation directly impact quality procedures and total quality management.

## The Federation of Oils, Seeds and Fats Associations Ltd.

The Federation of Oils, Seeds and Fats Associations (FOSFA), founded in 1971, is an international trade association based in London. First and foremost, FOSFA is a contract-issuing body. It is these contracts that help regulate trade and allow the parties involved to regulate the purchase and sale of commodities, including oils and fats. Also, FOSFA maintains technical standards, and is actively involved in their development. In addition, FOSFA operates a Member Analysts Scheme and a Member Superintendents Scheme; conducts trade education courses and seminars, and provides many other services of benefit to the industry.

A number of contract provisions have been introduced by FOSFA, including "Qualifications for all ships engaged in the ocean carriage and transhipment of oils and fats for edible and oleo-chemical use"; and, "Operational procedures for all ships engaged in the ocean carriage and transhipment of oils and fats for edible and oleo-chemical use." They also have a "List of acceptable previous cargoes," and a "List of banned immediate previous cargoes."

For many years, FOSFA has worked with similar associations to maintain the purity of oils, fats, and oleochemicals in all phases of transportation, handling, and storage. Their Council, Oils and Fats Section Committee, Technical Committee, and Shipowners Sub-committee have been proactive for many years in all matters relevant to the quality assurance of oils and fats.

## National Institute of Oilseed Products

The National Institute of Oilseed Products (NIOP), organized in 1934, is an international trade association based in Washington, DC. It has published a set of trading rules governing transactions among its members covering imported oilseeds and vegetable oils.

During early 1989, in response to industry and FDA concerns regarding prior cargo contamination, the NIOP adopted the following trading rules as a protocol for standard operating practices concerning cleaning, loading, surveying, and transporting edible fats and oils: Rule 5.10—NIOP rule for surveying of inbound/outbound parcels of vegetable oils; Rule 5.11—Vessel pumping system standard for ships loading edible oils into tanks which previously contained acceptable prior cargoes; and Rule 5.12—Prior cargo listings: (a) Acceptable prior cargo lists No. 1 & 2; and, (b) Unacceptable prior cargo list. The FDA accepted the NIOP prior cargo lists, plus the cleaning, inspection, and vessel-operating procedures.

In addition to these rules, the NIOP Technical Committee implemented procedures to administer petitions to add substances to an acceptable list. The petition(s) to the Technical Committee to add a substance must include factual information regarding criteria that determine acceptability. The criteria, for which documented verification is required, include identity, carcinogenicity, toxicity, removability, and analyzability. The protocol established by the NIOP Technical Committee to administer the petition includes a review of the petition by the Institute of Shortening and Edible Oils (ISEO) Technical Committee prior to submission of the petition to the FDA.

Government intervention and regulation was forestalled by NIOP promulgating and adopting a set of procedural standards to ensure quality performance in the shipping of edible oils. It is important to note that standards developed by and for industry are far more easily understandable and sustainable than those imposed by external forces.

## Conclusions

Here are a few key points to consider which bear upon quality aspects of shipping and handling lauric oils and oleochemicals.

Total quality management (TQM) and the efforts to deal with quality improvement programs (QIP) have had an impact on the vegetable and animal oils and fats industry for many years. Indeed, many cultural changes relative to quality concerns have occurred around the globe as a result of efforts to implement quality programs to supply quality products.

Quality of service and performance should be the main thrust of a service organization's daily activities. It must be a primary commitment of management and a major pursuit of employees. Quality performance requires management commitment, employee involvement, good communication, and positive action. Quality assurance is dynamic and must be constantly reviewed to ensure that procedures and quality control are suitable for their intended purposes.

The best way to avoid a quality problem is to prevent it. Operating procedures and instructions must be implemented successfully to prevent nonconformance. The operating procedures must be practiced uniformly in order to assure quality handling of products at a high level of customer service. A service organization must look to its customers for strict delineation of their needs and must solicit customers' comments as to whether those needs are being satisfied.

Quality policy, management policy, and on-the-job-training should result in a set of disciplined actions designed to bring about continuous improvement in every activity of the company. Each action is objectively measured to determine its effectiveness in maintaining and improving quality. When necessary, further action can be taken to improve conformance with agreed upon requirements.

As previously stated, there are many attempts being made by the private sector and governments to assess, analyze, and standardize all forms of quality. Since industries know how to survive and be successful, it is both essential and timely that they become more proactive in their working relationships with their respective government(s). It is extremely important that they discuss and manage common issues. Industrial integrity should continue to influence the relative concerns of government legislators and regulators constructively. If the private sector does not voluntarily participate in public deliberations, it is then reasonable to assume that government intervention will follow.

For many years, owners of the products and service groups providing transportation and storage have vigorously pursued voluntary programs to constantly upgrade equipment to state-of-the-art safety, environmental, and health standards. Partnerships have been formed between these groups to work on QIP and TQM. The pursuit of quality has been a part of ongoing programs dedicated to finding better transportation and handling procedures. Performance does make a difference in the continuation of a relationship between the buyer and the supplier of services.

The service industries are highly regulated on the overall quality of their respective businesses. The management of these firms and their operational procedures are under intense scrutiny by government regulators and inspectors; by risk assessors; by third-party audits; and by customers for safety, environmental protection, and other exposures to liability. The quality of the entire firm is under constant review. The pursuit of quality must go hand-in-hand with these other requirements.

"Enomics" (environmental economics) will bring about dynamic changes in our market place. During the balance of this decade we will see specific ways in which environmental restructuring takes place that will change many of the ways we process, transport, handle, package, distribute, and sell goods. "Enomics" will have an impact on quality programs.

I have attempted to weave a thread through the maze of some proposed or recently adopted statutes, directives, regulations, and standards which will impact quality aspects of shipping and handling lauric oils and oleochemicals in the future. From my point of view, the effort behind adoption of a voluntary quality improvement program is simple and direct. It is education coupled with the motivation of all individuals involved to work together in a team effort to understand what quality means. It means a degree of excellence—and excellence must be explained in order to be applied. It is also an understanding of what commitment means, and that quality is a commitment. It is an educational process, and quality management is a mindset.

# Nonfood Uses of Lauric Oils

Robert J. McCoy

The Procter & Gamble Company, 1 Procter & Gamble Plaza, GO-255-9, Cincinnati, OH 45202, USA.

## Abstract

Nonfood uses of lauric oils exceed that of food uses in many markets and offer the greatest potential for future growth. Surfactants are the major opportunity to gain market share vs. hydrocarbon-based feedstocks. The extreme price volatility of lauric oils relative to competing raw materials, such as other natural fats and oils and petroleum hydrocarbons, inhibits interest and investment in expanded nonfood use. The expanding palm kernel oil production has dampened price volatility by increasing the volume base and reducing vulnerability to localized weather conditions. Also, the major increase in exported coconut oil from Indonesia when world market conditions warrant has dampened price fluctuation. Additional progress is needed to give users reasonable assurance of adequate long-term supplies without extreme price volatility. Adoption of known and evolving technology to improve production efficiency, for example the appropriate use of fertilizer on coconut trees, can increase productivity and supply as well as temper the effect of drought on crop yields. While environmental attributes of products made from renewable feedstocks may increase the use of natural fats and oils, at present there is no clear trend of a substantial shift in that direction.

---

This paper reviews selected economic factors in the nonfood markets for the lauric oils, coconut oil (CNO) and palm kernel oil (PKO), that can influence their future potential in world markets. Lauric oils are good raw materials for oleochemical and other nonfood uses. They have unique characteristics and dependable quality. As is well known, coconut and palm kernel oils are currently the only large volume, commercial sources of vegetable-derived $C_{12}$ and $C_{14}$ triglycerides. According to *Oil World*, they represent 5.5% of total world consumption of fats and oils, with CNO at 3.6% and PKO at 1.9% (1).

Separate discussions are warranted for nonfood versus food uses because nonfood uses have become the largest outlet in most markets, with a little over one-half of world lauric usage believed to be in nonfood outlets. In addition, nonfood applications probably offer greater potential than food applications for market growth, provided their competitiveness with alternate raw materials can be improved. There appear to be favorable developments in this area.There has been greater price volatility for lauric oils than for other natural fats and oils; for example, soybean oil, as shown by comparative frequency distribution of monthly average prices (Figs. 1 and 2). This is also true versus ethylene, a competing feedstock. Although there is reason to believe this pattern is moderating, this price variability creates problems for long-term planning in product formulation and capital investment and in managing day-to-day financial risks. Manufacturers of the lauric-based intermediates usually cannot increase their product prices to recover higher costs when competing against intermediates made from hydrocarbons or nonlauric fats.This commodity market price risk in lauric creates a special challenge, given the absence of major organized futures exchanges, such as the Chicago Board of Trade (CBOT), where the market-price risk of many commodities can be hedged.

This paper will concentrate on economic issues that can influence decisions affecting current and future consumption. Existing literature (2–4) and other papers in these proceedings provide detailed information on specific uses.

## Nonfood Uses of Agricultural Products—General

Given the limited potential for expanding the food demand for agricultural crops, there has been a desire to develop more nonfood uses for a number of years. This interest intensified during periods of large inventory buildup and relatively low crop prices; for example, in the 1930s, 1950s, and 1980s. Public research and organizational support increased, as happened most recently in the early to mid-1980s, when initiatives by the U.S. Secretary of Agriculture (5–7), building on earlier work in the Agricultural Research Service, led to a program to expedite the commercialization of promising research developments. Such programs may help develop new or specialized vegetable oil crops, including annual lauric crops, as well as new uses. However, interest usually declines when prices recover, especially when they spike to very high levels. The resulting uncertainty about future economics and supply reliability tends to discourage users from making major commitments in raw materials, processes, and formulations.

There have been questions about the continued long-term economic competitiveness of agricultural crops, for example, natural fats and oils versus hydrocarbons. However, agricultural production capability has historically exceeded expectations and demand. There are still many opportunities to expand output with currently available technology and resources, and research promises continued technological advances (8). Dennis Avery, a former U.S. State Department economist, has written extensively on this subject (9). A good example for coconut production is the World Bank's $122 million financing of the Philippines "Small Coconut Farms Development Project."

A major U.S. agribusiness consulting firm in Memphis recently performed an in-depth study of the global oilseed

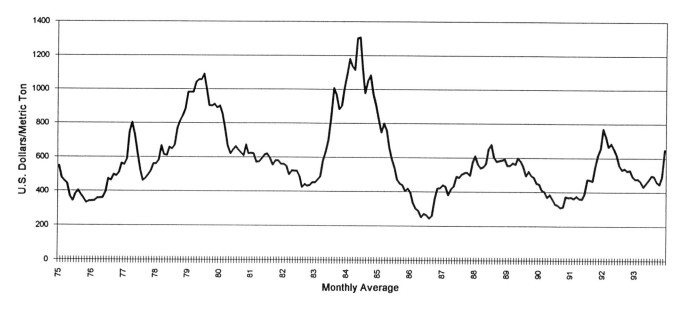

Fig. 1. Crude coconut oil prices CIF U.S. Ports. *Source:* Procter & Gamble..

industry for the next decade (10). The conclusion was that the production capacity of fats and oils can continue to supply the expanding world demand, with prices increasing less than the general price level.

## General Environmental Issues

Environmental issues have become an important factor in many nonfood applications during the past few years. There has been interest in "environmentally friendly" products—those that are typically viewed as being biodegradable or made from renewable, natural raw materials. If such a change occurred, it could, but not necessarily, create additional demand for certain agricultural products, including lauric oils.

"Life cycle inventories" are holistic studies that assess the total energy consumption and release of emissions to the environment as a result of a product's raw materials, processing, and logistics. A life cycle inventory conducted on surfactants by Pittinger *et al.* (11) has generally concluded that neither natural fat and oil derivatives nor petrochemical alternatives can be identified as the environmentally preferred option. Each has its own unique environmental profile that best serves to assist manufacturers in identifying areas for improvement in their chosen material rather than to claim environmental superiority. Such conclusions bring the formulation decision back to such factors as economics, consumer preference, and formulation compatibility.

## Nonfood Uses of Lauric Oils

Global nonfood consumption of laurics is currently estimated by one industry consensus to be between 50 and 60% of total lauric oil consumption of 4.6 million metric tons (MMT), the latter statistic from *Oil World* (1). With the exception of special confectionary fats, lauric oils have difficulty competing in large volume markets that generally use lower cost nonlauric oils, such as palm, soybean, and canola. These oils have the advantage of having highly mechanized agricultural production and processing systems versus the more labor-intensive systems for a crop such as coconut oil. (Palm kernel can have different economics, being a co- or by-product.) Also, the former are annual crops, and their production can be adjusted relatively quickly in response to market conditions.

## Current Applications

Before addressing the key economic issues, it may be helpful to recap the major nonfood uses of lauric oils briefly. Unfortunately, the lack of precise data makes it necessary to rely upon approximations.

*Soap.* It is estimated that about 600 thousand tons are used in soap. Where lauric oils are the high-cost fat versus tallow or other feedstocks, the economic justification is to provide lather. The market growth of traditional triglyceride-based sodium or potassium soaps varies from less than 1%/year in the United States to at least 10% in some developing countries. In Europe and the United States, formulations responding to consumer preferences for mildness and cosmetic attributes have tended to move away from pure soaps. The lauric oil content of the newer bars is generally 30–50% versus 15–40% in most traditional bars.

*Alcohol.* Medium-chain alcohols are believed to use about 1.75–2.0 MMT of lauric oils annually. This is the usage

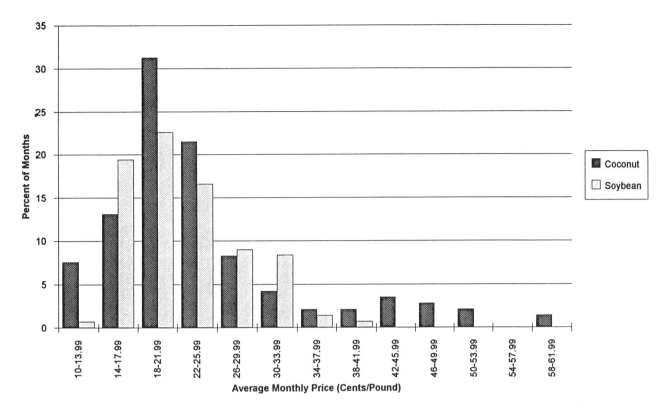

Fig. 2. Frequency distribution of coconut (CNO) and soybean oil (SBO) prices (CNO: CIF U.S. ports; SBO: cash Decatur) January 1982–December 1993. *Source:* Procter & Gamble.

class most vulnerable to hydrocarbon competition, yet it offers the greatest opportunity for expansion. Current market growth is believed to be 3–4%/year, about 40 thousand metric tons/year, equivalent to one new plant. There is increasing growth in sugar-based surfactants, such as alkyl polyglycosides.

*Fatty Acids and Methyl Esters.* The combined usage of lauric oils in fatty acids for nonsoap products and methyl esters for nonalcohol feedstocks is in the 100–200 thousand ton range. Methyl ester surfactant materials are very efficient, and they can use some of the lower cost triglyceride feedstocks (such as animal fat, palm oil products, and the long-chain fractions of lauric oils). These would mostly be in predominantly $C_{16}$–$C_{18}$-based formulations. While the world per capita production of animal fats is declining, there are high growth rates in palm oil production, an efficient way to produce $C_{16}$–$C_{18}$ if by-product sources are not adequate.

Methyl esters have been popular in China in the form of methyl ester sulfonates (MES), and may have large growth potential in the foreseeable future. In the Western world, glucose amides derived from methyl esters have been introduced in some dishwashing products.

## Market Potential

In 1991 Schirber (12) provided an interesting perspective of the size of the existing market supplied by petrochemicals, and, therefore, the size of the potential market for natural surfactants. A hypothetical change to all-natural products would require 80 new natural alcohol plants, each with a capacity of 40 thousand metric tons/year. It would also require about 4 MMT of natural oils—only slightly less than today's total world lauric oil production.

As is well publicized, recent additions to alcohol and methyl ester capacity are running well ahead of current usage rates, as discussed in Kuala Lumpur by Brunskill (13). The obvious challenge to market growth is to get end users to adopt the lauric based feedstocks.

## Economic Issues

The economic issues reviewed in this paper focus on the dynamics of lauric oil prices and their likely effect on consumption.

### Price Volatility

Historically, lauric oil prices have been among the more volatile of agricultural commodity prices. The causes of this volatility are well known: relatively inelastic demand in the short run, and the largely uncontrollable swings in production. With tree crops, the lead time to increase production is multiyear versus 1 year or less for annual crops. Once production comes on stream, it cannot be interrupted (except to leave the crop unharvested) and has relatively low out-of-pocket costs to harvest and process.

**TABLE 1**
**Philippine Copra Production (Thousand Metric Tons)**

| Calendar Year | Production | Change vs. Previous Year Quantity | Percent |
|---|---|---|---|
| 1980 | 2143 | — | — |
| 1981 | 2335 | +192 | +9 |
| 1982 | 2170 | −165 | −3 |
| 1983 | 2025 | −135 | −6 |
| 1984 | 1335 | −690 | −5 |
| 1985 | 1874 | +539 | +40 |
| 1986 | 2739 | +865 | +46 |
| 1987 | 2419 | −320 | −12 |
| 1988 | 1746 | −673 | −28 |
| 1989 | 1710 | −36 | −2 |
| 1990 | 2345 | +635 | +37 |
| 1991 | 1950 | −395 | +17 |
| 1992 | 1860 | −90 | −5 |
| 1993 est. | 1860 | 0 | 0 |

*Source: Oil World.*

Obviously, it is important for the cost of a given material, with value adjustments for any differences in performance, to be competitive with alternate raw materials. A simplistic analysis may only look at the average price over many years. However, this approach ignores the price volatility that has been a deterrent for many current and, more importantly, potential users of lauric oils during the past few decades. Some examples of the effect are as follows:

- Current users look for substitutions when prices are high. Some users leave the market permanently, giving up their ability to return by making changes in equipment or product attributes. An example is a switch to synthetic $C_{12}$–$C_{14}$ alcohol. More damaging long-term is the actual reformulation of an end-product to nonlauric formulations. The price to buy back either of these markets is high.
- Potential new users are discouraged from entering the market with new formulations.
- Producers of alternative raw materials or processes are encouraged. Examples from history include: frost damage to Florida's orange crop triggered a popular artificially flavored orange drink named Tang; green coffee price spikes led to processing changes that produced more cups of coffee per pound of beans; and price spikes in coconut oil a few decades ago spurred the production and use of hydrocarbon-based $C_{12}$–$C_{14}$ surfactants.

On the supply side, prices can go too low to sustain proper production practices, for example, adequate fertilization, and investment to replace aging coconut trees with better varieties. A whipsaw effect develops as prices cycle from one unsustainable extreme to the other. To quote an old commodity-market adage, "High prices make low prices, and vice versa."

The psychological effect can magnify the actual impact. People dislike uncertainty; they want to have reasonable confidence in the future status of their business. Managers of publicly owned companies can be under pressure from the financial markets to produce predictable profits. There is even greater anxiety if there are artificial interventions in markets, for example, large-scale contract defaults or government-sanctioned export quotas. Business and investment managers appear willing to cope with natural disasters, such as drought, if they are not also faced with the threat of arbitrary interruptions. It has been demonstrated on numerous occasions that attempts to interfere with free markets are not successful.

## Supply and Demand Elasticity— Short vs. Long Term

Elasticities of supply and demand are a function of time as well as price. Producers and consumers may not respond instantaneously to changes in price, either up or down. Even higher prices will often be paid for a period of time, as lower prices might be accepted. However, over time, more and more people work to make changes. What initially appears to be inelastic becomes elastic. This long-term elasticity is probably larger than can be assessed by economists or academic analyses that typically overlook the missed opportunities that do not appear in the statistical data.

The production of most agricultural crops is inherently unpredictable in the short term because of weather, disease, and insects. The effect of drought on sharply reducing coconut production (and oil palm fruit) is well known (Table 1). Compounding the problem, however, is the surge of copra yields to above-normal levels following what was apparently an enforced resting period for the trees. This abnormally high output comes into the market when consumption has been reduced by the high prices triggered by the earlier production shortfall. Conversely, withdrawal of good production practices when prices are low can cause delayed cuts in production just when the cycle is in the tight supply phase.

Variable production causes world inventories to fluctuate widely, requiring physical storage of the copra, palm kernels, or the respective oils. Fortunately, lauric oils are relatively easy to store from the quality standpoint. This allows the "surplus" from a bumper crop year to be carried until consumption recovers or there is a crop shortfall. However, it is costly to store and finance. Oil purchased for only $300/metric ton can increase the cost-to-usage twofold over 4–5 years. It is obvious that some smoothing of the supply flow helps the logistic efficiency of both producers and users.

Outside the lauric oil complex, where prices are less volatile, annual oilseed crops serve to adjust the supply picture in the short to medium term. Acreage adjustments are made depending not only on total demand, but in response to the less controllable output of tree crops and the by- and co-product materials. This could be a model for lauric oils.

## Market Price Risk and Hedging

As stated earlier, price variability can be an even bigger economic issue when the manufacturer competes against a different class of raw materials. Lauric oil users can encounter a major competitive divergence if their competition uses hydrocarbons, such as ethylene, or nonlauric triglycerides, such as animal fat. By contrast, a cooking oil made from soybean oil that competes against a cottonseed oil product will have a crude oil cost that tracks fairly well in most years.

Lauric oils are one of the more difficult commodities to protect, that is, hedge, from changes in market value. During the past decade, prices have ranged from about $225 to about $1300/metric ton CIF Europe or the U.S. Until a few years ago, there was a large "dealer market" in crude lauric oils which provided producers and users a reasonably effective market to hedge modest quantities for several months. However, this activity has declined radically and now offers only marginal opportunities. There has been no large-volume, organized futures market to facilitate risk transfer as is available for the soybean commodities, corn, precious metals, currencies, or hydrocarbons.

The palm kernel oil futures market in the Kuala Lumpur Exchange is new and yet to develop the high level of needed trading liquidity. For perspective, the CBOT volume in soybeans was about 15 times the size of the United States and South American crops (14). The Kuala Lumpur Exchange's trade in palm oil was about equal to the size of the Malaysian palm oil crop. Call and put options provide additional risk-management tools for the major commodities that permit producers and users of some commodities to bracket their price risk, albeit at an "insurance" cost. The concepts behind these options may have applications to lauric oils—again both to protect the producers on short-term production as well as to help producers and manufacturers of products to build long-term markets.

Numerous attempts to develop "steady state" lauric oil prices have not been totally successful. With a flat price in a cyclical market, one side inevitably gains for a period of time while the other side falls behind. There does not appear to be enough staying power to ride through the entire price cycle and realize the benefits of better market penetration.

## New Market Developments

As mentioned earlier, there have been developments in recent years that appear to be tempering the volatility in lauric oil prices.

- The expansion in palm kernel oil production, in total volume, share of total lauric oil production, and geographic diversification, has provided protection against adverse weather and normal yield variation of coconuts. During the 5-year period, 1978–82, PKO constituted 17% of total world CNO/PKO production (Table 2). That share increased to 29% in the 1987–88 crop year and to 36% in 1992–93 and is projected to continue expanding.

**TABLE 2**
**Total World Production of Coconut Oil (CNO) and Palm Kernel Oil (PKO) (Thousand Metric Tons)**

| October/September, by Year | CNO | PKO | Total | PKO Share of Total |
|---|---|---|---|---|
| Avg. 1978–82 | 2718 | 545 | 3263 | 17% |
| 1987–88 | 2851 | 1144 | 3995 | 29% |
| 1992–93 | 2983 | 1655 | 4638 | 36% |

Source: Oil World.

- During the past decade, Indonesia trade policy has increasingly allowed world market prices to determine how much of its lauric oil production goes for internal food use versus exportation. When coconut oil prices are at significant premium to palm oil or palm olein, coconut oil is exported and palm oil products are used in cooking. When prices are low, coconut oil is used domestically. Coconut oil exports have ranged from 6 thousand tons in 1986 to 351 thousand tons in 1992 (Table 3).

This increased response to market incentives, in addition to increasing Indonesia's foreign exchange earnings, has helped stabilize the supply of lauric oils available to the customers of the lauric-producing countries. As cited by Brookins, president of World Perspectives, Inc. "The lower the level of trade ... the greater the price shock to the world from changes in production" (15).

Before leaving Indonesia, it is important to note an interesting interplay in the statistical supply–demand balances. An increase in Indonesian lauric oil exports generally occurs with a decrease in total world consumption (Table 4). The higher exports are largely in response to higher prices triggered by a decline in production outside Indonesia. The reduction in Indonesian internal consumption generally leads to a net decline in world usage.

As nonfood users see such adjustments occurring without a major squeeze in the market, they will have more confidence in making lauric oils a part of their long-term future.

**TABLE 3**
**Indonesian Exports of Coconut Oil (Thousand Metric Tons)**

| January/December | Exports |
|---|---|
| 1984 | 35 |
| 1985 | 192 |
| 1986 | 6 |
| 1987 | 118 |
| 1988 | 207 |
| 1989 | 192 |
| 1990 | 194 |
| 1991 | 198 |
| 1992 | 351 |
| 1993 est. | 260 |

Source: Oil World.

## Other Substitutions

There are other adjustment mechanisms in world markets that keep the supply and usage between the food and nonfood sectors in balance. One example has implications for the competition between $C_{12}$–$C_{14}$ and $C_{16}$–$C_{18}$ materials for surfactant feedstocks. As the demand for palm olein for cooking oil markets increases, more palm oil stearin (about one-third of the fractionation yield) is produced as a by-product. Typical of by-products, it will be used; that is, priced to clear the market. Some of the stearin can go into food, but the pattern to date has been for a substantial portion to go to nonfood use, such as a replacement for tallow in soap, fatty alcohols, methyl esters, and methyl ester sulfonates. It should be noted that unsaturated $C_{16}$–$C_{18}$ fatty acids from vegetable and animal triglycerides, or the long-chain fractions from lauric oils, can also substitute for lauric oils in some applications, such as bar soaps. These interactions of the production and consumption of natural fats and oils, are examples of how they are all in a closed system. These commodities represent one of the freest markets in the world today. They have enjoyed a steady growth in consumption of 2–3%/year.

## Coconut Oil vs. Palm Kernel Oil

There can be substitution between coconut oil and palm kernel oil. The criteria on which is used in a specific situation can be complex. The decision is not based simply on which has the lowest delivered price. The lower acid and saponification values of PKO can make it worth somewhat less than coconut oil. Conversely, PKO can offer extra value through color, odor, processability, or other attributes. Also, refined, bleached, deodorized (RBD) PKO is often available at a smaller cost upcharge than for RBD CNO, reflecting the preferential tax treatment of further processed forms of palm products.

Other formerly minor differences are becoming increasingly important. One example is the higher phenol content in coconut oil from copra dried over an open fire. The World Bank and some bilateral donors are supporting work on this issue through developing better methods of drying and processing copra. Incidentally, phenol is also seen at low levels in some PKO, apparently as a result of charring the kernels by excessive temperatures in the indirectly heated dryers.

If the oils are used in fractionated products, it is important to get the maximum value from the use or sale of each fraction. Chain-length balancing is critical. For example, the higher content of $C_8$–$C_{10}$ in CNO versus PKO can make coconut oil worth more when $C_8$–$C_{10}$ product values are high. Prices of $C_{-8}$–$C_{10}$ fatty acids have ranged from about $775–2000/metric ton during the past 10 years. The lows were in mid to late 1980s when there was a surplus of $C_8$–$C_{10}$ by-products. Part of the improved demand for $C_8$–$C_{10}$ since then has been from specialty lubricants.

## Conclusion

In summary, lauric oils are good quality products with special characteristics. There is potential to expand nonfood

TABLE 4

Interaction of Coconut Oil (CNO) Production, Exports, and Consumption, with Palm Oil Prices October/September, 1991–92 vs. 1990–91 (thousand metric tons)

|  | October/September | | 1991–92 vs. 1990–91 | |
| --- | --- | --- | --- | --- |
|  | 1990–91 | 1991–92 | (Thousand Tons) | Percent |
| World CNO production excluding Indonesia[a] | 2323 | 2087 | −236 | −10% |
| CNO price vs palm oil price (Crude, CIF Rotterdam, U.S. $/metric ton)[b] | +$35 | +$272 | +$187 | +534% |
| Indonesian CNO exports[c] | 151 | 332 | +181 | +120% |
| Indonesian internal CNO consumption[d] | 682 | 433 | −249 | −37% |
| Indonesian internal palm oil consumption[e] | 1203 | 1558 | +355 | +30% |
| Total world CNO consumption[f] | 3302 | 2840 | −462 | −14% |
| Indonesian CNO production[g] | 843 | 758 | −85 | −10% |

[a]CNO production outside Indonesia declined.
[b]CNO prices versus palm oil increased.
[c,d,e]Indonesia increased CNO exports (c) and reduced internal consumption (d), replacing with palm oil (e).
[f,g]This all occurred even though total world CNO consumption declined (f) and Indonesia CNO production was down (g).
Source: Oil World Annual 1993.

uses, for example, the potential to gain market share versus petrochemical feedstocks if there is reasonable confidence in supply and price stability.

Uncertainty of both price and supply availability may be the biggest challenge for both producers and consumers. More stability should expand consumption over time. A time horizon of perhaps 5–10 years is needed.This may be about the minimum to justify investment in new production, to develop new formulations, design processes, build plants, and to establish a position in product markets.

The transition toward a more diverse and market responsive supply base, in the form of larger palm kernel oil production and Indonesian coconut oil exports, has been a major improvement. Additional progress is needed, such as smoothing the new Philippine coconut production. An annual crop to help temper the production cycle should encourage commitment to more usage. The development of hedging tools to permit transfer of the market price risk would also be helpful. As with any business producing a quality product, there needs to be continual effort to improve the efficiencies in production and processing.

## References

1. *Oil World,* ISTA Mielke GmbH, Hamburg, Germany (Weekly publication and various annual summaries).
2. Knaut, J., and H.J. Richtler, in *Proceedings of the World Conference on Oleochemicals into the Twenty-first Century*, edited by T.H. Applewhite, The American Oil Chemists' Society, Champaign, IL, 1991, pp. 27–38.
3. Kaufman, A.J., and R.J. Ruebusch, in *Proceedings of the World Conference on Oleochemicals into the Twenty-first Century,* edited by T.H. Applewhite, The American Oil Chemists' Society, Champaign, IL, 1991, pp. 10–25.
4. Edwards, I.S., *Southeast Asia Oleochemicals: Growth and Production.* Proceedings of the Annual Convention of the National Institute of Oilseed Products, Tucson, AR, 1993.
5. Sampson, R.L., *New Farm and Forest Products: Responses to the Challenges and Opportunities Facing American Agriculture,* Report of The New Farm and Forest Products Task Force to The Secretary, U.S. Department of Agriculture, Washington, D.C., June 25, 1987.
6. Princen, L.H., *Econ. Botany 36*:302 (1982).
7. *Industrial Uses of Agricultural Materials: Situation and Outlook Report.* Economic Research Services, ERS-NASS, 341 Victory Dr., Herndon, VA 20070.
8. McCoy, R.J., *Fats and Oils—Renewable Resources for the Chemical Industry,* Proceedings of the Chemical Marketing Rescarch Association Meeting, February, 1989.
9. Avery, D.T., *Trends and Turmoil in World Oilseed Economics,* Proceedings of the Annual Convention of the National Institute of Oilseed Products, Rancho Mirage, California, March, 1988.
10. Sparks Companies, Inc. *The Future Structure of the Global Oilseeds Industry,* (A proprietary, multiclient study) March, 1992.
11. Pittinger, C.A., J.S. Sellers, D.C. Janzen, T.M. Rothgab, and M.L. Hunnicutt, *J. Am. Oil Chem. Soc. 70* (1993).
12. Schirber, C.A., *INFORM 2*:1062 (1991).
13. Brunskill, A., *International Trade in Oleochemicals: Issues and Prospects,* PIPOC Conference, Kuala Lumpur, September, 1993.
14. *The Knight Ridder CRB Commodity Yearbook 1993;* John Wiley & Sons, Inc. New York, 1993.
15. Brookins, C., *A World Agricultural Trade Perspective: World Agriculture at the Crossroads, Asian Agricultural Production, May 1990;* World Perspectives, Inc., Washington, D.C.

# Discussion

The Chairperson commented briefly on the earlier meeting with the President and the concerns for improved productivity in the coconut complex. He particularly stressed the problems of improved quality and reduced price volatility.

Continuing the discussion on quality, there was a question regarding the rights customers have to access the records of storage companies if oil had deteriorated in storage. It was noted that the customer does have access to the records and samples. However, with respect to legal actions, it was stressed that discussions and reliance on insurance plans were better routes than litigation.

In another area, there was a question as to what the upper limit of oil yield is with a screw press. With palm kernel oil, the upper limit was said to be 42%. If higher pressures are attempted, there is a risk of heat damage to the products.

There was a question as to why there is a trend in the United States to buying refined, bleached, and deodorized oils in place of crude, because there are no real problems in processing crude oils. The response was that this is strictly a case of economics.

As for price volatility, there was a question as to how much volatility was acceptable. It was noted that there is not a precise range, but it is mainly a matter of timing of the price cycles.

# Identifying New Sources of Coconut Oil

William G. Padolina

University of the Philippines at Los Baños; Department of Science and Technology, DOST Compound, Bicutan, Taguig, Metro Manila, Philippines

## Abstract

The Philippines still ranks as one of the countries with the largest land area planted to coconut, with 3 million hectares planted to 300 million coconut trees. Coconut oil and other coconut products still rank among the top ten export products of the Philippines. However, the productivity of coconut farms has declined, and there is now a vigorous effort to revitalize the Philippine coconut industry. Thus, it is important when replacing senile coconut trees with new planting materials that superior or elite breeds be introduced in order to improve the competitiveness of coconut oil in the world market. This paper will discuss the recent advances made towards the identification of the elite lines, especially in terms of the oil yield and the quality of the oil as reflected in the fatty acid profile.

---

The importance of the coconut tree for edible and nonedible applications has elicited renewed interest in the search for new cultivars. This paper describes the work that has been conducted towards the breeding and selection of elite cultivars with superior characteristics such as high oil yields and improved oil quality. Coconut oil remains the main export item of the Philippines, therefore, this paper also focuses on the characteristics of the oil.

## Objectives of Coconut Breeding

Coconut-breeding work in the Philippines has always been guided by the following objectives (1): increased production of nuts and copra at different levels of soil fertility, improved quality of oil and increased protein content, and improved resistance to environmental stress and outbreak of pests and diseases. The institutions which are active in coconut-breeding work are the University of the Philippines at Los Baños, Visayas State College of Agriculture, Philippine Coconut Authority, Twin Rivers Research Center, and Matling Plantation. Their breeding work has been oriented towards the search for cultivars which are precocious and give high yields under a wide range of environment and management levels (Carlos, 1983, unpublished). As in any coconut-breeding program, the main activities are identification and collection of cultivars, evaluation and selection of cultivars, and seed production (Carlos, 1983, and Santos, private communication, 1984).

## Genetic Resources in Coconut

Sangalang, (private communication 1986), and Nair (2) identified common varieties used in coconut breeding: *Typica*—tall palms which are the most commonly cultivated in the coconut-producing countries. They grow about 30 m tall, have comparatively long prebearing ages, and are normally cross-pollinated to produce fruits which are medium to large in size; *Javanica*—dwarf palms, small-to-large seeded, and early bearing cultivars, usually with medium-sized fruits; and *Nana*—dwarf palms, short stature, smaller nuts of varying colors (green, yellow, and orange), prolific, and early bearing cultivars, usually planted as ornamentals or as a source of fresh nuts for refreshments. Following the procedure cited earlier, the breeding materials are derived from these varieties and hybridized.

The two biggest coconut genetic collections in Asia are located in the Philippines and in India. Activity in coconut genetic collections, conservation, and improvement conducted by the Philippine Coconut Authority during the last 15 years resulted in the collection of 82 different coconut ecotypes/varieties and populations from various sources (3). A review of these sources was made by Padolina (4) based on the unpublished work of Carlos in 1983.

The coconut germplasm acquisitions maintained at the Central Plantation Crops Research Institute, Kasaragod, India, are comprised of 41 indigenous and 86 exotic cultivars. The exotic collections consist of 72 talls, 12 dwarfs, one semitall, and one hybrid and come from 22 countries (2).

Romney and Dias (5) reported on coconut varieties in the Bahia State in Brazil. The tall varieties were identified as similar to the Jamaican tall and the dwarf varieties to the Nyior gading.

According to Carpio (6), the present system of coconut classification in the Philippines is based primarily on palm morphology and pollination characteristics. Other criteria that could be considered are mode of pollination, growth and maturation rate, composition of principal fruit components, yield, and other morphological characters (Santos, unpublished, 1983).

## Recent Advances in the Search for Elite Varieties

Nut components and copra yield are the most important characteristics used in assessing performance of planting materials (7). In fact, these two characteristics are critical in evaluating the oil yield from the materials.

**TABLE 1**
Weight of Whole Nut and Principal Component Character in Four Types of Coconut Genetic Material (9)

| Type | | Whole Nut (g) | Husk (g) | Shell (g) | Meat (g) | Water (g) | Copra (g) |
|---|---|---|---|---|---|---|---|
| Dwarf | Mean | 1271 | 412 | 171 | 411 | 281 | 227 |
| | Range | 792–1864 | 250–606 | 96–253 | 234–623 | 156–481 | 133–354 |
| | SD | 402 | 130 | 49 | 136 | 111 | 81 |
| | c.v. (%) | 32 | 32 | 29 | 33 | | |
| Tall | Mean | 1766 | 466 | 268 | 561 | 471 | 317 |
| | Range | 1158–2672 | 306–667 | 184–380 | 399–776 | 269–938 | 240–418 |
| | SD | 421 | 106 | 54 | 105 | 186 | 50 |
| | c.v. (%) | 24 | 23 | 20 | 19 | 39 | 16 |
| Dwarf × Tall | Mean | 1464 | 429 | 217 | 470 | 348 | 269 |
| | Range | 1107–2052 | 323–528 | 163–297 | 341–646 | 198–582 | 209–349 |
| | SD | 288 | 53 | 37 | 99 | 132 | 44 |
| | c.v. (%) | 20 | 12 | 17 | 21 | 38 | 16 |
| Dwarf × Dwarf | Mean | 1414 | 420 | 191 | 462 | 341 | 242 |
| | Range | 1354–1475 | 416–424 | 182–203 | 443–480 | 313–368 | 231–252 |
| | SD | 86 | 6 | 15 | 26 | 39 | 15 |
| | c.v. (%) | 6 | 1 | 8 | 6 | 12 | 6 |

*Abbreviation:* c.v., coefficient of variance.

**TABLE 2**
Performance of Coconut Hybrids Released in India (11)

| Hybrid | Parentage | Nut Yield Palm/yr. No. | Copra Yield Mean/Nut (g) | Copra Yield Mean/Palm (kg) | Copra/ha (t) | Oil Content (%) | Released by |
|---|---|---|---|---|---|---|---|
| Chandra Sankara | COD × WCT | 116 | 215 | 24.94 | 4.36 | 68 | CPCRI |
| Kera Sankara | WCT × COD | 108 | 187 | 20.20 | 3.54 | 68 | CPCRI |
| Chandra Laksha | LO × COD | 109 | 195 | 21.26 | 3.72 | 69 | CPCRI |
| Laksha Gabga (PHC-1) | LO × GB | 108 | 195 | 21.06 | 3.69 | 70 | KAU |
| VHC-1 | ECT × DG | 98 | 135 | 12.23 | 2.32 | 70 | TNAU |
| VHC-2 | ECT × DG | 107 | 152 | 16.26 | 2.85 | 69 | TNAU |
| Ananda Ganga (PHC-2) | ECT × MYD | 95 | 216 | 20.52 | 3.59 | 68 | KAU |
| Kera Ganga (PHC-3) | WCT × GB | 100 | 201 | 20.10 | 3.52 | 69 | KAU |
| WCT | | 80 | 176 | 14.08 | 2.46 | 68 | — |

*Abbreviations:* CPCRI, Central Plantation Crops Research Institute; KAU, Kerala Agricultural University; TNAU, Tamil Nadu University.

## TABLE 3
**Basic Features of Coconut Hybrids Recommended for Release in the Philippines (11)**

| Features | PCA 15-1 | PCA 15-2 | PCA 15-3 |
|---|---|---|---|
| Parentage | CAT x LAG | MRD x TAG | MRO x BAY |
| Age at first flowering (yrs.) 3–4 | 3–4 | 3–4 | 3–4 |
| No. of bunches/yr. | 14 | 15 | 14 |
| No. of nuts/palm/yr. | 81 | 75 | 63 |
| Copra/nut (g) | 266 | 296 | 277 |
| Copra/palm/yr. (kg) | 22.0 | 19.0 | 15.0 |
| Copra/ha/yr. (t) mean | 3.0 | 2.5 | 2.0 |
| Copra/ha/yr. (t) (1989) | 4.0 | 5.2 | 4.9 |

Santos et al. claim that the weight of copra per nut is heritable from parent trees. Since actual copra making is tedious and time consuming, conversion factors have been used to estimate copra recovery. It was found that the weight of the fresh meat or split nut is a more reliable basis for estimating copra yield (8).

Carpio used protein and isoenzyme electrophoresis of pollen and coconut meat, pollen morphology by SEM, and amino-acid analysis of pollen and coconut meat to detect variation among populations. Significant differences were not observed, but pollen-protein electrophoresis provided some discrimination and the peroxidase isoenzyme analysis showed some variation among populations (6).

In addition to nut characteristics and high yields, precocity, resistance to pests and diseases, long economic lifespan, good root/shoot ratio to guarantee stability during typhoons and drought, and high harvest index are desired characteristics (9). Louis and Chopra (10) reviewed the use of selection indices in coconut palm.

Table 1 summarizes the variations in nut characteristics encountered during the handling of different types of genetic materials in the Philippines. The tall and dwarf varieties show distinct variation. Furthermore, interpopulation variation is higher among dwarf material than in tall varieties. Santos (9) also notes that the intrapopulation variation in number of bunches and nuts per harvest is high in certain dwarfs, such as the orange, yellow, and green dwarfs and *catigan*. It was also found that it was difficult to ascertain genetic differences among local talls. Tall populations cross-pollinate more often than dwarfs, and show a higher viability in nut yield and in the number of palms with harvestable nuts when compared to dwarfs and hybrids. In general, talls are superior to dwarfs both in quantity and quality of copra per nut, but not in total nut production per palm (9).

Also, it was found that the number of nuts has more influence on total copra than copra per nut (9). Thus, increasing the number of nuts is a preferred breeding strategy to attain increased oil yield. Early flowering materials are expected to be potentially high yielders, and a high copra yield is expected from plants with higher numbers of leaves, bunches, and fruits set at harvest (9).

## Hybrids

The early reproduction maturity and higher yield of coconut hybrids from the cross of a dwarf parent and a tall parent as experienced in the Philippines confirms the observations of other breeders (9). Since the discovery of hybrid vigor in coconuts in 1937, many hybrids have been released for cultivation in different coconut-growing countries, a review of which was published by Dhamodaran et al. (11).

The vast potential of coconut hybrids to increase productivity has been demonstrated in India, the Philippines, and the Ivory Coast (Tables 2–4). Indian

## TABLE 4
**Performance of Ivory Coast Hybrids (mean of 9–12 years' data [11])**

| Hybrid Combination | Name | Bunches/yr. | Nuts/Tree, per yr. | Copra/Nut (g) | Copra Tree/yr. (kg) | Copra/ha/yr. (t) |
|---|---|---|---|---|---|---|
| WAT | | 11.7 | 55 | 235 | 12.8 | 1.74 |
| MYD x PYT | PB-122 | 13.8 | 104 | 253 | 26.3 | 3.57 |
| MRD x PYT | PB-132 | 14.0 | 95 | 282 | 26.7 | 3.63 |
| MYD x WAT | PB-121 | 14.5 | 104 | 247 | 25.8 | 3.50 |
| WAT x VITT | PB-214 | 13.6 | 114 | 209 | 24.0 | 3.26 |

**TABLE 5**
Fatty Acid (FA Distribution in the Oils from Four Philippine Coconut Varieties (13)

| FA | Niño | Golden | Laguna | Zamboanga |
|---|---|---|---|---|
| $C_6$ | 0.70 | 0.58 | 0.60 | 0.55 |
| $C_8$ | 9.84 | 8.83 | 7.65 | 10.97 |
| $C_{10}$ | 6.64 | 5.97 | 5.33 | 6.65 |
| $C_{12}$ | 50.28 | 51.50 | 47.34 | 52.67 |
| $C_{14}$ | 17.87 | 17.54 | 19.92 | 16.09 |
| $C_{16}$ | 7.03 | 6.93 | 9.12 | 5.52 |
| $C_{18}$ | 3.37 | 4.39 | 5.05 | 3.69 |
| $C_{18:1}$ | 3.94 | 3.70 | 4.41 | 3.45 |
| $C_{18:2}$ | 0.34 | 0.56 | 0.59 | 0.41 |

hybrids could yield from 2.32–4.38 t copra/ha/yr. In 1989, the hybrids released by the Philippine Coconut Authority achieved yields of 4.0–5.2 t copra/ha/yr, and the Ivory Coast hybrids yielded from 1.74–3.63 t copra/ha/yr. Dhamodaran et al. (11) further report that Indonesian hybrids yielded approximately 3 t copra/ha/yr.

In another review, it was observed that it has been difficult to effect population improvement in the yield of coconut by the present methods of breeding (12). Coconut is a highly cross-pollinated crop, has a long prebearing period, and takes a long time to attain yield stability.

## Recent Advances in Identifying Elite Coconut-Planting Materials

Having laid the basis and the general observations for coconut-breeding work, the exploitation of indigenous genetic resources is an important approach to the generation of elite lines. However, indigenous genetic resources cannot be utilized unless they are well characterized. Although much of the information is available on nut characteristics and copra yield, there is an increasing need to include more detailed analyses including the fatty acid profile of the oil, the triglyceride profile of the oil, protein and carbohydrate content of the meat, and other chemical characteristics.

Banzon and Resurreccion (13) analyzed the fatty acid composition of oils from four "types" of Philippine coconuts—coco niño, golden coconut, Laguna, and Zamboanga (Table 5). The Zamboanga type had the highest amounts of $C_8$, $C_{10}$, and $C_{12}$ fatty acids but had the lowest $C_{14}$, $C_6$, and $C_{18:1}$. The Laguna type had the highest amounts of $C_{14}$, $C_{16}$, $C_{18}$, $C_{18:1}$, and $C_{18:2}$ and the lowest in $C_8$, $C_{10}$, and $C_{12}$.

Rossell et al. (14) analyzed coconut-oil samples from the Philippines, Papua New Guinea, Vanuatu, North Sulawesi, and Sri Lanka (Table 6). They noted that the fatty acid compositions of the coconut-oil samples showed small variations with sample origin, with the main difference being relatively lower levels of unsaturated fats in the two samples of oil extracted from desiccated coconut from Sri Lanka. Desiccated coconut meat no longer contains the testa or paring; the oil from which contains high levels of unsaturated fatty acids. The highest lauric acid content was found in the Sri Lankan samples and the lowest from North Sulawesi. High levels of the medium-chain fatty acids $C_8$ and $C_{10}$ were found in the samples from Vanuatu.

The triglyceride carbon numbers were also analyzed by Rossell et al. (14) and are shown in Table 7. The predominant triglyceride carbon numbers were $C_{36}$, $C_{38}$, and $C_{40}$, which comprised around 51.6% of all the triglycerides.

In what appears to be the most extensive work in the proximate and fatty analyses of coconut cultivars, Del Rosario et al. (15) provided information by analyzing 32 coconut cultivars and hybrids, of both Philippine and exot-

**TABLE 6**
Distribution of Fatty Acid Composition with Sample Origin for Coconut Oil (14)

| Fatty Composition (wt%) Origin | # of Samples | $C_6$ | $C_8$ | $C_{10}$ | $C_{12}$ | $C_{14}$ | $C_{16}$ | $C_{18:0}$ | $C_{18:1}$ | $C_{18:2}$ | $C_{20:0}$ | $C_{20:1}$ |
|---|---|---|---|---|---|---|---|---|---|---|---|---|
| Philippines | 11 | 0.4–0.6 | 7.1–8.3 | 6.2–6.6 | 46.2–43.7 | 18.0–19.2 | 8.3–9.5 | 2.3–3.2 | 5.6–7.1 | 1.3–2.1 | 0.1–1.2 | $t^b$–0.1 |
| Papua, New Guinea | 4 | 0.4–0.6 | 6.9–7.5 | 6.4–6.8 | 47.1–50.3 | 17.8–18.1 | 8.3–9.3 | 2.3–3.0 | 5.8–6.3 | 1.4–1.6 | 0.1 | $t^b$ |
| Vanuatu | 5 | 0.5–0.6 | 7.3–9.4 | 6.6–7.8 | 47.1–48.4 | 16.8–17.9 | 7.7–9.1 | 2.3–2.9 | 5.1–6.5 | 1.4–1.6 | $t^b$–0.2 | $t^b$–0.2 |
| North Sulawesi | 1 | 0.5 | 7.3 | 6.3 | 45.9 | 18.1 | 9.7 | 2.7 | 7.4 | 1.9 | 0.1 | 0.1 |
| Sri Lanka | $2^a$ | 0.4–0.5 | 7.3–7.4 | 5.9–6.2 | 18.6–19.6 | 18.6–19.6 | 7.0–7.9 | 2.9 | 4.5–5.3 | 0.6–0.9 | 0.1 | $t^b$ |
| Overall range | | 0.4–0.6 | 6.2–7.8 | 6.2–7.8 | 45.9–50.3 | 16.8–19.2 | 7.7–9.7 | 2.3–3.2 | 5.4–7.4 | 1.3–2.1 | $t^b$–0.2 | $t^b$–0.2 |
| Mean (21 samples) | | 0.5 | 6.7 | 6.7 | 47.5 | 18.1 | 8.8 | 2.6 | 6.2 | 1.6 | 0.1 | $t^b$ |

[a]These results relate to dessicated coconut and were not used in the calculation of ranges and means. All other oil samples were extracted from samples of commercial copra.
[b]Trace (unquantified amount 0.05%). Traces of $C_{18:3}$ were found in all origins ranging up to 0.2% in one Vanuatu sample.

**TABLE 7**
**Triglyceride Carbon Number Composition (wt%) of Coconut Oil (14)**

| Triglyceride Carbon Number | Coconut Oil from Copra of All Origins | |
|---|---|---|
| | Range (%) | Mean (%) |
| $C_{28}$ | 0.7–1.0 | 0.8 |
| $C_{30}$ | 2.8–4.1 | 3.4 |
| $C_{32}$ | 11.5–14.4 | 12.9 |
| $C_{34}$ | 15.6–17.6 | 16.5 |
| $C_{36}$ | 18.3–19.8 | 18.8 |
| $C_{38}$ | 15.1–17.7 | 16.3 |
| $C_{40}$ | 9.2–11.1 | 10.2 |
| $C_{42}$ | 6.5–8.0 | 7.3 |
| $C_{44}$ | 3.6–4.6 | 4.2 |
| $C_{46}$ | 2.1–3.0 | 2.6 |
| $C_{48}$ | 1.6–2.6 | 2.3 |
| $C_{50}$ | 0.8–2.0 | 1.7 |
| $C_{52}$ | 0.4–2.0 | 1.6 |
| $C_{54}$ | 0.1–1.5 | 1.2 |

ic origin (Tables 8–10). Protein content varied from 5.11–11.03%, with a mean of 7.03%. Fat content ranged from 42.46–68.39%, with a mean of 59.80%. Crude fiber content varied from 1.88–19.97%, and ash values had a range of 1.55–4.40% with a mean of 2.36%. Nitrogen-free extract had a low of 14.62% to a high of 38.27% with an average of 23.89%. Total free sugar analysis showed a range from 1.92–17.84% with 0.03–1.68% representing reducing sugars (Table 9).

The fatty acid composition of the different coconut cultivars and hybrids showed a lauric acid range of 40.98–58.09% (Table 10). It was likewise observed that as the amount of lauric acid increased, the levels of the other fatty acids dropped. In the medium-chain fatty acids, $C_6$ had a range of 0.71–2.89%; $C_8$ from 5.35–13.65%; and $C_{10}$ from 4.40–8.91%. Myristic acid levels ($C_{14}$) ranged from 12.88–22.79%, and palmitic acid ($C_{16}$) levels ranged from 5.47–11.97%. Stearic acid ($C_{18}$) had a low of 0.31 and a high of 9.25%. Oleic acid ($C_{18:1}$) ranged from 2.84–9.66%, and linoleic ($C_{18:2}$) from 0.38–2.32% (15). The variability observed here could have potential application in genetic engineering.

It must be noted however, with regard to the work of Del Rosario et al., that sampling was a bit limited, and it may be too early to derive correlations. However, there are interesting trends that should be studied further. Additionally, a more comprehensive fruit-component analysis, including triglycerides must be undertaken.

Gruezo (16) studied eight "wild-type" coconut populations in the Philippine provinces of Eastern Samar and Surigao del Norte. He found two morphologically distinct forms, namely, large-fruited and small-fruited. These seednuts were believed to have originated from the South Pacific, such as the Palau Islands.

## Recent Developments in Coconut Tissue Culture

Rillo (17) and Padolina (4) reviewed work on tissue and embryo culturing of the coconut, and noted the efforts exerted on what seems to be a recalcitrant plant. In the Philippines, there are four research groups active in coconut tissue-culture research, because of the need to produce the large amounts of planting material for the replanting program. A technique to collect immature inflorescence nondestructively for tissue-culture purposes was developed by Rillo (18).

In February 1989, the Philippine-German Project on Tissue Culture was initiated in the Albay Research Center of the Philippine Coconut Authority. This work was being conducted in cooperation with British researchers from Wye College who have been working on this problem since 1981. Some interesting results have been obtained, but plant regeneration has not been achieved.

Ebert et al. (19) reported on the first results of the Philippine-German Project on Coconut Tissue Culture. Good calloid formation has been routinely obtained with immature embryos and inflorescence explants, which were made to develop into embryoids. Shoot development as well as root formation has been reported.

It has recently been reported that in 1991, a CIRAD–ORSTOM research team obtained somatic coconut palm embryos from a number of different organs (young leaves, inflorescence) and for several different varieties. The Montpellier researchers are now improving the process for larger-scale operations (20).

## Biotechnology and the Coconut

The feasibility of producing an economically important fatty acid in a transgenic plant has now been demonstrated by Voelker et al. (21). Using plant genetic engineering, the research group headed by Voelker in Calgene has induced laurate production in the common wall cress (*Arabidopsis thaliana*), a weed that normally produces only long-chain fatty acids. A 12:0-acyl-carrier protein thioesterase (BTE) was cloned, resulting in the accumulation of lauric acid in the storage triacylglycerols. They also reported, but did not elaborate, on the transformation of rapeseed with the napin-BTC gene which also resulted in the accumulation of laurate. No further information on the extent of production of laurate in rapeseed was reported.

Schwitzer (22) discussed the application of biotechnology in the extraction and modification of coconut oil. Enzymes could be used to maximize the extraction of oil from fresh coconut meat. Likewise, the oil could be modi-

## TABLE 8
**Proximate Composition (% Dry Basis) of Some Coconut Cultivars and Hybrids (15)**

| | Proximate Analysis[a,b] | | | | |
|---|---|---|---|---|---|
| Cultivars | Ash | Crude Fat | Crude Protein | Crude Fiber | NFE |
| 1  Macapuno | 3.41[b] | 56.39[m] | 6.8[hij] | 4.68[kl] | 28.71[ed] |
| 2  Tacunan | 1.94[mn] | 66.30[cd] | 5.2[pq] | 3.63[lmn] | 22.92[g] |
| 3  Catigan | 1.73[p] | 52.74[d] | 8.98[d] | 6.64[hij] | 21.91[c] |
| 4  Bago-Oshiro | 1.65[q] | 52.95[p] | 10.45[ab] | 6.31[ij] | 28.55[cde] |
| 5  West African tall | 1.88[no] | 59.34[jk] | 6.72[hijk] | 9.09[ef] | 22.97[g] |
| 6  OD × (Tambolilid × Macapuno) | 3.06[d] | 58.74[k] | 5.46[nopq] | 10.20[de] | 22.55[g] |
| 7  San Ramon × YD | 2.56[f] | 60.67[i] | 5.33[opq] | 14.57[b] | 16.87[hij] |
| 8  Concepcion tall × OD | 2.30[ij] | 65.19[e] | 5.11[q] | 11.11[d] | 16.29[ijk] |
| 9  OD × (Laguna tall × Cocono) | 2.13[k] | 68.08[ab] | 5.26[opq] | 8.90[fg] | 15.63[jk] |
| 10 Rabanuel GD × OD | 2.54[f] | 57.43[l] | 5.39[gh] | 7.90[gh] | 26.74[def] |
| 11 Lanao GD × OD | 4.40[a] | 43.45[r] | 6.08[lm] | 7.80[ghi] | 38.27[a] |
| 12 Lanao GD × (San Ramon × Coconino) | 2.24[ij] | 60.16[ij] | 6.24[klm] | 5.33[jk] | 26.03[f] |
| 13 San Ramon × Pascual GD | 2.00[lm] | 63.97[fg] | 6.41[ijkl] | 4.47[klm] | 23.15[g] |
| 14 Pascual GD × OD | 2.2[j] | 62.11[h] | 7.54[efg] | 4.70[kl] | 23.43[g] |
| 15 Pascual GD × YD | 2.30[ij] | 65.14[e] | 5.73[mnop] | 3.89[klmn] | 22.94[g] |
| 16 Lanao GD × San Ramon | 3,140 | 55.44[n] | 8.66[d] | 8.66[d] | 28.44[cde] |
| 17 YD × Laguna Tall | 2.23[ij] | 63.33[g] | 6.29[jkl] | 6.29[jkl] | 25.56[f] |
| 18 YD × San Ramon | 2.04[l] | 57.61[l] | 7.67[ef] | 7.67[ef] | 26.48[ef] |
| 19 Lupisan × OD | 2.40[gh] | 56.98[lm] | 11.63[a] | 3.53[lmn] | 26.06[f] |
| 20 YD × (Makapuno × Ordinary Tambolilid) | 2.58[f] | 50.48[g] | 7.86[e] | 19.97[a] | 18.91[g] |
| 21 Laguna tall | 3.01[d] | 67.25[bc] | 6.16[klm] | 5.31[jk] | 18.27[hi] |
| 22 Tacunan × Bago-Oshiro | 1.55[r] | 57.11[lm] | 6.17[klm] | 6.87[hi] | 28.29[cde] |
| 23 Variety 109 | 1.81[o] | 54.52[o] | 9.55[c] | 4.10[klmn] | 30.02[c] |
| 24 Dua-Dua | 2.83[e] | 60.88[i] | 6.10[lm] | 4.25[klm] | 25.96[f] |
| 25 Nigeria | 2.31[i] | 60.45[i] | 6.88[hi] | 3.79[klmn] | 26.57[def] |
| 26 Tanganyika | 2.60[f] | 68.30[a] | 7.02[gh] | 7.46[hi] | 14.62[k] |
| 27 Variety 21 | 2.03[l] | 66.57[cd] | 10.29[b] | 4.02[klmn] | 14.92[jk] |
| 28 Variety 107 | 2.35[hi] | 66.53[cd] | 5.80[mno] | 7.28[hi] | 18.04[hi] |
| 29 Laguna Tall × Raganuel GD | 1.82[o] | 65.60[de] | 10.29[b] | 3.72[lmn] | 18.57[h] |
| 30 Catigan × Bago-Oshiro | 1.86[o] | 42.46[s] | 7.97[e] | 12.85[c] | 34.86[b] |
| 31 P91C | 2.44[g] | 64.65[ef] | 7.24[fgh] | 3.09[mno] | 22.58[q] |
| 32 Markham | 2.01[lm] | 68.39[a] | 5.97[lmn] | 1.88[o] | 22.08[fg] |
| Mean | 2.36 | 59.80 | 7.03 | 6.72 | 23.89 |

[a]Means are the average of three determinations.
[b]Means followed by the same letter are not significantly different at P = 0.05 (DMRT).
*Abbreviations:* Orange Dwarf, OD; Yellow Dwarf, YD; Green Dwarf, GD.

fied by lipase enzymes through hydrolysis and *trans*esterification. These methods, if developed fully, could be harnessed to produce medium-chain triglycerides by exploiting enzyme specificity.

Schwitzer also reports the potential of producing coconut-oil substitutes by microorganisms like *Candida utilis, Rhodotorula,* and *Entomophthera coronata*. The oil produced by *E. coronata* seems to be the closest to the coconut oil. Other microbially mediated transformations of coconut oil have been reviewed by Padolina (23).

## Conclusion

This review is by no means comprehensive as it attempts to describe recent advances in the production of coconut oil. There is wide variability in the coconut palms which may be harnessed to produce elite lines suitable for specific commercial purposes. Modern techniques of genetic manipulation may be valuable in future breeding work.

**TABLE 9**
**Sugar Content of Some Coconut Cultivars and Hybrids (15)**

| Cultivars | Total Sugar[a,b] (g/100g meat) | Reducing Sugar[a,b] (g/100g total sugar) |
|---|---|---|
| 1 Macapuno | 11.95[ab] | 0.42[hijklmn] |
| 2 Tacunan | 6.16[gh] | 0.62[efghil] |
| 3 Catigan | 7.10[fg] | 0.06[fghi] |
| 4 Bago-Oshiro | 3.92[i] | 0.2[bop] |
| 5 West African tall | 5.96[h] | 0.21[klmno] |
| 6 OD (Tambolilid × Macapuno) | 9.89[de] | 0.85[cdefg] |
| 7 San Ramon × YD | 9.74[de] | 0.62[defgh] |
| 8 Concepcion tall × OD | 7.13[f] | 0.97[c] |
| 9 OD × (Laguan tall × Coconino) | 6.98[fg] | 0.62[defgh] |
| 10 Rabanuel GD × OD | 11.29[bc] | 0.68[cdefgh] |
| 11 Lanao GD × OD | 7.06[fg] | 1.36[b] |
| 12 Lanao GD × (San Ramon × Coconino) | 2.61[jkl] | 0.89[cdef] |
| 13 San Ramon × Pascual GD | 2.60[jkl] | 0.54[ghij] |
| 14 Pascual GD × OD | 3.14[ij] | 0.47[hijkl] |
| 15 Pascual GD × YD | 2.39[kl] | 0.46[hijklm] |
| 16 Lanao GD × San Ramon | 3.28[ij] | 0.9[cde] |
| 17 YD × Laguna Tall | 6.60[fgh] | 0.37[hijklmn] |
| 18 YD × San Ramon | 3.58[i] | 0.50[hijk] |
| 19 Lupisan × OD | 9.67[e] | |
| 20 YD (Macapuno × Tambulilid) | 10.30[de] | 0.20[klmno] |
| 21 Laguna Tall | 3.41[i] | 0.37[hijklmn] |
| 22 Tacunan × Bago-Oshiro | 6.30[gh] | 0.48[hijk] |
| 23 Variety 109 | 11.97[ab] | 0.83[cdefg] |
| 24 Dua-Dua | 17.84[m] | 0.16[lmno] |
| 25 Nigeria | 10.45[de] | 9.47[hijkl] |
| 26 Tanganyika | 12.25[a] | 0.15[mno] |
| 27 Variety 21 | 10.56[cd] | 0.22[klmno] |
| 28 Variety 107 | 8.17[o] | 1.63[ab] |
| 29 Catigan × Bago-Oshiro | 1.92[l] | 0.03[q] |
| 30 Laguna Tall × Rabanuel GD | 14.72[n] | 0.12[o] |
| 31 P91C | 12.71[a] | 0.30[ijklmno] |
| 32 Markham | 10.15[de] | 1.68[a] |

[a]Results are the average of three determinations.
[b]Means followed by the same letter are not significantly different at P < 0.05 (DMRT).
*Abbreviations:* Orange Dwarf, OD; Yellow Dwarf, YD; Green Dwarf, GD.

# References

1. Bernardo, F.A., Coconut breeding: objectives, methods, strategies and policies. *Proceedings of the National Coconut Research Symposium*, Tacloban City, Philippines, (1975).
2. Nair, M.K., *Indian Coconut J.* August: 17 (1992).
3. Santos, G.A. *Phil. J. Coconut Stud. 15*:16 (1988).
4. Padolina, W.G., *J. Am. Oil Chem. Soc. 62*:206 (1985).
5. Romney, D.H., and B.C. Dias, *Phil. J. Coconut Stud. 6*:30 (1981).
6. Carpio, C.B., *Phil. J. Bio. 11*:319 (1982).
7. Windart, W., and F. Rognon, *Oleagineux 33*:231 (1978).
8. Santos, G.A., S.B. Cano, and M.C. Ilagan, *Phil. J. Coconut Stud. 6*:34 (1981).
9. Santos, G.A., *Phil. J. Coconut Stud. 11*:24 (1986).
10. Louis, I.H., and V.L. Chopra, *Phil. J. Coconut Stud. 14*:10 (1990).
11. Dhamodaran, S., V.C. Viraktamath, and R.D. Iyer, *Indian Coconut J.* May:8 (1992).
12. Satyabalan, K., *Indian Coconut J.* November: 33 (1991).
13. Banzon, J., and A.P. Resurreccion, *Phil. J. Coconut Stud. 4*:1 (1979).
14. Rossell, J.B., B. King, and M.J. Downes, *J. Am. Oil Chem. Soc. 62*:221 (1985).
15. Del Rosario, R.R., C.M. Malijan, R.A. Fuentes, and M.R.S. Clavero, *Phil. Agriculturist 72*:147 (1989).
16. Gruezo, W. Sm., *Phil. J. Coconut Stud. 15*:6 (1990).

**TABLE 10**
Fatty Acid Composition of Some Coconut Varieties (wt % [15])

| Cultivars | $C_6$ | $C_8$ | $C_{10}$ | $C_{12}$ | $C_{14}$ | $C_{16}$ | $C_{18}$ | $C_{18:1}$ | $C_{18:2}$ |
|---|---|---|---|---|---|---|---|---|---|
| 1  Macapuno | 1.86 | 8.82 | 7.32 | 50.27 | 13.87 | 8.54 | 2.57 | 5.94 | 0.82 |
| 2  Tacuna | 2.51 | 7.43 | 6.26 | 48.58 | 18.59 | 9.48 | 2.38 | 3.72 | 0.42 |
| 3  Catigan | 1.02 | 6.13 | 5.69 | 51.28 | 19.63 | 9.12 | 2.71 | 3.80 | 0.65 |
| 4  Bago-Oshiro | 0.76 | 8.54 | 6.92 | 50.53 | 13.61 | 10.00 | 2.96 | 4.31 | 0.60 |
| 5  West African Tall | 1.68 | 8.78 | 5.76 | 47.52 | 19.84 | 7.32 | 2.68 | 4.10 | 7.32 |
| 6  OD × (Tambolilid × Macapuno) | 0.94 | 7.73 | 5.27 | 40.98 | 16.48 | 10.57 | 9.25 | 7.51 | 1.27 |
| 7  San Ramon × YD | 1.16 | 6.13 | 5.38 | 45.25 | 20.26 | 11.97 | 3.56 | 5.33 | 0.97 |
| 8  Concepcion Tall × OD | 1.13 | 7.87 | 6.72 | 51.86 | 18.00 | 7.73 | 2.78 | 3.98 | 0.54 |
| 9  OD (Laguna Tall × Coconino) | 1.22 | 7.82 | 5.72 | 52.93 | 15.33 | 7.76 | 1.71 | 6.82 | 1.20 |
| 10 Rabanuel GD × OD | 0.82 | 5.99 | 5.83 | 42.10 | 18.99 | 8.68 | 2.41 | 4.71 | 0.49 |
| 11 Lanao GD × OD | 0.88 | 6.30 | 5.05 | 53.23 | 18.90 | 7.74 | 2.81 | 4.63 | 0.62 |
| 12 Lanao GD × (San Ramon × Coconino) | 1.27 | 6.80 | 6.36 | 49.28 | 17.67 | 8.80 | 2.22 | 6.64 | 0.95 |
| 13 San Ramon × Pascual GD | 1.08 | 6.46 | 5.71 | 52.30 | 19.69 | 9.32 | 1.71 | 2.84 | 0.88 |
| 14 Pascual GD × OD | 1.25 | 5.35 | 8.11 | 58.09 | 15.82 | 7.78 | 2.49 | 5.39 | 0.72 |
| 15 Pascual GD × YD | 1.62 | 6.92 | 6.34 | 48.53 | 17.84 | 9.27 | 2.50 | 5.91 | 1.04 |
| 16 Lanao GD × San Ramon | 1.19 | 7.45 | 5.48 | 49.43 | 17.30 | 7.99 | 2.72 | 7.45 | 0.99 |
| 17 YD × Laguna Tall | 1.18 | 9.51 | 5.12 | 51.63 | 14.69 | 8.49 | 2.81 | 4.73 | 0.83 |
| 18 YD × San Ramon | 1.16 | 7.57 | 6.22 | 50.48 | 19.22 | 8.34 | 2.29 | 4.02 | 0.70 |
| 19 Lupison × OD | 1.34 | 8.48 | 6.20 | 50.97 | 16.72 | 7.71 | 2.24 | 0.07 | 0.65 |
| 20 YD (Macapuno × Tambolilid) | 1.88 | 8.99 | 6.17 | 57.52 | 12.88 | 5.78 | 1.91 | 4.06 | 0.80 |
| 21 Laguna Tall | 1.45 | 9.36 | 6.89 | 51.61 | 15.98 | 7.18 | 1.99 | 4.61 | 0.97 |
| 22 Tacunan × Bago-Oshiro | 0.94 | 7.67 | 6.98 | 51.49 | 16.69 | 8.64 | 3.94 | 3.26 | 0.38 |
| 23 Variety 109 | 1.45 | 8.27 | 7.10 | 49.43 | 17.03 | 8.42 | 2.18 | 5.25 | 0.61 |
| 24 Dua-Dua | 2.89 | 13.65 | 4.40 | 41.18 | 14.23 | 11.75 | 0.31 | 9.66 | 1.92 |
| 25 Nigeria | 1.27 | 7.83 | 6.49 | 54.84 | 15.72 | 7.08 | 1.74 | 4.33 | 0.70 |
| 26 Tanganyika | 1.98 | 7.70 | 6.92 | 54.64 | 16.24 | 6.31 | 1.94 | 3.81 | 0.47 |
| 27 Variety 21 | 1.52 | 10.21 | 8.91 | 51.51 | 15.20 | 5.47 | 2.21 | 3.98 | 1.01 |
| 28 Variety 107 | 1.14 | 6.07 | 4.93 | 58.09 | 14.55 | 6.97 | 2.73 | 5.06 | 0.46 |
| 29 Catigan × Bago-Oshiro | 0.71 | 6.81 | 5.74 | 45.81 | 22.79 | 9.64 | 2.06 | 5.06 | 0.46 |
| 30 Laguna Tall × Rabanuel GD | 1.45 | 9.36 | 6.89 | 51.61 | 15.93 | 7.18 | 1.99 | 4.61 | 0.97 |
| 31 P91C | 1.17 | 6.41 | 6.19 | 52.17 | 17.37 | 8.55 | 2.80 | 4.78 | 0.56 |
| 32 Markham | 1.07 | 8.10 | 7.52 | 53.73 | 15.91 | 7.06 | 2.31 | 3.64 | 0.68 |
| Mean | 1.35 | 7.77 | 6.25 | 50.56 | 17.00 | 8.37 | 2.59 | 5.00 | 0.83 |

*Abbreviations:* Orange Dwarf, OD; Yellow Dwarf, YD; Green Dwarf, GD.

17. Rillo, E.P., *Coconuts Today.* December:113 (1989).
18. Rillo, E.P., *Phil. J. Coconut Stud. 14*:16 (1989).
19. Ebert, A.W., E.P. Rillo, O.D. Orense, M.B.B. Areya, and C.A. Cueto, *Phil. J. Coconut Stud. 16*:12 (1991).
20. Anon., *Coconut Oleagineux 47*:327 (1992).
21. Voelker, T.A., A.C. Worrell, L. Anderson, J. Bleibaum, C. Fan, D.J. Hawkins, S. Radke, and H.M. Davies, *Science 257*:72 (1992).
22. Schwitzer, M.K., *Coconuts Today* April:35 (1984).
23. Padolina, W.G., *Coconuts Today* September:35 (1983).

# Potential Sources of Lauric Oils for the Oleochemical Industry

N. Rajanaidu and B. S. Jalani

Palm Oil Institute of Malaysia (PORIM), P. O. Box 10620, 50720 Kuala Lumpur, Malaysia.

## Abstract

Coconut (*Cocos nucifera*) copra and oil palm (*Elaeis guineensis*) kernel are the traditional sources of lauric oils. *Elaeis oleifera*, *Cuphea*, and *Orbignya* also have the potential to produce lauric oils. These are found in tropical and temperature regions of the world. Reports indicate that it is also possible to produce lauric acid in rapeseed using transformation techniques.

Oil palm has the advantage of producing both mesocarp and palm kernel oils. The latter has physical and chemical properties similar to coconut oil. Cultivation of the oil palm has been increasing for the past two decades in Malaysia. Thus, palm and palm kernel oils have become important sources of raw materials for the oleochemical industry.

At present, mesocarp content and mesocarp oil are emphasized in oil palm breeding at the expense of the kernel. However, an economic analysis shows that breeding for the highest possible content of kernel in oil palm fruits yields the highest rate of return from the oil palm. The kernel/bunch content of current oil palm-planting material is 4–8%. The natural variation for kernel/bunch in oil palm germ plasm ranges from 6–12% at family mean levels. Over the past 20 years, PORIM has assembled the largest oil palm germ plasm collection in the world. Some of the palms collected in Africa give kernel/bunch (K/B) > 15%. The prospects are bright to double the kernel content of future oil palm planting material. Kernel/bunch is highly heritable and maternally inherited.

There is an increasing demand for lauric acid ($C_{12}$) by the oleochemical industry. Lauric acid is an important raw material in the manufacture of detergents. Coconut (*Cocos nucifera*) copra and oil palm (*Elaeis guineensis*) kernels are traditional sources of lauric oils. These are cultivated in the tropical regions of the world. New sources of lauric oils are being investigated in temperate and tropical countries. *Cuphea* species, an annual that can be grown in a temperate zone climate, are one of the candidates (1). The tropical species Babassu or *Orbignya* spp (2) and *Elaeis oleifera* (3) are the other sources of lauric oil that presently are available to the oleochemical industry.

There is great interest in the application of recombinant DNA technology to oil crops to alter fatty acid composition (4). In one study, a $C_{12:0}$ preferring acyl-ACP thioesterase gene was isolated from the California bay plant (5). When the gene was transferred into rapeseed under the control of seed-specific promoters, there was significant accumulation of laurate in the seeds of the transformed plants (6).

The oil palm bunch consists of 500–1500 fruits. Oil palm fruit consists of three main parts, mesocarp, shell, and kernel. An increase in one component will generally result in a reduction in other components. The objective of this paper is to outline the level of natural genetic variation for kernel content in oil palm and the likely yields possible. We also outline the breeding methods for higher kernel content, and its impact on economic returns.

## Oil Palm Genetic Resources and Variation for Palm Kernel

There are three fruit forms of oil palm. They are the thick-shelled *dura* ($Sh^+ Sh^+$), the unshelled *pisifera* ($Sh^- Sh^-$) and their hybrid, the thin-shelled *tenera* ($Sh^+ Sh^-$). The inheritance of shell-thickness was discovered by Beirnaert and Venderweyen in 1941 (7) in the Belgian Congo. Since 1960, all the commercial oil palm-planting material has been *tenera* (also known as DxP). By exploiting a single gene for shell-thickness, the oil/bunch ratio was increased by 30%, from 16–18% in *duras* to 22–26% in *teneras*. A further 12–30% increase in oil yield is expected in clones; largely by exploiting the highly heritable trait, mesocarp oil/bunch. Oil palm breeders have been selecting for mesocarp at the expense of shell and kernels.

A study was carried out to investigate the amount of kernel variation in oil palms. The Palm Oil Research Institute of Malaysia (PORIM) has assembled a large collection of oil palm germ plasm for the past 20 years.

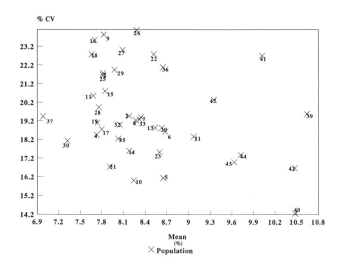

Fig. 1. Distribution of variations for kernel/bunch in oil palm populations.

**TABLE 1**
Anova Kernel to Bunch

| Source | DF | SS | MS |
|---|---|---|---|
| Replications (R) | 1 | 2.4789 | 2.4789 |
| Populations (P) | 39 | 1357.9740 | 34.8198[a] |
| Family within populations (F) | 160 | 1004.4080 | 6.2776[a] |
| P × R | 39 | 110.1055 | 2.8232 |
| F × R | 160 | 381.9461 | 2.3872 |
| Palms | 1644 | 3939.0110 | 2.3960 |

[a]Significant at < 1%.

**TABLE 2**
Variation between Family Means for Kernel to Bunch in Current Oil Palm-Breeding Material

| Trial No. | No. of Families | Overall Mean | Family Means | |
|---|---|---|---|---|
| | | | Min. | Max. |
| 0.180 | 58 | 5.81 | 4.23 | 7.29 |
| 0.189 | 93 | 6.03 | 4.18 | 8.11 |
| 0.190 | 15 | 5.95 | 5.15 | 7.40 |

Collections were made in West Africa and Central-South America (8). The first collection, made in Nigeria in 1973, was screened for kernel/bunch (K/B) (fruit/bunch × kernel/fruit). Forty-five populations were sampled at various sites distributed throughout the oil palm-growing areas. The K/B in these populations ranged from 7–10% for 45 samples[1]. These population differences are highly significant (Table 1). At the family mean level, the K/B varied from 6–12% for 185 samples with a mean value of 8.42[1]. The K/B in the current breeding material is 4–8% at the family mean level (Table 2). It is clear that the level of variation for K/B in the genetic collections is much higher. The distribution of variation for K/B in oil palm populations collected in Nigeria is shown in Fig. 1. The populations sampled in the northern part of Nigeria, a drier area, have a higher level of kernel content than populations sampled in the southern region of Nigeria. For individual palms, the K/B ranged from 2.6–15.3% (Fig. 2). The palms with high K/B can be bred with tissue culture techniques to fix the trait.

## Kernel Content and Economic Returns

It was indicated previously that breeders have been emphasizing high levels of mesocarp oil in breeding programs. Hartley (9) indicated that oil palm breeders should attempt to increase the kernel content in oil palm bunches. After the introduction of weevils in Malaysia (10) the K/B, on average, has increased from 5 to 7%. Using the 1992 palm oil and kernel prices, we have computed gross income. Figure 3 shows the prices of palm oil and palm kernel from 1980–1992. The price difference between these commodities has been consistent over the years. It clearly shows that income increases when the kernel content in oil palm bunches is higher (Table 3).

We have also computed the gross income by considering the maximum level of K/B at the population mean level of 10%. At this level, the combined income for palm oil and kernel was U.S.$ 11,520 per 100 metric tons (MT) of FFB fresh fruit bunches compared to U.S.$ 11,280 and U.S.$ 11,211 at the 7 and 5% K/B levels, respectively. It shows that better economic returns are possible by maximizing the genetically feasible level of K/B in the oil palm.

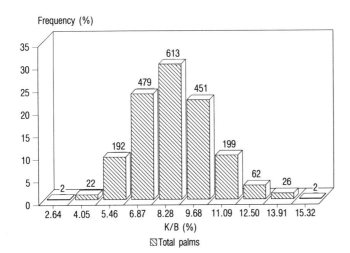

Fig. 2. Variation of palm level for kernel/bunch (%).

[1]Detailed data is available from the authors.

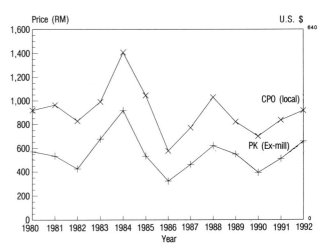

Fig. 3. Average annual prices of oil palm and palm kernel in Malaysia.

### TABLE 3
Theoretical Returns (U.S. $) from 100 MT Bunches Containing Fruits with 5, 7, and 10% Kernel/Bunch (KB)

|  | % K/B | | |
|---|---|---|---|
|  | 5 | 7 | 10 |
|  | (MT per 100 MT bunches) | | |
| Fruit to bunch | 65 | 65 | 65 |
| Mesocarp/Fruit | 82.31 | 79.24 | 74.62 |
| Shell/Fruit | 10.00 | 10.00 | 10.00 |
| Kernel/Fruit | 7.69 | 10.76 | 15.38 |
| Oil/Bunch | 26.75 | 25.75 | 24.25 |
|  | Returns (U.S. $) | | |
| Oil @ 366.60/MT | 9806.55 | 9439.95 | 8890.05 |
| Kernel @ 263.00/MT | 1315.00 | 1841.00 | 2630.00 |
| Total | 11,121.55 | 11,280.95 | 11,520.05 |

## Breeding for Higher Levels of Kernel Content in Oil Palm

The oil palm-planting material, DxP (*tenera*), is produced by crossing *dura* (female) with *pisifera* (male). The kernel content in oil palm fruits is maternally inherited. The *dura* mother palms with more than 16% K/F are crossed with *pisifera* which are known for their general combining ability, such as, Avros *pisiferas* (11). The inheritance of kernel size in the oil palm is illustrated in Fig. 4. A number of studies have shown that the heritability estimates for K/B and K/F are high in oil palm (12).

## Yield Potential of Oil Palm

Oil palm is the most productive oil-bearing plant species known; 1 hectare of oil palm in good growing conditions produces an average 4.5 MT of oil (ha/yr), 0.50 MT of kernel oil, and 0.45 MT of palm kernel cake. This is almost three times the oil yield of coconut and more than 10 times that of soybean (Table 4) (13).

The national average yield figures for Malaysia are given in Fig. 5. The national palm oil yield is 3.5–4 MT (ha/yr) and about 1 MT (ha/yr) kernel oil.

Table 5 shows the potential palm oil and palm kernel yields under extremely good growing conditions, 25 MT fresh fruit bunches (ha/yr). At 5% K/B, it is possible to realize 6.7 MT of palm oil and 1.25 MT of kernel. If the K/B is raised to 10%, the palm oil (ha/yr) drops to 6.1 MT, but the kernel (ha/yr) increases to 2.5 MT.

### TABLE 4
Yield of Different Oil Crops in kg Oil per Hectare per Year (13)

| Species | Oil (kg/ha/yr) |
|---|---|
| Oil palm | 2500–4000 |
| Coconut | 600–1500 |
| Olive | 500–1000 |
| Rapeseed | 600–1000 |
| Sunflower | 280–700 |
| Groundnut | 340–440 |
| Soybean | 300–450 |

## Conclusions

It is known that the kernel content in an oil palm bunch can be increased from 5 to 10%. There are oil palm families with >10% K/B in our oil palm germ plasm collections. The K/B can be fixed readily in oil palm-planting material since this characteristic is maternally inherited. It is also possible to multiply outstanding palms with high K/B by using tissue culture methods (14). If there is an indication for strong, long-term demand for lauric oils

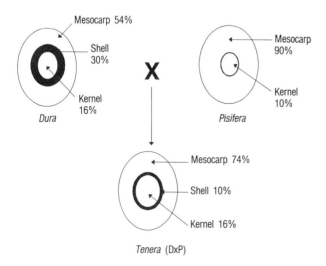

Fig. 4. Inheritance of kernel in oil palm.

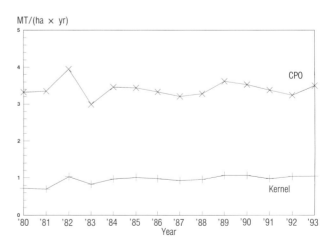

Fig. 5. Coconut palm oil (CPO) and palm kernel production.

**TABLE 5**
**Potential Oil and Kernel Yield at 25 MT Fresh Fruit Bunches (ha/yr)**

| Kernel/bunch | 5% | 7% | 10% |
|---|---|---|---|
| Oil yield (MT/ha) | 6.7 | 6.4 | 6.1 |
| Kernel yield (MT/ha) | 1.2 | 1.75 | 2.5 |
| Coconut (copra) | 1–5 (ha/yr) | | |

from users, it is possible for the oil palm breeders to embark on the production of planting materials with high kernel content.

## Acknowledgments

We thank the Director-General of PORIM for permission to publish this paper.

## References

1. Hirsinger, F., New Annual Oil Crops in *Oil Crops of the World*. Eds. G. Robblen, R.K. Downey, and A. Ashri. McGraw-Hill, New York, 1989, pp. 518–532.
2. Balick, M.J., Neotropical Oil Palms in *Oil Crops of the World*. Eds. G. Robblen, R.K. Downey, and A. Ashri. McGraw-Hill, New York, 1989, pp. 505–517.
3. Rajanaidu, N. *Plant Genetic Resources Newsletter*, FAO, Rome, 1982.
4. Battey, J.F., K.M. Schmid, and J.B. Ohlrogge, *Trends Biotechnol. 1*:122 (1989).
5. Voelker, T.A., A.C. Worrell, L. Anderson, J. Bleimbaum, C. Fan, D.J. Hawkins, S. Radke, and H.M. Davies, *Science 257*:72 (1992).
6. Voelker, T.A., A.C. Worrell, L. Anderson, J. Bleimbaum, C. Fan, D.J. Hawkins, and H.M. Davies, Engineering Laurate Production in Oilseeds. *Proc. 1992 Miami Biotechnology Winter Symposium*, 1992. 102.
7. Beirnaert, A., and R. Venderweyen, *Contribution a l'etude genetique et biometrique des varietes d'*Elaeis guineensis, *Jacq.* Publication INEAC, Serie Scientifique 27 (1941).
8. Rajanaidu, N., and V. Rao, Oil Palm Genetic Collections: Their Performance and Use of the Industry. *Proc. of 1987 International Oil Palm/Palm Oil Conference–Agriculture*, PORIM, Kuala Lumpur, Malaysia, 1988, pp. 59–85.
9. Hartley, C.W.S. The Oil Palm. *Tropical Agric. Series*, Longman, 1988.
10. Syed, R.A., *Bull. Ent. Res. 69*:213 (1979).
11. Rajanaidu, N., V. Rao, and A.H. Hassan, Progress of Serdang Elmina and Serdang Avenue Deli *Dura*-Breeding Populations in *Proc. of Workshop on Progress of Oil Palm-Breeding Populations*, PORIM, Kuala Lumpur, Malaysia, 1990, pp. 70–80.
12. Hardon, J.J., V. Rao, and N. Rajanaidu, A Review of Oil Palm Breeding in *Plant Breeding I*, ed. G.E. Russel, Butterworths, London, 1985, pp. 139–163.
13. Robbelen, G., Mutation breeding for quality improvement–A case study for oilseed crops. *Mutation Breeding Review*. Joint FAO/IAEA Division of Nuclear Techniques in Food and Agriculture. No. 6, pp. 1–44.
14. Jones, L.H., *Oil Palm News 17*:1 (1974).

# The Development and Commercialization of High-Lauric Rapeseed Oil

Andrew Baum

Calgene, Inc., 1920 Fifth St., Davis, CA 95616, USA.

## Abstract

Biotechnology is a powerful new tool for the development of crop varieties with superior agronomic and functional characteristics. In particular, genetic engineering allows for the development of crops that cannot be developed through conventional plant breeding because it allows for the introduction of genes from any organism into the target crop. In contrast, plant breeding is limited to introducing genes from sexually compatible species.

Since the early 1980s, scientists have been applying this technology to plants with the aim of developing oilseed varieties with modified oil compositions. Among the products currently under development are rapeseed varieties with elevated levels of lauric oil. First generation products with lauric/myristic acid contents of nearly 50% have been developed (the balance of the fatty acids being primarily $C_{18}$ unsaturates) and are undergoing their fourth season of field testing in the southeastern and midwestern regions of the United States. These high-lauric varieties are agronomically competitive with nongenetically engineered rapeseed, making it possible to produce high-lauric rapeseed oil at prices close to commodity prices of canola or soybean oil. In the future, second generation rapeseed varieties containing 70% lauric acid will be available.

First generation high-lauric rapeseed varieties could be commercialized as early as 1995 in the United States. Regulatory efforts, large-scale field trials, and planting scale-ups are all underway. The extent to which these first lauric oil varieties will be commercialized will depend in large part upon the relationship of lauric oil prices to rapeseed-oil prices. Second generation products could have advantages over current lauric oils but will still be part of the current lauric oil complex. Projecting the long-term relationship between these oils is problematic at best.

It is not clear whether or not high-lauric rapeseed oil will be supportive of or detrimental to existing lauric oil suppliers. While a new source of lauric oil could have a negative impact on lauric oil pricing, it could also improve supply stability and in turn facilitate the development of new lauric oil applications and increase lauric oil demand.

The focus of this paper will be on the development and commercialization of a high-laurate rapeseed oil product (HLRO) developed using genetic engineering. I will first explain the process by which the HLRO product has been developed in the context of similar technologies. I will then describe how the HLRO prototype has been developed and present a status report on where that product is in the development process. I will conclude by giving some thoughts on what the impact of an HLRO product might be on world lauric markets.

## Biotechnology in Context

One definition of biotechnology is: "the use of living organisms to make commercial products" puts some of the new biotechnology tools in context. Recently one of the newest plant biotechnology tools, genetic engineering, has been confused with plant biotechnology itself. Plant biotechnology is not new. Humans have been engaged in plant biotechnology since they stopped living in caves and began improving crops for cultivation. Genetic engineering is just another tool for accomplishing this end, albeit a very powerful one.

Plant biotechnology consists of three techniques: plant breeding, cell biology, and recombinant DNA or genetic engineering technology. Each one of these technologies offers advantages for today's agricultural industry. Recombinant DNA technology provides new opportunities for expansion of the industry by leveraging the more traditional plant-breeding and cell biology techniques.

What advantages does genetic engineering offer over more conventional technology? First and most importantly, genetic engineering technology allows for the development of products that cannot be developed through cell biology or classical plant-breeding techniques. This is possible because genetic engineering allows for the movement of genes from any organism: plant, animal, or bacteria, into the target crop. Second, genetic engineering leads to shortened product-development cycles in comparison to plant breeding or cell biology. This might seem like an odd statement to make given that the plant genetic engineering industry began in earnest in 1981, and we are only now introducing the first products to the market. However, much of the first half of the 1980s was spent in developing the basic tools and techniques required to make genetic engineering technology possible. Now that those tools and techniques are in place, I think we will begin to see a steady flow of products into the market place, with new product development times taking from 3 to 5 years from initiation of the product development process to first commercial release. Finally, genetic engineering allows for a level of proprietary protection that is not available with other plant biotechnology techniques. It is possible to get composition of matter patents for genetically engineered plants and genetically engineered plant products. This is a much stronger level of protection than the plant-variety protection that is available for the results of plant breeding. Such protection is critical to the biotechnology industry because the investments in the technology are such that unless the developer of a genetically engineered plant product has proprietary protection, the investments will not generate an appropriate return.

## Plant Genetic Engineering

Given that genetic engineering is a powerful technology, how does the process actually work? Genetic engineering can be seen as consisting of five sequential steps: protein purification, gene isolation, plant cell transformation, plant cell regeneration, and plant breeding. In the case of rapeseed the last three steps: transformation, regeneration, and plant breeding are routine for several groups throughout the world. Transformation refers to the insertion of a foreign gene into a plant cell. In the case of rapeseed this process has become relatively routine; in fact patents have been awarded on this technology. Regeneration refers to the process in which the transformed plant cell is grown through cell-culture techniques into a whole plant. Plant breeding refers to the process by which the genetically engineered plant is converted into an agronomically competitive cultivar. As you will see, agronomic competitiveness will be key to the successful commercialization of high-laurate rapeseed.

Since the process of putting a foreign gene into plant cells and then growing those cells into agronomically competitive varieties is relatively routine for many plant species, the challenge for the plant genetic engineering industry is the identification and isolation of the genes responsible for economically valuable traits. This is typically done via a two-step process. First, the enzyme or protein responsible for the desired change in a plant is purified using conventional biochemical techniques. This is often a difficult, time-consuming, and expensive process. However, once the protein or enzyme has been purified, isolating the gene responsible for that protein is straightforward.

In particular, the purification of the proteins responsible for the biosynthesis of oils in plants is often problematic. However, it is possible to purify the proteins and then isolate the genes responsible for specific changes in the fatty acid composition in rapeseed and other oil seeds.

## Plant-Oil Biosynthesis

The ability to modify the chemical composition of rape and other oils is based on the fact that the biosynthetic pathway by which plants make oils is well characterized and the basic biosynthetic pathway is conserved across different plant species. That is, basically all plants use the same mechanism to produce seed oils. The implications of this fact are profound, for it allows genetic engineers to move the gene responsible for a specific fatty acid composition in one plant, say a tropical weed, into an agronomically competitive species and allow the economic production of that fatty acid.

The basis of plant-oil biosynthesis is an elongation process in which the plants build fatty acids two molecules at a time. During this elongation process, the growing fatty acid is attached to an enzyme called acyl carrier protein (ACP). Once the ACP is cleaved from the fatty acid, elongation ceases and the fatty acid is attached to a glycerol moiety and made into one of the triglycerides which constitute the vast majority of vegetable oils. The enzymes responsible for cleaving the ACP from the elongating fatty acid are called thioesterases. Different plants have different thioesterases that recognize different chain lengths and degrees of unsaturation. In temperate crops, such as soybeans, rapeseed, and sunflower, thioesterases recognize 16 and 18 carbon fatty acids of different degrees of unsaturation. In tropical plants, like coconut and palm kernel, there are thioesterases that specifically recognize 8, 10, 12, and 14 carbon fatty acids (Fig. 1).

In developing high-laurate rapeseed, our scientists looked for a plant that produced large quantities of lauric acid in its seed. The plant that they chose to work with was the California Bay Laurel tree, which grows wild in California and has as much as 60% laurate in its seed oil. Using seeds from this tree, our scientists purified the specific thioesterase responsible for the production of laurate in the Bay Laurel tree and then proceeded to isolate the gene responsible for the production of that enzyme. Once the enzyme was purified and the gene isolated, development of a high-laurate rapeseed plant was relatively straightforward using the transformation/regeneration and plant-breeding techniques described previously. The impact of the Bay Laurel's $C_{12}$ thioesterase on the chemical composition of rapeseed oil was dramatic. Normal rapeseed has no lauric acid at all, and consists entirely of $C_{16}$ and $C_{18}$ fatty acids. Transgenic rapeseed, on the other hand, has significant amounts of lauric acid, and our best lines now have as much as 45% laurate by weight in the seed oil. It is interesting to note that in addition to the 45% laurate produced, our best plants also produce myristic acid (Fig. 2). We have found that the ratio of lauric to myristic acids in our transgenic plants is typically 10 to 1.

The development of HLRO is a dramatic example of the power of genetic engineering. We were able to fundamentally change the chemical composition of rapeseed oil by engineering the rapeseed plant to produce a fatty acid it had never produced before.

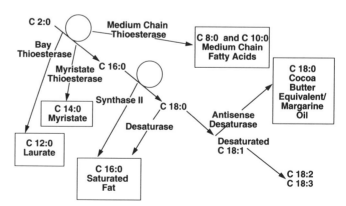

**Fig. 1.** Rapeseed oil modification.

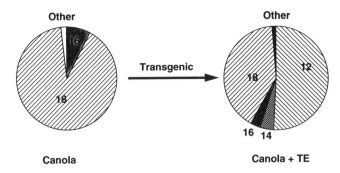

Fig. 2. High-laurate canola.

## High-Lauric Oil Product Status

Our high-laurate rapeseed plants are now in their fourth season of field trials. Experience to date has shown that the high-laurate plants are agronomically indistinguishable from regular rapeseed varieties and are quite competitive in terms of yield and other agronomic characteristics with those varieties. This is an important fact. In order for high-laurate rapeseed to be produced economically, the high-laurate rapeseed varieties must be agronomically competitive with regular varieties, so that it is not necessary to pay significant premiums to the farmer to grow the crop.

With the product development work largely complete, two other key hurdles must be overcome before HLRO can begin to be commercialized. First, regulatory approval for the genetically engineered product must be obtained. In the case of the United States, the products will need to be cleared for environmental safety and food use. With respect to food use, the Food and Drug Administration (FDA) policy in place is very rational and reasonable. In addition to demonstrating that the oil is safe for food use, we also need to demonstrate to the FDA that the meal by-product of HLRO is safe when fed to animals. These safety studies are well underway, and we do not see this as being a rate-limiting step to commercialization. It must also be demonstrated to the United States Department of Agriculture (USDA) that the high-laurate rapeseed crops will not be an environmental threat by creating some new type of weed. We submitted the required documentation addressing this issue to the USDA in November 1993 and hope to have formal approval in calendar year 1994. The issues in Europe are essentially the same, although it should be noted that the regulatory situation in Europe appears to be much less settled and will be much more of an issue than it is in the United States, where the regulatory process has already been defined.

In addition to obtaining regulatory approval, it will be necessary to have an integrated production infrastructure in place to ensure that the HLRO product can be produced on a reasonable cost basis. Obviously, the first step of the production process is the development of the high-laurate varieties themselves. Once the varieties have been developed, we will contract with farmers for the production of high-laurate rapeseed-oil varieties, and buy back their entire production on an agreed-upon basis at the time the crop is planted. Once the crop has been produced, we will then consolidate the crop in a series of grain elevators throughout the production territory, arrange for the seed to be crushed on a toll basis at a third party crushing facility, sell the resulting meal as a commodity, and then sell HLRO into the lauric oil markets. Although conceptually quite simple, the actual implementation of this process is quite complex, as Fig. 3 demonstrates. Our company has been working for the last 7 years to develop this infrastructure in the Southeastern and Northern tier states of the United States, and we are now in a position to produce HLRO (and other identity-preserved rapeseed oils) at a cost roughly comparable to that of canola oil.

When will our HLRO enter commercial markets? We plan to commercialize this product in the fall of 1995 if it is economically viable to do so. Plants with greater than 45% $C_{12}$ and $C_{14}$ are currently in their fourth season of field trials, and we are confident that these varieties are agronomically competitive. The regulatory approval process is well underway. The production infrastructure for HLRO is in place, and we believe that we will be able to produce hundreds of millions of pounds by the mid-1990s, if market conditions warrant such production.

## Successive Products

A second generation of high-laurate products which will have $C_{12}$ compositions in excess of 70 or even 80% are under development and scheduled for release in 1997 or 1998. In addition to HLRO, we are developing a series of related products as well. These products include rapeseed varieties with $C_8$ and/or $C_{10}$ compositions in excess of 40% and a set of rapeseed-oil products with $C_8$ and $C_{10}$ compositions in excess of 70 or even 80%. These products are scheduled for introduction in the last 3 years of the 1990s.

## Impact of High-Lauric Rapeseed Oil on Lauric Markets

It can be argued that the production of HLRO could present a threat to existing producers of lauric oils. While this is certainly a defensible view, it is not a view to which we subscribe. In fact, we believe that HLRO will benefit both the consumers and producers of lauric oils. There are several reasons for this belief. First, total HLRO production will be small versus total world output. It would take nearly 2 million acres of high-laurate rapeseed production to amount to even 10% of the total lauric oil production in the world today. Total North American (including Canada) production of rapeseed is only 10 million acres. Second, a new source of lauric oil will dampen price volatility, the effect of which could spur the development of new lauric uses and applications by giving lauric consumers confidence that both supply and price will be more predictable and reliable. Finally, it should be noted that HLRO will only be economically competitive when the price of lauric oils exceeds that of temperate oils like soybean and canola

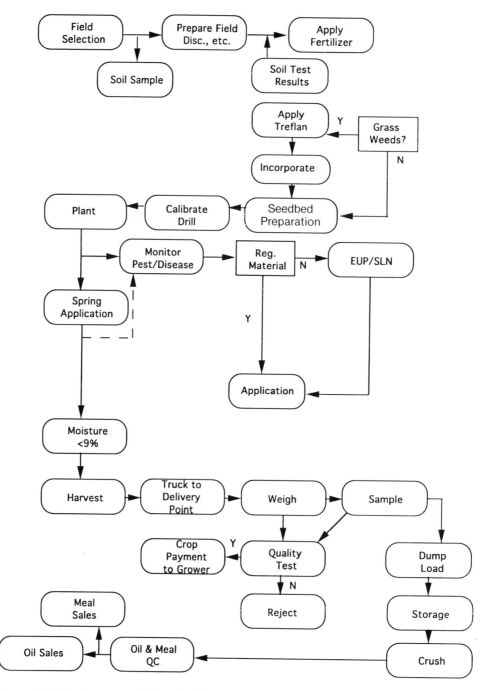

**Fig. 3.** High-lauric rapeseed-oil production.

oils, since the basis of HLRO cost will be that of other commodity oils produced in the temperate oil complex. The corollary of this fact is that HLRO will become more of a factor in those years when lauric supplies become tight and prices rise, creating a significant margin between temperate and lauric oil prices.

## Conclusion

In summary, using genetic engineering techniques, we have developed rapeseed-oil varieties with more than 45% laurate ($C_{12}$) and myristate ($C_{14}$) in the seed oil. These products are well advanced in the product-development process and could be commercialized as early as 1995. A second generation product with a laurate content in excess of 70 or 80% will be launched in the mid-1990s, as will be products with high concentrations of capric ($C_{10}$) and caprylic ($C_8$) acids. The impact of these oils on the existing lauric oil markets is difficult to predict given the relative volatility of those markets. However, it is our belief that the net benefit will be positive by dampening supply and price volatility which could actually lead to an expansion to the uses of lauric oil in the years to come.

# Discussion

With respect to new sources of lauric oils, there was considerable discussion about fungal sources and tissue cultures of the coconut: the latter has not yielded any nuts to date, while the former is some work in the early 1980s that was not pursued. There also was an interest in the variation of fatty acid composition and of aflatoxin in the copra that appears to be dependent on the region of the copra's origin.

As for the development of rapeseed varieties with a high lauric acid content, there was discussion dealing with the marketing of such an oil if it appears to be unhealthy. Also, there was concern that the "green" movement might be opposed to such developments, but this was not perceived to be a problem if the public perceives the value of the end-products. When asked if high myristic acid oils could be developed, the speaker acknowledged that they could, but he could foresee no particular markets for myristic acid. When questioned, the speaker could foresee no problems with the regulatory groups for approval of the nonedible oils, although the meals may pose a problem. He also predicted that, at least for nonedible use, the high-lauric varieties would have an attractive subsidy in the European Economic Community. As for the varieties being winter or spring types, it was noted that at present the spring type are in use but the winter varieties are under development.

In summarizing this Session, one chairperson went into considerable detail regarding the significance of the meeting with President Ramos. He noted five key points which were

1. The need to limit the middlemen by establishing farmer coops;
2. The establishment of a research center to be funded by Japan similar to the Palm Oil Research Institute of Malaysia;
3. The revitalization of the coconut industry under the Philippines 2000 plan;
4. Execution of the AOCS-IAC idea to foster better communication; and
5. The role of this President as a "hands-on" type of manager who is fostering policies that encourage foreign investments.

# Fractionation of Lauric-Based Fatty Acids Achieving Consistent Product Quality

Kee Pee Ho and Madhev Bala Subramaniam

Palm-Oleo Sdn-Bhd., Lot 1245 Kundang Estate, Rawang Selangor, 48000, Malaysia.

## Abstract

The oleochemical industry in Malaysia is slightly more than a decade old. During this period, splitting capacity has reached a figure of about 500,000 metric tons (MT)/yr, the main raw materials being based on palm and palm kernel oil. By the end of 1995, the splitting capacity has been targeted for 750,000 MT/yr.

The changing trends in the demand for industrial oleochemical products have resulted in a need to be able to produce a variety of products, both distilled and pure fractions. As consumers find substitutes for raw materials, the industry can remain viable only if it can maintain or improve quality while being competitive.

A variety of fractionated products from palm oil and palm kernel oil is produced in Malaysia. The main fractionated products from palm kernel oil are caprylic, capric, lauric, and myristic acids. The fractionated products must be of high purity and be competitive as equivalent substitutes. Where once 92% purity of a fraction was demanded, present day expectations are in the 98–99% range. Even this sometimes is insufficient. There are further quality requirements, such as good heat stability, low color, low peroxide value and unsaponifiable matter, odor, low metal content, to list a few. The concept of high quality is now intertwined with consumer requirements. The production of such products requires stringent control of raw materials, proper storage and handling facilities, selection of the type of processing facilities, and tight in-process quality control including proper handling of products in transit. This paper discusses some methods to achieve the increasing quality demands by consumers.

The past decade has seen the emergence of a new force in the oleochemical industry. Where the domain was once in the West and Far East, the center of oleochemical processing has slowly concentrated in the ASEAN regions. The emergence of Malaysia, Indonesia, and the Philippines has changed the scenarios of the sourcing and processing of raw materials. The rapid emergence of the palm oil industry has, to a certain extent, compelled these countries to search for new areas of utilization, and it is only natural that this would include the oleochemical industry.

The capacity to produce palm kernel and coconut oils have been increasing steadily over the years. In Malaysia alone the palm kernel oil industry is at about 900,000 metric tons (MT)/yr. This has contributed to the emergence of the capacity to produce raw materials rich in lauric components. Table 1 shows the typical composition of palm, palm kernel, and coconut oils in relation to beef tallow. Currently, the fat-splitting capacity in Malaysia stands at about 500,000 MT/yr, where the raw material is almost exclusively based on palm or palm kernel oil. Indonesia will be a major producer well within the next decade.

The mode of processing is usually customer driven, as to whether products are fractionated to their pure components or just into distilled forms. Even then, purity is no longer the single issue in today's customer-oriented society. Other demands, which include color, stability, Iodine Value (IV) and odor, have been included as partial requirements for the industry. More attention is paid to the selection of raw materials right up to the storage and delivery of the final product, as this will have a strong bearing on product quality. Competition is now more evident as more suppliers try to attain a higher quality level in their finished products to capture market niches.

## Raw Material Control and Storage

Current trading rules are based on local specifications. In Malaysia, the governing authority is the Palm Oil Refiners and Licensing Authority (PORLA). The oil trade categorizes raw materials according to titer, color, free fatty acids, moisture, and insoluble and unsaponifiable matter. These qualities were not intended for the fat-splitting industries. In addition to these specifications, other specifications, such as rancidity, peroxide value, and the heat history and storage condition of the oil before its use normally are

**TABLE 1**
**Typical Composition of Oils**

| Fatty Acid | Coconut | Palm Oil | Palm Kernel Oil | Soybean |
|---|---|---|---|---|
| C8:0  | 6.5  | —    | 2.5  | —    |
| C10:0 | 6.0  | —    | 5.0  | —    |
| C12:0 | 49.5 | —    | 48.5 | —    |
| C14:0 | 19.5 | 1.5  | 17.5 | —    |
| C16:0 | 8.5  | 4.2  | 16.5 | 11.5 |
| C18:0 | 2.0  | 41.8 | 1.5  | 3.5  |
| C18:1 | 6.0  | 43.0 | 12.5 | 23.1 |
| C18:2 | 1.5  | 9.5  | 2.0  | 55.8 |
| C18:3 | —    | —    | —    | 6.1  |

included. These are qualities which affect the oils during splitting and in their processing. Not only will they affect the yield, they will also affect the color and quality of the fatty acids produced and their subsequent processing.

The immediate affect on yield can be seen because, if we were to use refined, bleached, and deodorized (RBD) oils instead of crude oils, the glycerine and fatty acid yields could be increased by about 5%. But more importantly, the quality of the fatty acids is improved. The ensuing product will have a better color and tolerance to heat. At times when dealing with higher quality feedstocks containing unsaturated fatty acids, antioxidants may be necessary to minimize the formation of peroxides.

In addition to these features, consideration must be paid to the storage conditions and materials of construction. Generally, for raw materials, it is advisable to maintain a storage temperature about 10°C higher than its titer to maintain pumpability. Too high a temperature will result in oxidation. Similarly, heating too rapidly under atmospheric conditions results in product deterioration. The rate of heating should be slow and evenly distributed with small temperature differences between each strata. For this reason, mixers are necessary to ensure even distribution of heat and homogeneity of composition. Generally raw materials such as oils are not very corrosive and are stored in mild steel vessels.

## Processing of Lauric-Based Fatty Acids

Once split, the fatty acids are stored in stainless steel tanks and are ready to be used in further processing. Ideally, they will be processed quickly or stored in nitrogen-blanketed tanks to avoid peroxide formation.

The methods of distillation and fractionation have changed considerably over the last few decades (1,2). Bubble-cap trays are now obsolete and can be found only in undergraduate textbooks. More and more processors are converting to structured packings. The conversion increases efficiency because distillation of heat-sensitive materials requires good vacuum with a low pressure differential within the column. In a typical situation, a column with a height equivalent to 40 theoretical plates will suffer a pressure drop of about 8 mbar (top to bottom) as opposed to a bubble-cap column, where it can be as high as 50–60 mbar. The immediate effect will be an increase in the bottom distillation temperature. The difference between the two distilling temperatures can be as high as 30°C. This indicates that in columns, steam is no longer required for partial vapor-pressure reduction.

Other virtues of structured packing are its high efficiency (bubble-cap trays have efficiencies in the range of 50–85%) and fast through-put per theoretical plate. This results in minimal decomposition of the fatty acids, as contact time is now far less. With such packings, decomposition drops to less than 1% during distillation.

Another important aspect in achieving good product quality is the vacuum system. Normal distillation is done at approximately 10 mbar. The salient feature here is not to oversize the vacuum system, but rather to minimize leaks into the column. For this reason, flanges are almost always avoided, and pumps are hermetically sealed rather than those with conventional mechanical seals. A good system is one with minimal leakage. Operating at temperatures in excess of 200°C, leakage will certainly result in the oxidation of the fatty acids as excess air is introduced into the columns. The most visible impact of these approaches is the subsequent quality of the product. Stability, in both heat and alkaline conditions, will improve if distillation temperatures are kept low and there is minimal degradation of the fatty acids.

Selection of the reboiler is vital for the fractionation of fatty acids. Generally, there have been many publications on various aspects of the design of such reboilers. The falling film type is more commonly used as the heat transfer efficiency is much higher, and pressure drops are quite low. Its sizing is important. While fractionating fatty acids, the temperature difference between the heating medium and the fatty acid must be kept low, preferably about 30°C, to prevent cracking at elevated temperatures. Also, the bottom reflux should be arranged so that the contact time is sufficient to induce distillation only, and that there is sufficient wetting of the reboiler. Hence, the determination of the overall heat transfer rate is important to calculate the heat transfer area required to achieve distillation.

This subject would not be complete without discussing energy conservation and construction materials. If properly designed, the condensers can be used to generate sufficient steam to provide general heating and for use as the motive steam in vacuum generation. Materials used in such processes suffer from thermal stress and corrosion. For this reason, the general specified material is SS316L with a minimum molybdenum content of 2.5% to withstand chemical corrosion. For the evaporator, SS317L generally is the material of choice.

## Towards Achieving Product Quality

Market requirements have changed tremendously over the years. Where once standard purity was considered 92%, customer acceptance today stands at a minimum of 98%. In addition, end-users are now seeking suppliers with ISO certification as part of their requirement to achieve consistent quality towards total quality management. Other requirements are usually added. For example, an ethylene *bis*-steariamide (EBS) user will demand a product with good heat stability, whereas a metallic soap maker will have a requirement for good alkaline stability. There should be no odor as well. The production of such high purity and quality products with good economy has changed the pattern of operations. These requirements pose new challenges to the producer. There is now a requirement to achieve product purity and quality consistently and economically. The package of operations begins with the raw materials and extends to the finished goods.

In achieving these goals, the fractionation plant must be in stable running condition. Data, such as temperature, pressure, reflux and feed rates, must be readily available and easy for the operator to evaluate and take remedial actions. It is for this reason that the Distributed Control System (DCS) has gained rapid recognition in our industry. Data is continuously updated and important loops are continually tuned by means of an auto tuner, which can readily be installed into more-advanced DCS systems. Temperature profiles at each structured packing layer normally are advisable, and feed control should be cascaded to level control. At steady state, the temperatures profiles become constant as a steady reflux and product flow at each stream is maintained. The product is initially removed based on material balance and composition of the feedstock. Quality control is achieved by continuous laboratory analysis at predetermined intervals. A typical fractionation plant will consist of three columns with a distiller. During a run based on a single pass of palm kernel fatty acids, it is possible to obtain pure fractions of lauric and myristic acids of up to 99% minimum purity with a top fraction of caproic, caprylic, and capric acids with a bottom fraction of a mixture of palmitic, stearic, and oleic acids. If all of the above conditions are achieved, it is possible to meet the product quality and purity requirements of most buyers. At times it might be necessary to introduce an antioxidant, such as BHT plus citric acid and other additives depending on the end-user application and requirements.

## Final Product Storage and Finishing

The intermediate storage of the fractionated fatty acids is an important area, as it will affect final product quality. As purity is now at a minimum of 99%, these tanks, which are mainly stainless steel in construction, are dedicated to prevent the problem of contamination and are insulated. All pipe lines associated with such tanks are dedicated, as well. The pumps should also be constructed of stainless steel, and there should be a polishing filter associated with each pump.

Storage is an important area. Ideally, the products should be shipped to the end-user as soon as possible. The storage time should be at a minimum. Tanks should be designed for nitrogen blanketing. For this purpose, the tanks require breather/vent valves and rupture discs to protect the tanks and the products during filling and discharge. The nitrogen valve must be calibrated to the operating pressure, with pressure switches linked to annunciators for an early warning alarm. Tank temperatures should be kept 5–10°C above the product titer to maintain fluidity and prevent thermal damage. Heating should be slow and based on a hot water recirculation system rather than on low-pressure steam.

Transportation of products from the factory should be under nitrogen blanketing, as well. Tankers should be built for bottom loading rather than top loading. They should have the fixtures and safety features required for nitrogen blanketing.

Most plants generally have the capability to either flake or bead solids or to drum finished products. This will help to protect the product once it has been produced. The quicker products are consumed, the less the chance of possible contamination and deterioration.

Ultimately, the end-user will view the product according to the quality and the finishing of the product. Packing and drumming operations must be tidy and organized. This is especially true when there are multiple products being packed or drummed. Care must be taken to avoid mix-ups. Bags and drums must be marked neatly and well cared for to avoid tears and scratches. During transport, care in handling from the packing room to the warehouse and ultimately to the containers is important to ensure customer satisfaction. There is no point in having the best of products in badly finished packaging.

## Conclusion

As producers, we must be responsive to customer requirements. The first impression is the most vital, and this comes from the way the goods are packed, handled, and stacked when delivered. Excellent packaging, coupled with the capability to produce products of good quality and purity, consistently and economically, will ensure that the targeted goals can be met. It must be emphasized that to attain such a standard not only must the equipment be good, but we must have a workforce that is well trained and achieve the desired results. This philosophy towards total quality management requires a concerted effort from both the management and the workforce. Training and understanding of processing and packing is part of the requirement towards achieving better quality.

## References

1. Stage, H. and K. Bose, *Chem. Age of India, 25*:8 (1974).
2. Stage, H. and K. Bose, *Chem. Engng. World, 11* (1976).

# The Fractionation of Lauric Oil Components for Cocoa Butter Substitutes

Soon Wong

Atlanto Sdn Bhd, 25B Jalan ss 15/8A, Subang Jaya, 47500 Petaling Jaya, Selangor, Malaysia.

## Abstract

In the early 1960s there was an economic need to find a replacement fat for cocoa butter. Hydrogenation, interesterification, and fractionation were investigated. In recent years, enzymatic interesterification has provided a new approach. However, conventional processes are still well-rooted in a number of factories largely for economic reasons. Physical fractionation and hydrogenation are attractive alternatives to produce a cocoa butter substitute (CBS) from palm kernel oil in particular. Other lauric oils, such as coconut oil and tucum oil, have been tried to a much lesser extent. There are practical limitations when using hydrogenated natural fats without going through a physical separation cycle. Several commercial products and their relationships to the characteristics of physically separated lauric oils also are highlighted. Some of these conventional technologies have been improvised. The quality and efficiency of production in Malaysia has improved greatly. New technology offers a high degree of automation without sacrificing the yield or quality, and much of the technology now is offered commercially.

---

The rise in the price of cocoa butter many years ago stimulated remarkable interest in developing compatible fats to replace cocoa butter. Much of the commercial research was shrouded in secrecy; largely because of the extremely attractive price premium of cocoa butter substitutes over conventional, commodity-type oils and fats. Today the main lauric and nonlauric oils and fats that have been evaluated are as follows: palm kernel oil (PKO), coconut oil (CNO), and tucum oil. The main lauric oil processed through fractionation and hydrogenation to make cocoa butter substitute (CBS) fat is palm kernel oil. Natural nonlauric oils and fats (for making CBE) that requires fractionation are palm midfractions, sal fat, and shea nut oil. Polyunsaturated oils that are processed through selective hydrogenation and fractionation for making cocoa butter extender (CBX) fats include soybean oil, cottonseed oil, sunflowerseed oil, and safflowerseed oil.

## Production of Specialty Fats with Desirable Enrichment Through Fractionation

The technological approaches over the years have gradually evolved from simple hydrogenation, interesterification, and fractionation techniques to more sophisticated biotechnological processing (1,2). Gene cloning of *Cuphea* for CBS as well as rapeseed and palm for CBE cannot be ruled out, as they offer very attractive alternatives. Extensive research in these directions currently is going on in various companies and research institutions. The current "conventional" dry physical fractionation technology using Alfa Laval, Tirtiaux, Oiltek, De-Smet and EMI plants cannot effectively produce a good lauric stearin fraction for use in high quality CBS. Basically there are five types of such physical separation technologies which can perform the "selective fractionation" to produce acceptable lauric fractions. They are physical separation by conventional labor-intensive hydraulic press process; physical separation by an automated press process (3–11), physical separation by molecular still distillation techniques, physical separation by detergent fractionation, and physical separation by solvent fractionation. Although these last three processes are physical techniques in theory, in practice they involve chemicals and intense heating which are not too popular because of chemical pollution and the high energy costs of production. The advantages and disadvantages of these five physical separation processes are summarized as follows: Hydraulic pressing has a cheap investment for quick return on investment (ROI), low utility and maintenance costs, and very good quality product, but it is fairly labor intensive. An automated press system offers a very good quality product, fairly low utility costs, automated press operation, and can be used for producing other types of specialty fats; however, it is more expensive than the conventional press process, hence, it has a longer ROI. Molecular distillation can lead to a very good quality product, be used for various grades of CBS and fractions, is a very automated operation, but it uses expensive equipment and it cannot be used for CBX and CBE fats production. By using detergent separation fairly acceptable quality can be produced, the system is automated, but there are lower yields of a lauric stearin fraction when compared to other processes, higher maintenance costs, and chemical pollution in wastewater. Also, chemical residues in CBS cannot be ruled out entirely, and the plant cannot be used to produce other specialty fats. By using solvent fractionation with acetone, a very good quality stearin fraction can be obtained, and one can use the same plant for other specialty fats. It is a highly automated process, but it is an expensive investment and has a very long ROI with high production costs.

The physical separation of lauric oils by hydraulic pressing to produce these specialty fats has received the most attention. This is largely because of the much lower capital investment and a larger market share of CBS products. The types of lauric oils and the economic viability that need to be assessed are as follows: Tucum oil that has a yield of stearin of 65% (± 5%) is only available in small commercial quantities. Palm kernel oil with a yield of stearin of 40% (± 5%) is abundantly available. Coconut oil that yields stearin at 15% (± 5%) is abundantly available.

The quality, yield, and availability of source oils are important considerations for investment strategies for lauric oil based CBS products. The above yield factors are based on mechanical press separation with no acetone or detergents used in the fractionation process. The quality of the fractionated, coconut stearin is somewhat inferior because of the wider spread of fat triglycerides. However, this quality deviation can be improved by careful fractionation.

## Materials and Methods

### Fractionation

Tucum fat, palm kernel oil, and coconut oil were allowed to crystallize slowly to 8°C below the slip melting point of the respective fats. Thereafter, they were wrapped in nylon fabric for high pressure fractionation (pressing) on a hydraulic press. The oil portion (the olein) was collected in a gutter and the fat portion (the stearin) was unwrapped from the nylon cloth.

### Preparation

The fatty acid methyl esters were prepared by following the Palm Oil Research Institute of Malaysia's (PORIM) method. The iodine value (IV) analysis was done by the Wijs Method. The slip melting point was done on open ended standard glass capillary tubing with the fat held in the capillary glass tube for stabilization at 26°C for 40 h (IUPAC method).

### Instrumentation

The GLC instrument used was an HP 5890 model with modified adaptations to suit cold-hot, on-column triglyceride analysis. Every GLC run was preceded by automated background correction, a standard feature of the HP 5890 instrument. The SFC analysis was done on the Bruker Minispec pc 120s instrument in conjunction with accurate calibration standards from Bruker. Fat tempering at every stage of the analysis was done in thermostated water baths with an accuracy of ± 0.2°C.

## Results and Discussion

The lauric oils, in particular palm kernel oil [Malaysian origin (12)] and tucum oil (Brazilian origin) must be physically separated very accurately to produce an acceptable stearin fraction (13). To obtain this precise quality (Table 1) the separation process must be performed to exact stan-

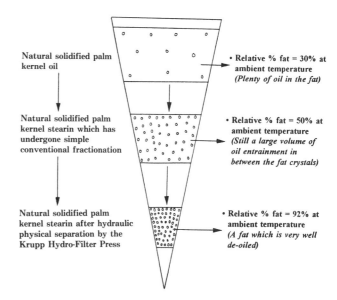

**Fig. 1.** Schematic representation of relative amounts of solids and liquids in various coating fats.

dards. This is where conventional fractionation techniques fail because the relative desirable triglycerides enrichment is not possible by conventional means. This is best illustrated diagrammatically as shown in (Fig. 1).

However, this is only a very simplified picture because merely enriching the solid fat alone does not necessarily give a usable fat of "The desirable confectionery interest" (14,15). The spread of solid triglycerides (at room temperature) within the enriched stearin must not be too diverse, which would otherwise broaden the melting range. Hence this renders the fat of lesser confectionery value. We can see this when comparing the physical separation of PKO stearin (Fig. 2) and CNO stearin, when it is fractionated by hydraulic physical separation. The illustration in Fig. 2 above is merely looking at the separation by mechanical press fractionation. In molecular still distillation however, it may be possible to give a cleaner cut to enhance molecular purity to produce excellent CBS from CNO.

In general the greater the degree of de-oiling (or de-olienation) the purer would be the molecular enrichment for the palm kernel stearin (PKS) solid triglycerides. Consequently, the solid character exhibited by the fractionated fat should be greater. This can be seen from the analyses in Tables 2 and 3.

### TABLE 1
**Properties of Tucum Oil and Its Fractions from Pressing**

| Oil Type | SFC % at Temperature (°C) | | | | | | |
|---|---|---|---|---|---|---|---|
| (Fraction) | 10 | 15 | 20 | 25 | 30 | 35 | 40 |
| Crude | 84.4 | 77.6 | 65.3 | 42 | 0 | 0 | 0 |
| Stearin | 93.4 | 91.3 | 86.0 | 71.8 | 32.0 | 0.4 | 0 |
| Olein | 66.3 | 54.8 | 36.9 | 7.1 | 0 | 0 | 0 |

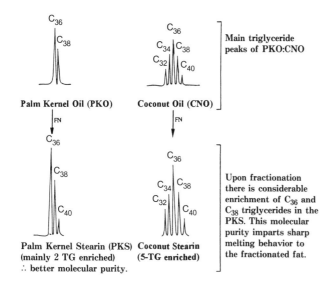

Fig. 2. Palm kernel oil stearin vs. coconut oil stearin. The other smaller TG peaks are ignored for relative comparison.

Normal hydrogenated (unfractionated) PKO (even hydrogenated to less than 0.5 IV) has the following inherent inadequacies as a good confectionery fat:

- It has an inadequate SFC body at certain desirable temperatures.
- It is too waxy when compounded into chocolate confectionery products.
- It has insufficient natural shrinkage for good mold release to minimize compound chocolate breakage in the production lines.
- It does not have the desirable cooling sensation of a good chocolate fat.

Only a well-fractionated PKS which is then hydrogenated to hydrogenated palm kernel stearin (HPKS) can exhibit

### TABLE 2
**Properties of Hydrogenated, Fractionated Palm Kernel Oil**

| Oil Type (Fraction) | SFC% at Temperature (°C) | | | | |
|---|---|---|---|---|---|
| | 10 | 20 | 30 | 35 | 40 |
| Hydrogenated Palm Kernel Oil (PK) (IV 0.5–2.5) | 96.2 | 85.6 | 34.7 | 13.8 | 6.4 |
| Hydrogenated PK Stearin[a] (IV 0.5) | 97.1 | 97.2 | 40.3 | 3.0 | 0 |
| Hydrogenated PK Stearin[b] (IV 0.5) | 98.4 | 97.7 | 53.2 | 4.9 | 0 |

[a]Original IV 7.2; acceptable molding fat.
[b]Original IV 5.7; acceptable molding fat.

### TABLE 3
**Characteristics of Highy Enriched Crude Palm Kernel Oil Stearin**

| Iodine Value | 8.0 | 7.0 | 6.5 | 5.5 |
|---|---|---|---|---|
| | FAME (%) | | | |
| $C_{8:0}$ | 3.6 | 3.4 | 3.6 | 3.1 |
| $C_{10:0}$ | 2.9 | 3.0 | 2.9 | 1.6 |
| $C_{12:0}$ | 56.0 | 56.8 | 57.1 | 58.9 |
| $C_{14:0}$ | 19.9 | 20.3 | 20.3 | 21.0 |
| $C_{16:0}$ | 7.8 | 7.7 | 7.6 | 7.8 |

these confectionery features (16). Hence, the physical separation techniques today have been very well developed to cater to this need as can be seen from some commercial products marketed around the world (Table 2).

Palm kernel oil stearin produced by mechanical press fractionation gives a better "solid fat" with increased solid fat body contribution of higher purity than a similar coconut oil stearin. This is because CNO has a large amount of shorter chain fatty triglycerides, imparting residual liquidity to the fractionated fat (Fig. 3).

Krupp Maschinentechnik of Germany has developed a commercial scale hydropress for automated mechanical fractionation of specialty fats. This process has overcome the rather labor intensive method of conventional lauric oil hydraulic press fractionation. There are many engineering advantages of this system as follows:

- The hydropress can be cooled to the temperature of fractionation.
- The thickness of the fat chamber can be adjusted to the desired thickness to suit different quality requirements.
- The engineering design enables it to handle lauric as well as nonlauric specialty oils and fats.
- There is absolutely no manual handling of the oil cakes during the fractionation process.
- There is no cloth blinding as in the conventional hydraulic press process, as the oil contact area is made up of stainless steel.
- It has excellent consistency right through the whole press cake surface to produce high-quality CBE fats equivalent to those produced by the acetone plants.

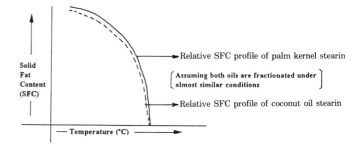

Fig. 3. SFC of PK stearin and coconut oil stearin.

- The press can be run on preset programmed pressure build-up settings for different types of specialty oils and fats.

This hydropress has been designed with great attention to detail to produce not only specialty fats but also specialty olein (e.g., palm olein with IV up to 72 and cloud points as low as −4°C). This has not been possible with any other current dry fractionation technologies available.

It is predicted that dry physical separation for lauric and nonlauric oils will be dominant into the twenty-first century. The chemical and solvent physical fractionation processes may not be competitive. Furthermore, they will become unpopular as emerging consumer groups worldwide for natural dry physical separation rather than risk having chemical and acetone residue in their chocolates.

## Acknowledgments

The author would like to thank Tan Boon Keng (Lam Soon Bhd, Malaysia), T. Willner and V. Stempel (Krupp Maschinentechnik GmbH, Germany) for reviewing this paper.

## References

1. Macrae, A.R., and R.C. Hammond, *Biotechnol. Gen. Eng. Rev. 3*:193 (1986).
2. Matsuo, T., et al., *Japanese Patent 1, 139*:154 (1983).
3. Willner, T., et al., Proc. *PORIM International Palm Oil Development Conferences, Malaysia* September 5–9, 1989.
4. Willner, T., et al., *Food Sci. Technol. 12*:586 (1989).
5. Willner, T., W. Sitzmann, and E. Münch, *Edible Fats and Oils Processing: Basic Principles and Modern Practices*, D.R. Erickson, ed., 1990, pp. 239–245.
6. Willner, T., W. Sitzmann, and E.W. Münch, Paper presented at the 81st Annual Meeting of the American Oil Chemists' Society, Baltimore, April 22–25, 1990.
7. Willner, T., and W. Sitzmann, *Fat Sci. Technol. 93*:S598 (1990).
8. Willner, T. and W. Sitzmann, *Seminar on Fats and Oils Processing in dry fractionation, Helsingor,* October 10–11, 1990.
9. Willner, T., W. Sitzmann, and K. Weber, *Fat Sci. Technol. 93*:588 (1991).
10. Willner, T., W. Sitzmann, and K. Weber, *Proceedings on Oils and Fats in the Nineties, Fredricia, Denmark* 23–26, 3 (1992).
11. Willner, T., W. Sitzmann, and K. Weber, *Seminar Oils and Fats Technology of Karlshamns Engineering, Sweden,* January 21, 1992.
12. Siew, W.L., and K.G. Berger, *PORIM Technology Series,* No. 6, Malaysia (1981).
13. Soon, W., *Specialty Fats vs. Cocoa Butter,* p. 522, 1991.
14. Kalanithi, N., and Y.S. Salmi, *PORIM Research Report* 00084, Malaysia.
15. Kalanithi, H., Y.S. Salmi, and Nor Rosnah, Md., *PORIM Research Report* 00128b, Malaysia (1987).
16. Flingoh, C.H. Oh, and Zukarinah Kamaruddin *Elaeis, PORIM, Malaysia 1,2,* 03–108 (1989).

# The Production of Fatty Alcohols and Their Amino Derivatives from Coco Fatty Acid Methyl Esters

Morio Matsuda, Masamitsu Horio, Kiyoshi Tsukada, Koushiroh Sotoya, Hiroshi Abe, and Rikio Tsushima

KAO Corporation, Chemicals and Materials Research Laboratory, 1334 Minato, Wakayama City 640, Japan.

## Abstract

Fatty alcohols from natural feedstocks, such as fatty acid methyl esters or fatty acids, have been produced by catalytic hydrogenation at high temperatures and pressures. Copper–chromite catalyst has been widely used for this reaction for years. Because of its chromium component, copper–chromite catalyst has environmental problems in preparation, handling, and disposal. Therefore, we have developed a nonchromium catalyst for the hydrogenation of fatty acid methyl esters to produce the corresponding alcohols.

For economic reasons, we chose copper metal as the main catalyst component and iron oxide as the cocatalyst. This catalyst showed the same high activity as copper–chromite, but the activity of the recovered catalyst dropped because of iron oxide reduction. To prevent reduction of iron oxide, we added aluminum oxide as the third catalyst component. We finally developed the double-promoted copper–iron–aluminum oxide catalyst to catalyze the hydrogenation of methyl esters.

The filterability, activity, and selectivity of the catalyst are important for industrial use. We investigated preparation methods to improve filterability. In this paper, we present the performance of the catalyst in the hydrogenation of coco fatty acid methyl esters regarding activity, selectivity, and filterability. Another catalyst is also introduced for the amination of alcohols prepared from methyl esters.

Long-chain fatty tertiary amines, especially N,N-dimethyl alkyl amines (DMAA), are very important intermediates for producing cationic surfactants, amphoteric surfactants, and amine oxides. There are many industrial processes that produce DMAA. Catalytic amination of fatty alcohol is the most economical and simplest process to produce fatty tertiary amines. In this process, amination catalysts are important. The abilities required of the catalyst in this reaction, the mechanism of fatty alcohol amination, and the performance of the newly developed amination catalyst are presented in this paper.

---

Aliphatic alcohols produced from coconut, palm kernel and palm oils (hereafter called higher alcohols) are used widely as source materials for various industrial agents, such as synthetic detergents, surfactants, plasticizers, and additives for cosmetics and synthetic resins. Recently, higher alcohols produced from natural oils and fats have been praised highly because of the renewable nature of the source; therefore the materials are now being produced extensively worldwide. As a result, production has now exceeded that of synthetic alcohols. Figure 1 and Table 1 show the manufacturing processes and technologies (1). Presently, the liquid phase suspension method using Cu–Cr oxide catalyst (2,3) is the mainstay. So far, no substitute for Cu–Cr oxide catalyst has been developed. However, more stringent regulations on the manufacture, usage, and disposal of this catalyst are expected in the future, due to chromium toxicity. The development of a new catalyst using no chromium therefore is desired strongly.

To satisfy this need, we began to develop alternate catalysts. While researching a catalyst using copper as the major constituent, we investigated the stability of an activator using various metals, and developed a catalyst composed of Cu–Fe–Al oxide. For an industrial catalyst to be economical in a liquid phase suspension process, various

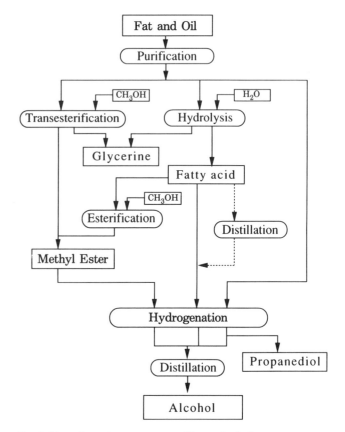

**Fig. 1.** Manufacturing processes of fatty alcohols.

## TABLE 1
### Technical Methods for Fatty Alcohol Manufacture

| Process | Raw Materials | Catalyst | Reaction Conditions |
|---|---|---|---|
| Fluidized bed | Methyl Esters | Powdered Cu–Cr | 250–300°C 200–300 atm |
|  | Fatty acids | Powdered Cu–Cr | 300–320°C 300≈330 atm |
| Fixed bed | Methyl esters | Molded Cu–Cr Molded Cu–Zn Molded Cu–Si | 200–250°C 50–300 atm |

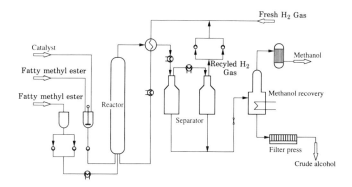

Fig. 3. High-pressure fatty alcohol process.

problems including cost and disposal of catalyst must be solved aside from satisfying the required functions activity, selectivity, and filtration. The Cu–Fe–Al oxide catalyst developed has satisfied all these conditions.

Aside from developing a new catalyst, we have also succeeded in the industrial production of tertiary amines directly from higher alcohols. Higher alcohol amine derivatives (hereafter called amine derivatives) prepared from oils and fats, such as tallow, coconut oil, and palm oil, are important value-added products for the manufacture of softening agents, antistatic agents, gasoline additives, cosmetic bases, pesticides, and detergents. Higher functional and quality amine derivatives produced at lower cost are in great demand resulting in increased research to develop new amines and production technologies (4).

As shown in Fig. 2, several manufacturing methods are used to produce tertiary amines. For high quality products, the most economical and simplest method is to convert higher alcohols directly into amino compounds. With this method, the selectivity, yield, or quality of the resulting amines varies greatly depending on reaction conditions, such as pressure, temperature, and amount of catalyst. Therefore, it is important to develop a catalyst which demonstrates high activity and high selectivity at low temperature and low pressure. The development of Cu–Ni catalyst satisfies all these conditions.

## Development of Cu–Fe–Al Catalyst

Figure 3 shows an example of a higher alcohol-manufacturing facility (5). Heated methyl ester is the starting material. The methyl ester, a catalyst methyl ester mixture, and heated hydrogen are continuously fed into the bottom of the reaction vessel. The methyl ester is hydrogenated while passing through the reaction vessel. A filter or centrifuge is used to separate the reaction product into alcohol and catalyst, and part of catalyst is reused.

The reaction vessel is not equipped with a stirring apparatus. The catalyst is mixed by the flow of methyl ester and hydrogen. As a result, the catalyst used in the liquid phase suspension must have optimal particle diameter, density, specific gravity, and strength in addition to general properties, such as activity, selectivity, durability, and filterability. Since the plants presently used in the industrial production of higher alcohols have optimized their facilities for Cu–Cr catalyst, the physical properties of a catalyst should approximate those of the Cu–Cr catalyst, making the development of a substitute catalyst more difficult.

### Catalyst Evaluation Methods

Using a 0.5 l stirred type autoclave, methyl ester was hydrogenated. During the reaction, sampling was conducted periodically. Figure 4 shows the results of the oils and fats and gas–liquid chromatography (GLC) analyses. The reaction may be divided into an initial stage which is a zero-order reaction and a latter stage which is a first-order reaction. In the initial stage, alcohol is generated from the methyl ester, and an ester wax is generated by an ester interchange of the methyl ester and generated alcohol. In the latter, alcohol is generated from the ester wax. We used the zero-order rate constant for catalyst screening, and the zero-order and first-order rate constant for comparison of catalyst optimization.

According to the results of GLC analysis of samples taken during the reaction, the representative by-products are hydrocarbons, dialkylethers, and alkylmethylethers produced by the dehydrogenation of alcohol. We used the

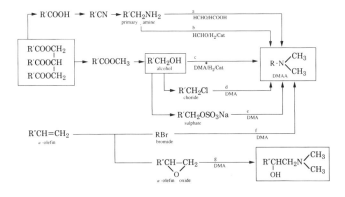

Fig. 2. Manufacturing processes of DMAA. a, Leuckart method; b, reductive methylation method; c, alcohol method; d, chloride method; e, sulfate method; f, bromide method; g, oxyamine method; and *, $(CH_3)_2NH$.

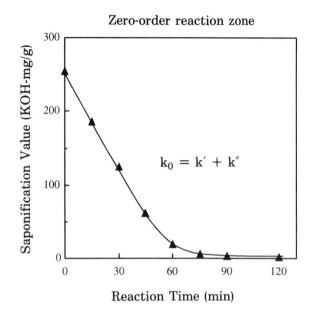

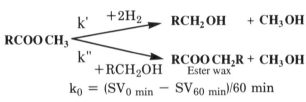

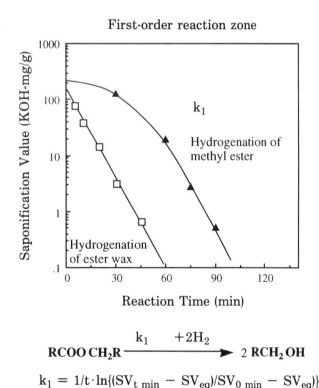

**Fig. 4.** Reaction pathway and definition of rate constant. Reaction conditions: Temperature, 250°C; Pressure, 150 atm; Catalyst, Cu–Cr–O (2.5 wt%).

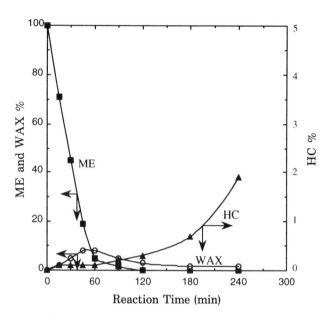

**Fig. 5.** The correlation between ME, WAX, HC and reaction time. Reaction conditions: Temperature, 250°C; Pressure, 150 atm; Catalyst, Cu–Cr–O (2.5 wt%).

amount of hydrocarbons generated as an indicator of selectivity (Figs. 5 and 6).

We started catalyst development by evaluating and analyzing the Cu–Cr catalyst. Copper is essential for the activity and selectivity of the catalyst to hydrogenate methyl esters. The Cu–Cr catalyst is assumed to be stabilized when the Cu forms a spinel structure with Cr to demonstrate and maintain its functions. Therefore, we focused on the trivalent metals which might form a spinal structure with Cu as a substitute for Cr.

Using Cu as an activator and trivalent metals, such as Fe, Al, Nb, La, and Bi, we produced and evaluated catalysts. As shown in Fig. 7, Fe and Al demonstrated initial activity equal to or higher than that of Cr. When X-ray diffraction (XRD) analysis was conducted on Fe, which

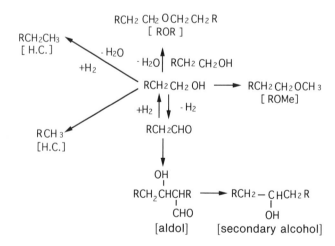

**Fig. 6.** Side reactions of the hydration process.

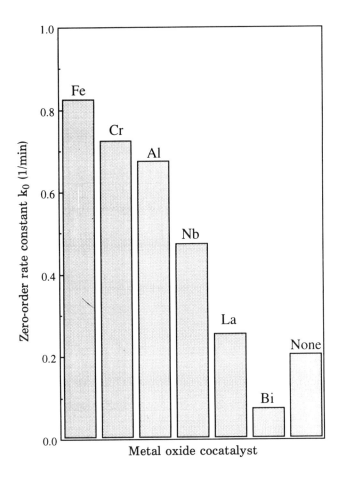

Fig. 7. Activities of copper-based binary oxide catalysts. Catalyst Composition, CuO:MO = 70:30 (weight ratio); Preparation, coprecipitation; Calcination, 450°C/2 h. Reaction conditions: Temperature, 250°C; Pressure, 150 atm; Catalyst, 5 wt%.

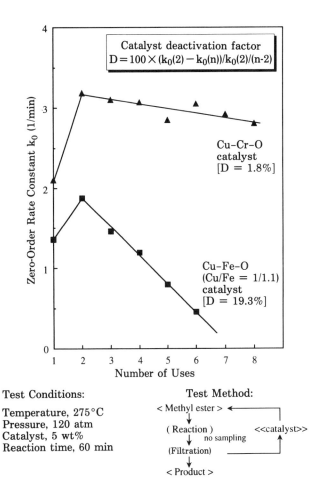

Fig. 8. Catalyst reuse test.

demonstrated the best activity, a spinel structure of $CuFe_2O_4$ was confirmed. Since this result indicated stabilized functions similar to those of Cu–Cr, it was decided to use Fe in combination with Cu for further study.

To determine catalyst function stability, the entire amount of catalyst was collected by filtration after one use. The recovered catalyst was added to methyl ester and reacted under the same conditions. Compared with the Cu–Cr catalyst, the activity of this Cu–Fe catalyst was greatly reduced (Fig. 8). After six uses, the Cu–Fe catalyst was analyzed with XRD, and the generation of α-Fe was discovered to be the result of Fe reduction under high temperature and pressure. The Cu–Fe spinel structure which had stabilized the Cu had broken down, and the activity was reduced by copper sintering, an increase in the particle diameter of Cu (Fig. 9).

We investigated the effect of a tertiary metal to prevent Fe reduction. A correlation was observed between the amount of Al added and prevention of Fe reduction. When the amount of Al was increased, generation of α-Fe was inhibited and the sintering of Cu was prevented. When the Al:Fe ratio was greater than or equal to 1.0 in the bulk composition, the drop in activity was equal to that of the Cu–Cr catalyst (Figs. 10 and 11). We also found that the rate constant $k_1$ and selectivity were correlated to the amount of Al added, and that performance comparable to the Cu–Cr catalyst could be obtained with an aluminum to iron ratio of 1.2 or greater (Fig. 12).

As the next step, the surface Cu–Fe–Al composition of the catalyst and the electron status of Cu were analyzed using ESCA (X-ray photoelectronic spectrometer) and

TABLE 2
Surface and Bulk Composition of Catalyst

| Composition | Cu/Fe/Al = 1/1.1/1.2 | | Cu/Fe/Al = 1/1.1/0.3 | |
|---|---|---|---|---|
| | Surface | Bulk | Surface | Bulk |
| Cu | 6.2 | 30.3 | 26.2 | 41.3 |
| Fe | 2.0 | 33.6 | 47.7 | 46.1 |
| Al | 91.8 | 36.1 | 26.1 | 12.4 |

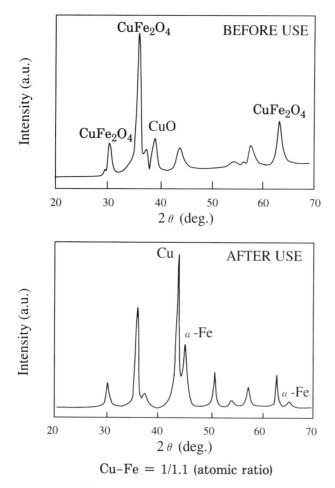

**Fig. 9.** X-ray diffraction pattern of Cu–Fe catalyst.

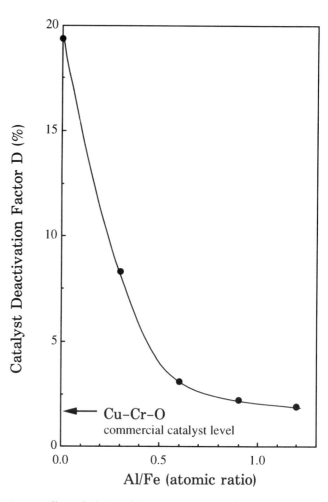

**Fig. 10.** Effect of $Al_2O_3$ addition to Cu–Fe catalyst. Reaction conditions: Temperature, 275°C; Pressure, 120 atm; Catalyst, 5 wt%.

compared to catalyst activity. Table 2 shows the bulk composition and surface composition. Because of the large differences in the bulk and surface compositions, a composition gradient was produced by ion-etching analysis with ESCA (Fig. 13). A considerable change in the composition occurred in the surface layer of the catalyst, which seemed to influence the activity, selectivity, and stability.

For industrial catalysts, filtration (including repeated usage) and particle distribution are very important. We confirmed that the new catalyst has functions and physical properties that approximate those of the Cu–Cr catalyst (Fig. 14).

## Manufacture of Tertiary Amine by the Alcohol Method

For some time, researchers have tried to develop a method to manufacture tertiary amines: *N,N*-dimethylalkylamines by directly converting higher alcohols into amines. Various studies were reported and patents were filed regarding catalyst and reaction mechanisms. However, due to insufficient activity and selectivity of amine compounds, the method has not been industrialized. Catalysts play an important role in the direct alcohol–amino reaction. If a catalyst with a high activity and high selectivity is developed, the manufacture of tertiary amines having versatile structures, (natural or synthetic, long alkyl chain, saturated or unsaturated, and straight chain or branched); improved yield; cost reduction; and improved quality, simple facility, gentle reaction conditions, and reduced amount of catalyst can be expected, making industrial production possible.

Many researchers have investigated mechanisms to turn higher alcohols into amines. A mechanism for the reaction of alcohol with DMA (dimethylamine) has been suggested (Fig. 15[6]). Alcohol is dehydrogenated into aldehyde, then intermediate reaction products are formed by the addition of aldehyde and amine, and the resultant products subjected to hydrogenolysis. The formation of the intermediate reaction products with the addition of aldehyde and amine occurs without the presence of a catalyst (7). The catalytic functions required are both dehydrogenation and hydrolysis.

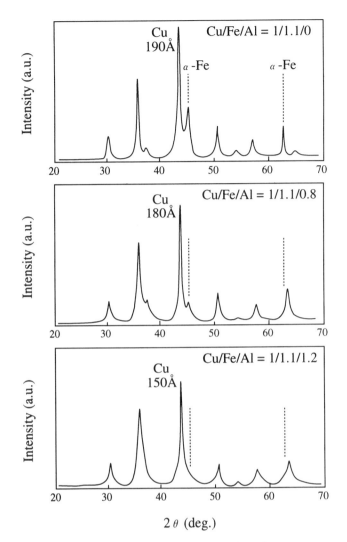

**Fig. 11.** X-ray diffraction pattern of Cu–Fe–Al catalyst.

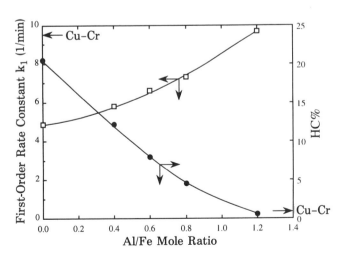

**Fig. 12.** Effect of protection of iron oxide from reduction by addition of $Al_2O_3$. Reaction conditions: Temperature, 275°C; Pressure, 250 atm; Time, 4 h; Catalyst, 2.5 wt%.

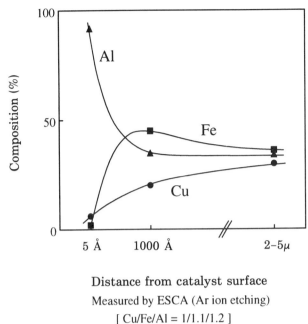

**Fig. 13.** Composition profile of the catalyst particles.

Catalyst-involved side reactions, [1] such as generation of hydrocarbons and carbon monoxide by decomposition of aldehyde, [2] generation of aldol by condensation of aldehyde, as well as monomethyl amine (MMA) and trimethyl amine (TMA) by disproportionation of DMA occur as well. The third side reaction is observed most often, and MMA produced by this reaction further reacts with 2 moles of aldehyde and generates a dialkyl (long-chain) tertiary amine. In

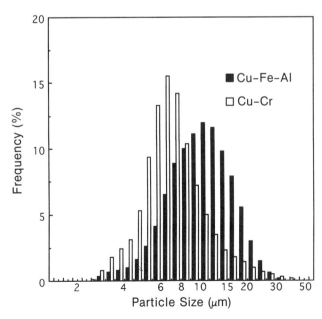

**Fig. 14.** Particle-size distribution of new catalyst.

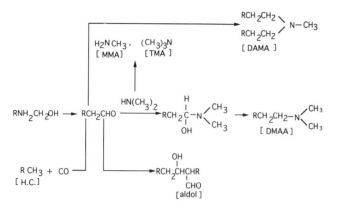

Fig. 15. The supported mechanism of direct amination of catalyst.

**TABLE 3**
**Composition and Performance of Direct Amination Catalyst[a]**

| Catalyst | Reaction Temperature (°C) | Amount of DMDA (%)[b] |
|---|---|---|
| Cu–Ni | 200 | 97.0 |
| Cu–Cr | 210 | 37.0 |
| Raney–Ni | 210 | 25.5 |
| Ni | 210 | unreacted |
| Pd/C, Ru/C | 210 | < 10 |

[a]Reaction conditions: Pressure: 1 atm; Catalyst: 1.0% vs. alcohol; Time: 5 h.
[b]DMDA: N,N-dimethyl dodecyl amine.

order to obtain desirable amines, the development of a catalyst which has well balanced dehydrogenation and hydrogenolysis functions and inhibits irregular formation of amine is necessary.

Copper-based catalysts are reported to have excellent activity and selectivity for the alcohol method of amine generation. However, a catalyst that depends only on copper for activity has a very slow reaction rate, and while large amounts of aldol and enamine are generated, little desirable amine is obtained under ordinary pressure and reaction temperatures of around 200°C. In view of the reaction mechanisms, such an outcome indicates a hydrolytic deficiency, and that the dehydrogenating and hydrolytic functions are not well balanced. Therefore, it is considered difficult to satisfy the activity and selectivity conditions required by the alcohol method with a single metal.

In order to balance the hydrolytic and dehydrogenating functions of copper, we tried combining copper and nickel. As shown in Fig. 16 and Table 3, the maximum activity was obtained by using Cu and Ni at a ratio of 4:1, and it demonstrated higher activity and selectivity compared to conventional Cu–Cr and Raney–Ni catalysts. Thus, demonstrating a marked combined effect of Cu and Ni.

In addition to research on the reaction mechanism, the development of the Cu–Ni composite catalyst relied heavily on analyses at the electronic, atomic, and molecular levels using up-to-date equipment (8,9). We analyzed the surface Cu–Ni composition and Ni valence in the Cu–Ni composite catalyst using ESCA and compared them to catalytic activity and selectivity. Figure 17 shows the bulk and surface metal compositions of the catalyst and the time

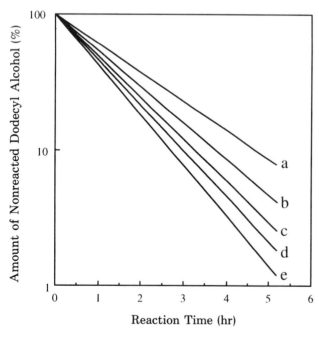

Fig. 16. The activities of supported Cu–Ni catalyst. Reaction conditions: Temperature, 200°C; Pressure, 1 atm. Cu/Ni bulk composition: a, 0.25/1; b, 1/1; c, 2/1; d, 4/1; e, 5/1.

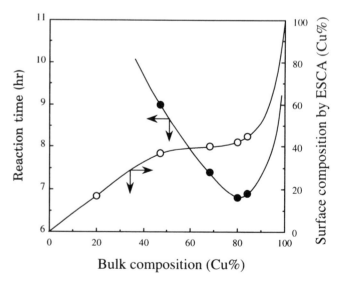

Fig. 17. The correlation between the catalyst composition and catalytic activity. Reaction conditions: Temperature, 200°C; Pressure, 1 atm; Alcohol content < 1.0%; Catalyst, 1.0 wt%.

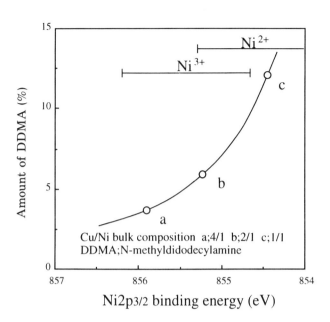

**Fig. 18.** The correlation between the binding energy of $Ni_2P_{3/2}$ and the amount of DDMA.

required for reaction. It was noted that the bulk composition of the catalyst is not necessarily identical to the surface composition, and that a slight difference in the surface composition changes the reaction activity considerably. As trivalent Ni increases in the catalyst surface, less *N*-methyl dialkyl amine is generated (Fig. 18).

## Conclusion

This paper has described the development of chromium-free catalyst for higher alcohol manufacture and of a catalyst for converting higher alcohols into tertiary amines. The development of an efficient catalyst will simplify the manufacturing process, reduce the cost, and improve the quality. In the future, more attention will be focused on catalysts and catalysts will be required to be more versatile. The development of new catalysts depends on basic research of the catalytic action at electronic, atomic, and molecular levels. A fundamental understanding of catalytic processes is important, and will produce more efficient catalysts than simple trial and error.

## References

1. Katoh, A., *Use of Palm and Palm Kernel Oil*, Saiwai Press, Tokyo, Japan, 1990, p. 219.
2. Adkins, H,. and K. Folks, *J. Am. Chem. Soc. 53*:1091 (1931).
3. Adkins, H., and K. Folkvs, *J. Am. Chem. Soc. 54*:1145 (1932).
4. Kreutzer, U.R., and D. Saika, *J. Jpn. Oil Chem. Soc.* (Yukagaku) *88* (1979).
5. Voeste, Th. *J. Am. Oil Chem. Soc. 61* (1984).
6. Baiker, A., and J. Kijenshi, *Catal. Rev. Sci. Eng. 27* (1985).
7. *Organikum*, VED Deutscher Verlag der Wissenschafen, Berlin, 1977.
8. Abe, H., and J. Okabe, *J. Jpn. Oil Chem. Soc.* (Yukagaku) *37*:519 (1988).
9. Abe, H., K. Hashiba, Y. Yokota, and K. Okabe, *J. Jpn. Oil Chem. Soc.* (Yukagaku) *37*:519 (1988).

# Catalytic Hydrogenation of Lauric Oils and Fatty Acids

R.S. Murthy

Natural Oleochemicals Sdn. Bhd., Plo 428, Jalan Besi Satu, Pasir Gudang Industrial Estate, 81700 Pasir Gudang, Johor, Malaysia.

## Abstract

Hydrogenation of lauric oils, such as palm kernel and coconut oils and their fatty acids, have attained great importance recently in view of the increased number of applications in industries, such as foods, cosmetics, and detergents. Hardening of coconut and palm kernel oils and their solid and liquid fractions presents opportunities to prepare products with steep melting properties which are valuable in foods. Stearic acid must have an iodine value (IV) of less than 0.5 for use in a variety of industrial applications. Hydrogenation of the oils and fatty acids involves three phases, the liquid substrate, the gaseous hydrogen, and the solid catalyst. The loop reactor helps to bring the three phases in close contact for efficient hydrogenation. The nature of the nickel catalyst surface and texture are presented. Experimental data obtained in both the laboratory stirred tank reactor and the commercial loop reactor are included for substrates, such as crude palm kernel fatty acids (CKFA) and crude palm kernel oil (CPKO). The loop reactor was found to be very efficient in reducing the IV at lower catalyst consumption.

Hydrogenation of palm kernel oil (PKO), coconut oil (CNO), and their fractions, olein and stearin, yields products with improved solid fat content (SFC) and steep melting point behavior rendering them very useful in applications such as confectionery fats and as cocoa butter substitutes (CBS). The lauric CBS is highly resistant to oxidative rancidity. The hydrogenated stearin is better suited as a cocoa butter extender (CBX) with a high compatibility with cocoa butter and a lower tendency to develop soapy flavors. The fractionated solid portion of the lauric oils has a very high SFC with a melting point that is still lower than body temperature (1,2).

While hydrogenated lauric oils are useful in the food industry, the fatty acids find useful applications in shampoos, shaving creams, and lubricants (3). Triple pressed stearic acids, which are blends of palmitic and stearic acids, with an IV < 0.5, are used in the cosmetic and toiletries industries. Stearic acid and blends of palmitic and stearic acids with low IV are produced by first splitting the triglyceride oils and then hydrogenating the fatty acids either as is or after distillation. The alternative is to hydrogenate the oil to the desired IV first and then split the hardened oil to obtain the fatty acids which subsequently can be distilled to free them from residual nickel (Fig. 1). The nickel consumption for triglyceride oils, such as crude palm kernel oil (CPKO) and palm stearin, is low since there is no catalyst poisoning due to the nickel soap formation that occurs when fatty acids are the feed stocks. This paper describes in detail the hydrogenation of the lauric oils and their fatty acids, the preparation and textural characteristics of the nickel catalyst, and the mode of working of the loop reactor. The experimental data obtained for the hydrogenation of CPKO, crude palm kernel fatty acids (CKFA), and blends of $C_{16}$–$C_{18}$ fatty acids for both the laboratory autoclave and the commercial loop reactor are included.

## Experimental

Catalytic hydrogenation of palm kernel oil, coconut oil, and their split fatty acids was carried out in our laboratory using a Parr-type stirred tank reactor and a commercially available nickel catalyst. Four hundred grams of the oil was used for the reaction. The substrate was placed under vacuum at 100°C to remove moisture and subsequently pressurized with hydrogen to 20 bar. The hydrogenation reaction was completed in 2 hr at 180°C.

The catalyst Nysofact 101 IQ (Engelhard) had a nickel content of 22% w/w and it had a pore size distribution of 12–25 Å (Fig. 2) suitable for complete hydrogenation of oils and fatty acids. The nickel-metal area of the catalyst

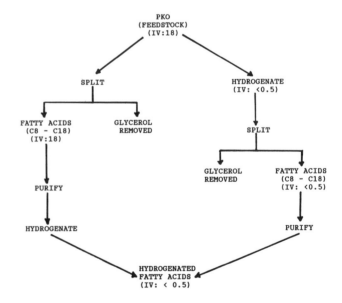

Fig. 1. Routes for production of hydrogenated fatty acids from palm kernel oil.

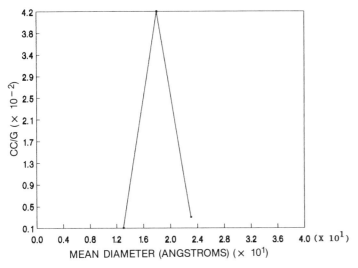

Fig. 2. Pore-size distribution of 101 IQ catalyst.

was 110 m²/g catalyst. The support for the catalyst was found to be silica as the defatted catalyst showed IR vibration at 1020 cm$^{-1}$.

Palm kernel and coconut oils have similar fatty acid compositions in the $C_{12}$ and $C_{14}$ range. The former has higher unsaturates (20%) as compared to the latter (10%), resulting in a higher IV for CPKO. However $C_8$ and $C_{10}$ fatty acid contents of CNO are double that of PKO (Tables 1 and 2).

Low IV can be achieved quickly in the loop reactor with lower catalyst consumption. The main features of the loop reactor are shown in Fig. 3. The reactor assembly consists of a high-pressure vessel with a nozzle protruding from the top. The nozzle introduces a jet of oil along with the catalyst particles into the reactor during which hydrogen gas is pumped into the system. Intimate mixing of the oil, gas, and catalyst occurs. The slurry from the reactor is pumped through a heat exchanger and back into the reactor with a centrifugal pump. Unreacted hydrogen from the headspace is recycled into the system. The reactants and products can be heated or cooled by passing them through the heat exchanger.

Hydrogenation of triglyceride oils and the fatty acids was carried out with a commercial dry-reduced, supported nickel catalyst. Engelhard Nysofact 101 IQ and Unichema

### TABLE 1
### Laboratory Hydrogenation of Crude Coconut Oil

| Analysis | Before Hydrogenation | After Hydrogenation (catalyst dosage (kg/t)) | | |
|---|---|---|---|---|
| | | 0.40 | 0.75 | 0.90 |
| IV | 9.51 | 3.44 | 0.71 | 0.49 |
| AV | 4.65 | 6.00 | 5.47 | 5.85 |
| MOI (%) | 0.057 | — | — | — |
| Color | 6.5R | 3.0R | 3.0R | 1.2R |
| (5 1/4" cell) | 20Y | 11Y | 13Y | 10Y |
| Fatty Acid Composition (%) | | | | |
| $C_6$ | 0.43 | 0.43 | 0.39 | 0.40 |
| $C_8$ | 5.22 | 5.22 | 4.97 | 4.96 |
| $C_{10}$ | 4.91 | 5.52 | 4.72 | 4.70 |
| $C_{12}$ | 46.74 | 45.76 | 45.68 | 45.70 |
| $C_{14}$ | 20.09 | 19.24 | 19.65 | 19.60 |
| $C_{16}$ | 10.39 | 10.50 | 10.62 | 10.60 |
| $C_{18}$ | 2.81 | 9.72 | 13.36 | 13.62 |
| $C_{18:1}$ | 7.68 | 3.47 | 0.46 | 0.32 |
| $C_{18:2}$ | 1.73 | 0.14 | 0.15 | 0.10 |

### TABLE 2
### Laboratory CPKO Hydrogenation

| Analysis | Before Hydrogenation | After Hydrogenation (catalyst dosage (kg/t)) | | |
|---|---|---|---|---|
| | | 0.1 | 0.4 | 1.0 |
| IV | 19.10 | 16.43 | 4.30 | 0.23 |
| AV | 7.72 | 7.18 | 7.59 | 7.59 |
| Color | 6.0R | 0.1R | 0R | 0R |
| (5 1/4" cell) | 60Y | 6Y | 1.0Y | 1.0Y |
| Fatty Acid Composition (%) | | | | |
| $C_6$ | 0.12 | 0.10 | 0.10 | 0.08 |
| $C_8$ | 2.39 | 2.55 | 2.46 | 2.48 |
| $C_{10}$ | 2.73 | 2.81 | 2.77 | 2.80 |
| $C_{12}$ | 50.82 | 48.47 | 48.74 | 48.42 |
| $C_{14}$ | 17.7 | 17.19 | 16.91 | 16.18 |
| $C_{16}$ | 8.43 | 8.87 | 8.61 | 8.19 |
| $C_{18}$ | 1.96 | 2.89 | 17.3 | 21.65 |
| $C_{18:1}$ | 14.76 | 16.66 | 2.92 | — |
| $C_{18:2}$ | 1.63 | 0.48 | 0.20 | 0.21 |

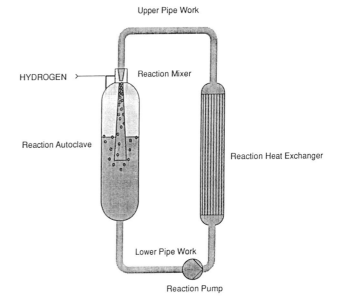

Fig. 3. Loop reactor assembly.

P9932 were reported to be very efficient for the complete hydrogenation of oils and fatty acids. These catalysts have optimal pore-size characteristics that facilitate the hydrogenation of both the large triglyceride molecules (15–20 Å breadthwise) and the fatty acids (5 Å). The preparation of the catalyst involves the use of inorganic supports, such as silica, for deposition of the nickel salt, such as basic nickel carbonate, at the precipitation stage (4–6). The silica source can be kieselguhr, synthetic silica, or ash obtained by burning rice husks. A hot solution of nickel is reacted with a sodium carbonate solution at 95–98°C in the presence of silica and kept under intense stirring. When the base [$Ni(OH)_2$ plus $Ni(CO)_3$] is precipitated on the silica support a pH 9–9.5 is maintained with a nickel to silica ratio of 2–4.

The precipitate is dried and ground to a fine powder which is activated by flowing dry hydrogen gas over it at 400–450°C. The activated nickel catalyst consists of $Ni-SiO_2$ particles 3–20 μm in diameter. Metallic nickel crystallite size is 40–55 Å, with a nickel metal area of 100–120 $m^2$/g catalyst. Optimal nickel-support interactions need to have the right degree of reduction (on the order of 70–80%). The degree of reduction is also dependent upon the extent of drying, the particle size of the basic nickel carbonate powder, the purity of the hydrogen, and the temperature. Promoters, such as molybdenum, help to enhance activity of nickel catalyst for fatty acids (7).

The carbonate to hydroxide ratio of the precipitate has a significant influence on the reducibility of the nickel in the final catalyst (8). An optimal ratio of carbonate to hydroxide is required for a high degree of reduction associated with high metal area. The pore-size distribution of nickel-silica catalyst can be varied to generate a catalyst with the desired pore size, suitable for large triglyceride molecules and fatty acids, when the nickel carbonate is precipitated, or at the high-temperature reduction stage, or both. For complete hydrogenation of triglycerides and fatty acids, high active catalyst with pore size of about 25 Å is needed.

Commercial nickel catalysts are supplied coated with hardened fat due to the pyrophoric nature of metallic nickel. The hydrogenation of triglycerides and fatty acid substrates involves the liquid substrate, the solid catalyst, and hydrogen. The manner in which these three are brought together controls the efficiency of hydrogenation. A high degree of mixing is achieved in reactors such as the loop reactor (9).

## Results and Discussion

### Hydrogenation Studies Using the Laboratory Autoclave

*Hydrogenation of CNO, PKO and Their Fatty Acids.* The results of hydrogenation of CNO and CPKO with different catalyst dosages are shown in Tables 1 and 2, respectively. Iodine values of the coconut oil could be reduced from 9.51 to 0.49 in 2 hr with 0.9 kg/t catalyst.

There was very little increase in the acid value (AV) from 4.65 to 5.85. There was considerable improvement in the color on hydrogenation. The hydrogenated product showed a color of 1.2R 10Y (in 5 1/4″ cell) as compared to 6.5R20Y for the feedstock. The fatty acid composition showed an increase of stearic acid content from 2.81 to 13.62, with almost complete elimination of oleic and linoleic acids at an IV of 0.49.

Crude palm kernel oil has a higher IV (19.10) in view of the high contents of oleic acid and linoleic acid, when compared to an IV of 9.51 for coconut oil. The results presented in Table 2 show the effects of catalyst dosage on IV reached in 2 hr. A final IV of 0.23 can be reached with a 1.0 kg/t catalyst dose without any enhancement in AV. The color improved significantly on hydrogenation with 6R60Y decreasing to 0R 1Y. The linoleic and oleic acids were completely hydrogenated to stearic acid at an IV of 0.23.

The IV of crude coconut fatty acids was reduced from 9.99 to 0.76 at a catalyst dosage of 0.9 kg/t (Table 3). The IV reduction was more pronounced with a catalyst dosage of 1.1 kg/t, resulting in a lower value of 0.48. The oil was more easily hydrogenated than the fatty acids.

The hydrogenation pattern for CKFA indicates that an IV of 19.67 could be lowered to less than 0.5 at a catalyst dose of 1.8 kg/t while having oleic and linoleic acids almost fully hydrogenated (Table 4). Even in this case, the catalyst consumption was significantly higher for the fatty acids than for the triglycerides.

*Hydrogenated Medium- and Long-Chain Palm Kernel Fatty Acids.* Hydrogenated, stripped palm kernel and coconut fatty acids ($C_{12}$–$C_{18}$ fatty acids) are very useful as soap feed stocks. The hydrogenation of these acids proceeds smoothly to convert oleic and linoleic acids com-

**TABLE 3**
**Hydrogenation of Crude Coconut Fatty Acid in the Laboratory**

| Analysis | Before Hydrogenation | After Hydrogenation (catalyst dosage (kg/t)) | | | |
|---|---|---|---|---|---|
| | | 0.40 | 0.75 | 0.90 | 1.10 |
| IV | 9.99 | 9.55 | 5.02 | 0.76 | 0.48 |
| Color | 20R | 30R | 20R | 17R | 8R |
| (5 1/4″ cell) | 52Y | 52Y | 40Y | 30Y | 30Y |
| Fatty Acid Composition (%) | | | | | |
| $C_6$ | 0.24 | 0.23 | 0.20 | 0.11 | 0.17 |
| $C_8$ | 6.03 | 5.93 | 5.32 | 5.10 | 5.36 |
| $C_{10}$ | 5.47 | 5.45 | 4.93 | 5.57 | 5.12 |
| $C_{12}$ | 49.18 | 46.31 | 47.99 | 47.41 | 47.47 |
| $C_{14}$ | 18.12 | 20.02 | 19.45 | 19.38 | 19.21 |
| $C_{16}$ | 9.04 | 10.06 | 10.04 | 9.74 | 9.89 |
| $C_{18}$ | 2.10 | 2.50 | 7.01 | 12.07 | 12.32 |
| $C_{18:1}$ | 7.60 | 7.76 | 4.72 | 0.46 | 0.17 |
| $C_{18:2}$ | 2.32 | 1.75 | 0.32 | 0.16 | 0.29 |

**TABLE 4**
**CKFA Hydrogenation Using Laboratory Autoclave**

| Analysis | Before Hydrogenation | After Hydrogenation (catalyst dosage (kg/t)) | | |
|---|---|---|---|---|
| | | 1.0 | 1.5 | 1.8 |
| IV | 19.67 | 1.92 | 0.65 | 0.48 |
| M & I | 0.051 | 0.038 | 0.035 | 0.025 |
| Fatty Acid Composition (%) | | | | |
| $C_6$ | 0.18 | 0.28 | 0.25 | 0.22 |
| $C_8$ | 2.23 | 2.71 | 2.48 | 2.38 |
| $C_{10}$ | 2.96 | 2.97 | 2.74 | 2.69 |
| $C_{12}$ | 46.40 | 47.38 | 46.17 | 45.72 |
| $C_{14}$ | 17.01 | 16.28 | 16.57 | 16.60 |
| $C_{16}$ | 9.23 | 8.44 | 8.83 | 8.90 |
| $C_{18}$ | 2.20 | 10.15 | 22.27 | 23.07 |
| $C_{18:1}$ | 17.38 | 1.61 | 0.46 | 0.17 |
| $C_{18:2}$ | 2.40 | 0.20 | 0.24 | 0.23 |

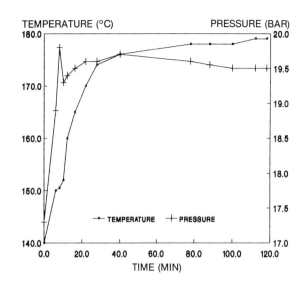

Fig. 4. Loop Reactor: Pressure and temperature profiles for a loop reactor with CKFA as the feedstock.

pletely into stearic acid when a product with IV < 0.5 is formed, the titer being 33–38°C and 28–32°C for the palm-kernel- and coconut-based acids, respectively. Crude palm kernel oil contains free fatty acids (FFA) in the range of 2–5% depending upon the storage conditions of the oil. The oil is freed from the FFA by distilling the acid during the deodorization process at the refining stage. The distillate has about 75% of the FFA in the range of $C_8$–$C_{18:1}$, with an IV of 17.5, and having an unsaponifiable matter concentration up to 1–2%. Hydrogenation of the fatty acid distillate coupled with appropriate purification steps offers opportunities to produce saturated fatty acids as feedstocks for soaps. However, hydrogenation is rendered difficult due to the presence of excessive unsaponifiables. A catalyst dose of 3 kg/t is required to reduce the IV from 17.5 to less than 1, eliminating the oleic and linoleic content almost completely at 180°C and a hydrogen pressure of 20 bar.

## Performance of Loop Reactor for Hydrogenation of Fatty Acids

*Crude Palm Kernel Fatty Acids.* A 12-ton commercial-size batch of CKFA was carried out using the loop reactor (Fig. 3). A low IV of 0.47 was achieved in 2 hr with 1.1 kg/t of the catalyst. Both oleic and linoleic acids were completely converted to stearic acid. The pressure and temperature profiles observed during the reaction in the loop reactor are shown in Fig. 4. After 5 min, the hydrogen pressure dropped from 19.7 bar to 19.3 due to the reaction and stabilized at that level until the reaction was completed. The temperature of the reaction increased from 140°C to 175°C in the first 30 min and stabilized at 175°C for 2 hr until the end of the reaction.

As the fatty acid reactant and the catalyst used for both laboratory study and commercial loop reactor are the same, a direct comparison was attempted (Fig. 5). Final IV achieved in 2 hr for the laboratory reactor were 2.70, 1.75, 0.70, and 0.50 for a catalyst dose of 0.8, 1.0, 1.55, and 1.8 kg/t of fatty acid, respectively. With the loop reactor, the required IV of 0.5 was obtained with a catalyst dose of only 1.1 kg/t compared to 1.8 kg/t in the case of the laboratory autoclave under similar pressure and temperature conditions.

*Oleic-Rich Long-Chain Fatty Acids.* As discussed, hydrogenation of palm kernel fatty acids yields a product with a very low IV. The unsaturated components, oleic and linoleic acids are present only at a 20% concentration in

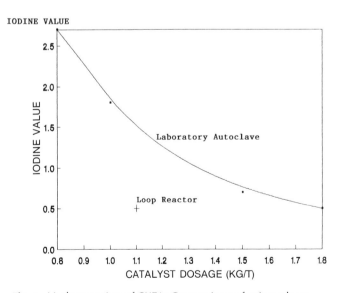

Fig. 5. Hydrogenation of CKFA: Comparison of using a loop reactor and laboratory autoclave to hydrogenate CKFA at 20 bar/180°C.

**TABLE 5**
**Hydrogenation of Oleic-Rich Fatty Acid in the Loop Reactor**

| Analysis | Before Hydrogenation | After Hydrogenation (2 hr) |
|---|---|---|
| IV | 67.20 | 0.71 |
| AV | 200.00 | 202.07 |
| Fatty Acid Composition (%) | | |
| $C_6$ | — | — |
| $C_8$ | — | — |
| $C_{10}$ | — | — |
| $C_{12}$ | 2.71 | 1.17 |
| $C_{14}$ | 1.04 | 0.63 |
| $C_{16}$ | 23.55 | 21.34 |
| $C_{18}$ | 8.58 | 76.20 |
| $C_{18:1}$ | 57.36 | — |
| $C_{18:2}$ | 6.77 | 0.66 |

CKFA; therefore if hydrogenation of the $C_{18:1}$- and $C_{18:2}$-rich feed is carried out, the volume to be handled can be reduced fivefold. Hydrogenation of such a mixture of $C_{16}$–$C_{18}$ fatty acids with an IV of 67 gave a product with an IV of 0.71 in the loop reactor and the material would be useful as triple pressed stearic acid. (Table 5 and Fig. 6). The pressure-temperature profiles observed in the reactor are shown in Fig. 7.

## Influence of Bleaching Earth

Bleaching earth is known to improve the hydrogenation activity of nickel catalyst due to adsorption of catalyst poisons, such as phosphorous and sulphur compounds, present in fatty acids (10). Therefore, studies were carried out in the laboratory using added bleaching earth (Tonsil optimum FF) at 0.5% level for hydrogenation of CKFA (at 180°C and hydrogen pressure of 20 bar). The results

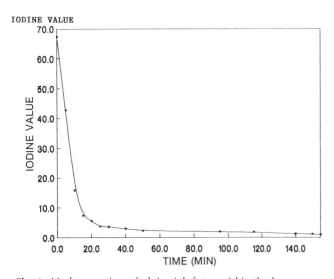

Fig. 6. Hydrogenation of oleic-rich fatty acid in the loop reactor.

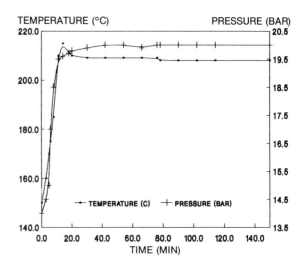

Fig. 7. Pressure and temperature profiles for hydrogenation of oleic-rich fatty acids in the loop reactor.

given in Table 6 show the reduction in catalyst consumption and a vast improvement in color of the hydrogenated product. A catalyst dose of 2.0 kg/t was required to achieve an IV of 0.5, and the same performance was possible with only 1.75 kg/t catalyst when bleaching earth was added at 0.5% (of oil). Such improvements have been observed even for CPKO. The phosphorous content of the oil is 6 ppm with a negligible quantity of sulphur. Therefore, the benefits due to bleaching earth for the oil and fatty acid hydrogenation are small. Other adsorbents, such as active carbon, synthetic silica, and zeolite molecular sieves, are worth considering for this reaction.

## Spent Nickel Catalyst from Fatty Acid Industry

*Reuse of Spent Nickel Catalyst.* The catalyst used in hydrogenation essentially consists of nickel (12–15%), inorganics (10%) due to the presence of the support and addition of filter aid, with the rest being the hardened oil.

**TABLE 6**
**Hydrogenation of CKFA: Influence of Bleaching Earth**

| Bleaching Earth: 0.5% w/w (tonsil optimum FF) | | | | |
|---|---|---|---|---|
| Catalyst Dose P9932 | 0 min (initial IV) (color) | IV after Hydrogenation | | |
| | | 30 min | 60 min | 90 min (color) |
| 2.0 kg/t | 18.11 (6.7R 70Y) (5 1/4" cell) | 10.52 | 4.33 | 0.50 (4.8R 16Y) |
| 1.75 kg/t | 18.11 | 15.98 | 7.17 | 0.74 (4.3R 12Y) |
| 1.75 kg/t + 0.5% BE | 18.11 | 12.53 | 6.53 | 0.46 (1.1R 10Y) |

The spent catalyst available from the hydrogenation of refined PKO or refined palm stearin has sufficient activity remaining that in conjunction with fresh catalyst it can be used for hydrogenation of the fatty acids, since the FFA present in the spent catalyst does not pose a problem. Our laboratory studies indicate that an IV of 0.4 is achieved for CKFA with the addition of 1.0 kg/t of spent catalyst (containing 12% Ni from refined palm stearin hydrogenation) along with 1.5 kg/t of fresh catalyst. For the same performance, fresh catalyst at the level 2.0 kg/t is required. The catalyst used in the fatty acid hydrogenation is not active, and therefore it is not reusable. The spent catalyst is reprocessed to recover nickel metal.

*Recovery of Nickel from Spent Catalyst.* Nickel is recovered from the spent catalyst by first removing the fat either as soap, by treating the catalyst with hot alkali; or burning off the fat at high temperature in a furnace. The spent catalyst (free of fat) consists of nickel (as oxide) and inorganics such as silica, alumina, magnesia, and ferric oxide. The mass is digested with hot sulfuric or nitric acid or a mixture of the two; nickel goes into solution as a salt along with soluble impurities, such as Al, Mg, and Fe. The insoluble silica is filtered off, and a pure nickel solution is recovered by selectively precipitating the rest of the impurities. The nickel solution is useful in electroplating or is reprocessed as catalyst for hydrogenation of vegetable oils and fatty acids through either the dry reduction route or the formate route.

## References

1. Grothues, B.G.M., *J. Am. Oil Chem. Soc. 62*:390 (1985).
2. Pease, J.J., *J. Am. Oil Chem. Soc. 62*:426 (1985).
3. Kaulistian, P., *J. Am. Oil Chem. Soc. 62*:431 (1985).
4. Coenen, J.W.E., in *Preparation of Catalysts II,* Elsevier, New York, 1979, p. 89.
5. Coenen, J.W.E., *J. Oil Technol. Assoc. India 16* (1969).
6. Srinivasan, G., R.S. Murthy, K.M. Vijaya Kumar, and P.V. Kamat, *J. Chem. Tech. Biotechnol. 30*:217 (1980).
7. Srinivasan, G., and R.S. Murthy, Indian Patent 157579, 1984.
8. Nitta, Y., T. Imanaka, and S. Teranishi, *J. Catalysis. 96*:429 (1985).
9. Buehler G., in *Fatty Acids in Industry,* edited by R.W. Johnson and E. Fritz, Marcel Dekker, Inc., New York, 1989, pp. 113–152.
10. Zschau, W., *J. Am. Oil Chem. Soc. 61*:214 (1984).

# Discussion

The Chairperson summarized the subject matter that was covered and a question was raised as to the prevention of breakdown and discoloration during distillation of coconut fatty acids. The response was that careful control of raw material quality is helpful, as is the control over heating by using high vacuum and by minimizing the temperature differentials in the still. After distillation, storage of products under nitrogen helps to prevent color and odor development.

There also were two questions regarding hydrogenation. The first dealt with the most economical size of reactors using the Buss loop. The second regarded the reuse of catalyst. In the first instance, information has been published on the Buss loop in earlier world conferences. In the second case, catalyst reuse is highly dependent on the process and products in question. There are no set rules, and each situation must be assessed for acceptability and economics of catalyst reuse.

# Chemical and Physical Properties of Palm Kernel Oil

C.L. Chong[a] and W.L. Siew

Chemistry and Technology Division, Palm Oil Research Institute of Malaysia (PORIM), P.O. Box 10620, 50720 Kuala Lumpur, Malaysia.

## Abstract

Physical and chemical properties of palm kernel oil have been established, including the density in air, viscosity, and fatty acid composition. The density in air of Malaysian palm kernel oil lies in the range of 0.9087–0.8809 g/mL for refined oil and in the range of 0.9077–0.8799 g/mL for the crude oil in the temperature range of 35°–75°C. The viscosity of palm kernel oil decreases from 40.41 cP to 10.43 cP for the refined oil and from 39.03 cP to 8.7 cP for the crude oil over the same temperature range.

Palm kernel oil contains mainly $C_{12}$ and $C_{14}$ fatty acids. The Wijs iodine value (IV) of Malaysian palm kernel oil varies normally from 16.2–19.2, with a mean of 17.8, but slightly higher values, up to 19.6, have been observed. A wide variation in oil composition has been observed in different sections of the kernel. Quality control data, monitored from 1981–1986, showed seasonal effects on the IV of palm kernel oil. The IV also is affected by kernel size. Data on the carbon number analysis and solid fat content of Malaysian palm kernel oil also are given.

The crystallization behavior of palm kernel oil has been studied by means of X-ray diffraction, differential scanning calorimetry, and optical microscopy. Palm kernel oil crystallizes in the β'-form, in the temperature range 0°–25°C, as is evidenced by the short spacings of the diffraction pattern obtained, and the needle-shaped crystals observed in optical microscopy.

Palm oil is obtained from the mesocarp of the palm fruit while palm kernel oil is extracted from the kernel of the fruit. Though both types of oils are obtained from the same fruit, they have very different chemical and physical properties, with palm oil being rich in $C_{16}$ and $C_{18}$ fatty acids while palm kernel oil is rich in the $C_{12}$ fatty acid. Hence palm kernel oil is often termed a lauric oil, together with coconut and barbassu oil. However, its chemical composition is similar to but broader in range than that of coconut oil, and so are its physical properties. Due to this, its areas of application are more versatile than those of coconut and other lauric oils. This paper deals with the observed chemical and physical properties of palm kernel oil, and the effect of kernel size on its iodine value (IV).

[a]To whom all correspondence should be addressed.

## Chemical Properties

### Fatty Acid and Triglyceride Composition (1)

Malaysian palm kernel oil shows a well defined and fairly uniform chemical composition as illustrated by its fatty acid composition shown in Table 1. From the table, it can be seen that the fatty acid chains present in palm kernel oil stretch from those containing six carbon atoms ($C_6$) to those with eighteen carbon atoms ($C_{18}$). The major fatty acids present are those of the lauric, myristic, palmitic, and oleic acids with lauric acid being the dominant fatty acid (48%), hence its designation as a lauric acid.

The triglyceride composition of Malaysian palm kernel oil is shown in Table 2 in the form of carbon number analysis. The carbon numbers of palm kernel oil triglycerides range from $C_{28}$–$C_{54}$ with $C_{36}$ and $C_{38}$ as the dominant triglycerides present. This is consistent with the results observed in its fatty acid composition.

### Iodine Value

Iodine value of Malaysian palm kernel oil lies in the range of 16.2–19.2 (2). The average IV is 17.8. These values compare well with those quoted by King et al. (3) There are samples of palm kernel oils having IV higher than those usually observed; during certain periods of the year, the IV of palm kernel oils can be observed to be higher than the mean. Generally, these trends or samples concern end-users, as high IV tend to affect fractionation processes and the yields of the palm kernel stearin obtained.

**TABLE 1**
**Fatty Acid Composition of Malaysian Palm Kernel Oil (1)**

| Composition | Mean (%) | SD | Observed Range (%) |
|---|---|---|---|
| $C_6$ | 0.3 | 0.07 | 0.1–0.5 |
| $C_8$ | 4.4 | 0.47 | 3.4–5.9 |
| $C_{10}$ | 3.7 | 0.24 | 3.3–4.4 |
| $C_{12}$ | 48.3 | 0.94 | 46.3–51.1 |
| $C_{14}$ | 15.6 | 0.33 | 14.3–16.8 |
| $C_{16}$ | 7.8 | 0.36 | 6.5–8.9 |
| $C_{18}$ | 2.0 | 0.19 | 1.6–2.6 |
| $C_{18:1}$ | 15.1 | 0.24 | 13.2–16.4 |
| $C_{18:2}$ | 2.7 | 0.24 | 2.2–3.4 |
| Others | 0.2 | 0.09 | trace–0.9 |

**TABLE 2**
**Triglyceride Composition of Malaysian Palm Kernel Oil**

| Composition | Mean (%) | SD | Observed Range (%) |
|---|---|---|---|
| $C_{28}$ | 0.55 | 0.27 | 0.1–1.9 |
| $C_{30}$ | 1.25 | 0.26 | 0.8–2.1 |
| $C_{32}$ | 6.34 | 0.54 | 5.6–6.8 |
| $C_{34}$ | 8.43 | 0.49 | 7.7–9.5 |
| $C_{36}$ | 23.33 | 1.48 | 19.1–26.2 |
| $C_{38}$ | 16.96 | 0.64 | 14.8–18.5 |
| $C_{40}$ | 9.79 | 0.36 | 9.3–10.8 |
| $C_{42}$ | 9.10 | 0.40 | 8.3–10.1 |
| $C_{44}$ | 6.56 | 0.33 | 5.9–7.2 |
| $C_{46}$ | 5.14 | 0.34 | 4.7–5.8 |
| $C_{48}$ | 5.79 | 0.53 | 4.8–6.9 |
| $C_{50}$ | 2.30 | 0.31 | 1.5–3.4 |
| $C_{52}$ | 2.17 | 0.34 | 1.7–3.3 |
| $C_{54}$ | 2.43 | 0.45 | 1.8–3.7 |

**TABLE 4**
**Percentage Size Distribution and Iodine Value of Kernels in Random Samples of 1 kg Batches**

| | Amount of kernels (%) | | |
|---|---|---|---|
| Size of Kernels | 1 | 2 | 3 |
| Large | 15.32 | 19.13 | 24.89 |
| Medium | 50.53 | 57.81 | 52.79 |
| Small | 34.05 | 23.06 | 22.32 |

| | IV | | |
|---|---|---|---|
| Size of Kernels | 1 | 2 | 3 |
| Large | 16.8 | 16.4 | 16.2 |
| Medium | 17.4 | 17.3 | 17.6 |
| Small | 18.2 | 18.4 | 18.9 |
| IV of overall batch (Wijs) | 17.7 | 17.5 | 17.7 |
| Calculated IV | 17.5 | 17.5 | 17.6 |

Large: >12.5 mm; Medium: 9.5–12.5 mm; Small: <9.5 mm.

It has been found that certain factors affect the IV of the kernel oil. In a study where kernels were sliced and divided into outer, inner, and center portions, oil extracted from the different sections showed IV ranging from 9 to 32 (Table 3). The highest IV was found in the outer layer of the kernel, and the lowest IV in the center of the kernel. This unusual characteristic of the palm kernel suggests that the size of kernels would affect the IV of the oil (Table 4). Palm kernels vary in size from less than 9.5 to those greater than 12.5 mm. The larger kernels produce oil of IV of 16.2–16.8 as compared to IV of 18.2–18.9 in smaller kernels.

Table 5, which shows kernels from three plantation mills, indicates that the IV is affected by the composition of the kernel and the amount of broken kernels. The IV of broken kernels is generally higher because some seepage of crude palm oil may have occurred if the nut was broken during the extraction process. Thus processing conditions and the size of the kernel are two main factors affecting the IV of kernel oil.

## Other Chemical Properties

The values in Table 6 show the saponification value and unsaponifiable matter for Malaysian palm kernel oils.

**TABLE 3**
**IV Distribution of the Oil in the Kernel**

| Fruits | Oil | IV (Wijs) |
|---|---|---|
| Unsterilized | Outer | 31.9 |
| Fruits | Middle | 15.1 |
| | Inner | 9.3 |
| | PKO | 18.9 |
| Sterilized | Outer | 26.6 |
| Fruits | Middle | 15.0 |
| | Inner | 9.7 |
| | PKO | 18.7 |
| Production | Outer | 20.3 |
| Kernel | Middle | 14.9 |
| Ex-mill | Inner | 13.2 |
| | PKO | 17.3 |
| | Shells | 23.4 |

**TABLE 5**
**Mean IV of Kernels (from 3 plantation mills)**

| Mill | Kernel Type | Amount (%) | IV |
|---|---|---|---|
| J | Small | 14.7 | 19.3 |
| | Medium | 40.6 | 18.7 |
| | Large | 11.0 | 17.7 |
| | Broken kernels | 32.6 | 18.1 |
| | Homogeneous samples | | 18.4 |
| TM | Small | 12.3 | 19.3 |
| | Medium | 42.0 | 18.0 |
| | Large | 16.2 | 17.8 |
| | Broken kernels | 27.5 | 18.0 |
| | Homogeneous Sample | | 18.3 |
| UB | Small | 18.1 | 19.3 |
| | Medium | 45.1 | 18.2 |
| | Large | 12.6 | 17.9 |
| | Broken kernels | 24.9 | 18.3 |
| | Homogeneous Sample | — | 18.3 |

Results from weekly samples collected from April 1987 to May 1988.

TABLE 6
Saponification Values and Unsaponifiable Matter

| Parameters | Mean Value | Range | SD |
|---|---|---|---|
| Saponification value | 245 | 243–249 | 1.4 |
| Unsaponifiable matter (%) | 0.3 | 0.1–0.8 | 0.16 |

Source: Ref. 2.

## Physical Properties

### Density in Air

The values in Table 7 show the density in air for crude palm kernel oil (CPKO) and refined, bleached and deodorized palm kernel oil (RBDPKO). These results are shown graphically in Fig. 1. From the results, it can be seen that the density of palm kernel oil, like that of other vegetable oils, decreased as the temperature increased. The temperature coefficient was 0.0007 g/mL. The free fatty acid content (FFA) of the RBDPKO was low (<0.1%) while that of the crude palm kernel oils used was between 2–4% as lauric acid. It can be seen that the density in air of RBDPKO was higher than that of the CPKO. This was due to the removal of some of the materials during refining.

### Viscosity

Viscosity is a measure of the internal friction of the oil molecules. The data in Table 8 gives the viscosity values for CPKO and RBDPKO. The results are shown graphically in Fig. 2. The results show that the viscosities of crude and refined palm kernel oils decreased as temperature increased. The viscosity of RBDPKO was observed to be slightly higher than that of CPKO at the same temperature.

### Refractive Index, Color and Slip Melting Point

The values observed for the refractive index, color and slip melting point for Malaysian crude palm kernel oils are as

TABLE 7
Density of Palm Kernel Oil

| Temperature (°C) | Density in Air (g/mL) | |
|---|---|---|
| | CPKO | RBDPKO |
| 35 | 0.9077 | 0.9087 |
| 40 | 0.9042 | 0.9052 |
| 45 | 0.9007 | 0.9017 |
| 50 | 0.8972 | 0.8983 |
| 55 | 0.8938 | 0.8948 |
| 60 | 0.8903 | 0.8913 |
| 65 | 0.8868 | 0.8879 |
| 70 | 0.8833 | 0.8844 |
| 75 | 0.8799 | 0.8809 |

TABLE 8
Viscosity of Palm Kernel Oils

| Temperature (°C) | Viscosity (cP) | |
|---|---|---|
| | CPKO | RBDPKO |
| 30 | 39.03 | 40.41 |
| 35 | 32.23 | 33.10 |
| 40 | 26.5 | 26.88 |
| 45 | 22.12 | 22.48 |
| 50 | 18.61 | 19.49 |
| 55 | 15.94 | 16.73 |
| 60 | 13.24 | 14.73 |
| 65 | 11.39 | 12.88 |
| 70 | 10.00 | 11.32 |
| 75 | 8.78 | 10.43 |

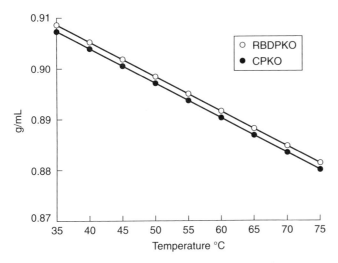

Fig. 1. Apparent Density in air (g/mL) of RBDPKO and CPKO.

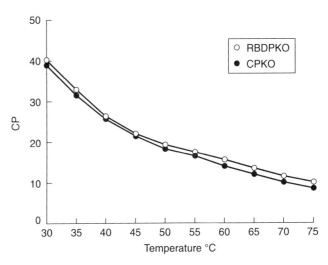

Fig. 2. Viscosity of RBDPKO and CPKO.

shown in Table 9. From the values in Table 9, it can be seen that crude palm kernel oil is dark yellow in color as compared to the reddish-orange color of crude palm oil. This lighter color is due to the absence of trace amounts of carotene present in the kernel oil (<10 ppm) as compared to that in crude palm oil (<500 ppm). After refining, the amount of carotene is further reduced, and the oil becomes a very light yellow in color.

The slip melting point of palm kernel oil is also observed to be much lower than that of palm oil (34.2°C). This is due to the shorter fatty acid chains present in palm kernel oil triglycerides as compared to those present in palm oil.

## Solid Fat Content

From its slip melting point, it can be seen that palm kernel oil is a semisolid oil at ambient temperature (<28°C) and a liquid oil above 28°C. The values in Table 10 give the typical solid fat content (SFC) of CPKO and RBDPKO as measured by wideline NMR. From Fig. 3, it can be seen that palm kernel oil contains a relatively high solids content at ambient temperature, with a sharp melting point below 30°C. This sharp melting point below body temperature (34°C) and its high solids content at ambient temperature renders palm kernel oil or its solid fractions suitable

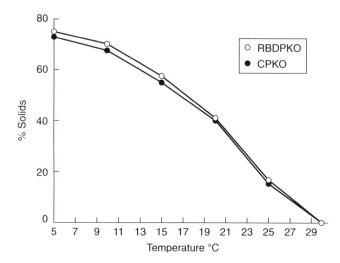

**Fig. 3.** SFC of RBD palm kernel oil and crude palm kernel oil.

**TABLE 9**
Refractive Index, Color and Slip Melting Point of Palm Kernel Oil

| Property | Mean Value | Range Observed | SD |
|---|---|---|---|
| Refractive index | 1.4509 | 1.4500–1.4518 | 0.0005 |
| Slip melting point (°C) | 27.3 | 25.9–28.0 | 0.33 |
| Color (Lovibond 5¼″ cell) | 5.5 red 50 yellow | 4.0 red–8.0 red 37 yellow–77 yellow | 0.9 red 0.7 yellow |
| Color (FAC) | — | 5–13 | 1.4 |
| Carotene content (ppm) (446 nm) | 7.6 | 4.3–11.8 | 1.5 |

Source: Ref. 2.

**TABLE 10**
SFC of Palm Kernel Oils by NMR

| Temperature (°C) | SFC (%) CPKO | RBDPKO |
|---|---|---|
| 5 | 73.1 | 74.4 |
| 10 | 67.5 | 68.6 |
| 15 | 56.5 | 58.1 |
| 20 | 40.0 | 41.1 |
| 25 | 17.2 | 18.2 |
| 30 | — | — |

for many edible applications, for example as a coating fat or a cocoa butter substitute (CBS) either with or without further modification, since the properties observed give the cooling sensation when melted in the mouth and produce other properties necessary for confectionary applications.

## Crystallization behavior

The crystallization behavior of palm kernel oil was investigated over the temperature range 0°–25°C. It was observed that over this temperature range, the palm kernel oil triglycerides tend to crystallize out in the β′-form, as evidenced by the two strong diffraction lines at 4.19 Å and 3.77 Å and the needle-shaped crystals observed under the optical microscope. This β′-form accounts for the difficulty in separating the occluded olein from the stearin, hence the high-pressure presses required to process palm kernel stearin.

## Thermal Behavior and Fractionation

The crystallization process can be monitored thermally by means of differential scanning calorimetry (DSC), both during the crystallization and melting phases of the process. Figure 4 shows a crystallization thermogram of palm kernel oil while Fig. 5 shows a melting thermogram. Compared to palm oil, the cooling thermogram of palm kernel oil shows similar olein and stearin peaks. However, unlike the palm oil-cooling thermogram, the two crystallization peaks are not well separated, but one merges into the other. This explains the high amount of occluded olein and the difficulty in separating the olein and stearin by normal processing equipment. From the melting thermogram, it can be seen that the melting thermogram of palm kernel is less complex than that of palm oil, with smaller shoul-

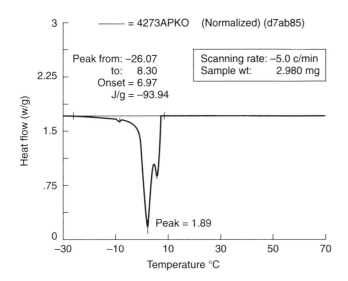

Fig. 4. Cooling thermogram of palm kernel oil.

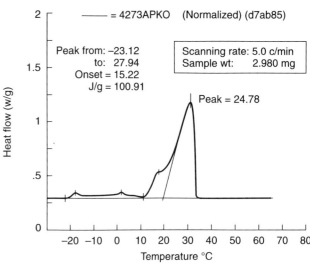

Fig. 5. Heating thermogram of palm kernel oil.

ders and fewer inflection points. This could be due to the more narrow chemical composition observed.

Since palm kernel oil can be crystallized at room temperature or below, commercial fractionation of palm kernel oil can be and is being carried out to produce palm kernel olein and palm kernel stearin with different chemical and physical properties. Palm kernel olein is used as a feedstock for the oleochemical industry as a primary lauric source, while palm kernel stearin is in high demand as a lauric fat or coating.

## Acknowledgments

Permission from the Director-General of PORIM to present this paper is gratefully acknowledged.

## References.

1. Siew, Wai Lin, *PORIM Bulletin 19*:19 (1986).
2. Siew, W.L., and K.G. Berger, *PORIM Technology* 6 (1981).
3. King, B., and A.Z. Stephonie, *Leatherhead Food R.A. Research Report* 559 (1986).

# Characteristics and Properties of Malaysian Palm Kernel-Based Specialty Fats

**Thin Sue Tang and Flingoh Chuan Ho Oh (deceased)**

Palm Oil Research Institute of Malaysia (PORIM), No. 6, Persiaran Institusi, 43650 Bandar Baru Bangi, Selangor Darul Ehsan, Malaysia.

## Abstract

A recent survey of Malaysian-produced palm kernel specialty fats showed that the palm kernel stearin and hydrogenated palm kernel stearin have consistent properties and compositions within narrow ranges. However, palm kernel olein and its hydrogenated products showed a slightly wider spread in iodine value and slip melting point. The stearins and hydrogenated stearins from pressing and detergent processes have comparable slip melting points and lauric acid contents. The solid fat contents (SFC) of the products are affected by the free fatty acid content. Hydrogenated palm kernel oil shows a fairly wide spread of composition and properties, and has comparatively high solids at 35°C making it less desirable as a cocoa butter substitute.

Palm kernel oil is an important co-product from the milling of the fruit of the palm, *Elaeis guineensis*. Its production in the world has been increasing steadily over the last 20 years, largely as a result of the successful cultivation of the oil palm and the down-stream processing of palm oil in Malaysia and Indonesia (Table 1). Today palm kernel oil and coconut oil are the main sources of lauric oil (Table 2), which is an important raw material for the oleochemical and the food industries.

Palm kernel oil, though obtained from the same tree as palm oil, has very different characteristics and fatty acid composition compared to the latter (Table 3 [1–3]). Crude palm oil is rich in carotenoids, which give it a distinctly reddish color; is high in palmitic (C16:0) and oleic (C18:1) acids; and is slow melting, with a slip point of about 33–39°C. Crude palm kernel oil is yellowish, with virtually no carotenoids. It is high in lauric (C12:0) and myristic

### TABLE 1
**World Production of Palm Kernel Oil (1000 MT)**

| Country | 1970 | 1980 | 1990 | 1991 | 1992 |
|---|---|---|---|---|---|
| Malaysia | 31 | 236 | 827 | 782 | 812 |
| Indonesia | 2 | 48 | 229 | 256 | 273 |
| Nigeria | 35 | 80 | 146 | 170 | 171 |
| Thailand | NA[a] | NA[a] | 21 | 22 | 25 |
| Niugini | NA[a] | NA[a] | 12 | 15 | 22 |
| Other African | NA[a] | NA[a] | 85 | 90 | 96 |
| South America | NA[a] | NA[a] | 52 | 55 | 63 |
| Others | 332 | 181 | 76 | 72 | 70 |
| Total | 400 | 545 | 1448 | 1462 | 1532 |

[a]NA: Not available.
Sources: *Oil World 1958–2007* and *Oil World Annual 1993*.

### TABLE 2
**Consumption (MMT) of Lauric Oils and their Percentage Share in the Total World Consumption**

| | Palm Kernel Oil | | Coconut Oil | |
|---|---|---|---|---|
| Year | Consumption | % Share[a] | Consumption | % Share[a] |
| 1960 | 0.42 | 1.42 | 1.88 | 6.44 |
| 1970 | 0.39 | 0.98 | 2.02 | 5.08 |
| 1980 | 0.61 | 1.07 | 2.58 | 4.52 |
| 1990 | 1.34 | 1.67 | 3.13 | 3.90 |
| 1992 | 1.54 | 1.78 | 3.12 | 3.90 |
| 1995(f)[b] | 1.89 | 1.97 | 3.14 | 3.25 |
| 2000(f)[b] | 2.59 | 2.29 | 3.15 | 2.75 |

[a]% Share is computed based on total world consumption of 17 major commercial oils and fats.
[b](f) forecast figure.
Source: Chow, C.S. *PORIM Report* G (183) 93.

### TABLE 3
**Characteristics of Palm Kernel Oil and Palm Oil (1–3)**

| | Palm Kernel Oil | Palm Oil |
|---|---|---|
| SMP (°C) | 25.9–28.0 | 33.0–39.0 |
| IV (Wijs) | 16.2–19.2 | 50.1–54.9 |
| Fatty Acid Composition (%) | | |
| C6:0 | 0.1–0.5 | — |
| C8:0 | 3.4–5.9 | — |
| C10:0 | 3.3–4.4 | — |
| C12:0 | 46.3–51.1 | 0.1–0.4 |
| C14:0 | 14.3–16.8 | 1.0–0.4 |
| C16:0 | 6.5–8.9 | 40.9–47.5 |
| C18:0 | 1.6–2.6 | 3.8–4.8 |
| C18:1 | 13.2–16.4 | 36.4–41.2 |
| C18:2 | 2.2–3.4 | 9.2–11.6 |
| Other | T–0.9 | 0.3–1.3 |

**TABLE 4**
Exports (1000 MT) and the Average FOB Price (RM)[a] of RBD Palm Kernel Specialty Fats from Malaysia for 1985–1993

| Products | 1985 | | 1990 | | 1991 | | 1992 | | 1993[b] | |
|---|---|---|---|---|---|---|---|---|---|---|
| | MT | RM | MT | RM | MT | RM | MT | RM | MT | RM |
| PK[c] Olein | 4.9 | 1368 | 41.1 | 828 | 53.3 | 914 | 51.0 | 1294 | 39.0 | 1043 |
| HPK[d] Olein | 0.01 | 2452 | 9.0 | 1468 | 11.3 | 1487 | 11.7 | 1709 | 8.5 | 1574 |
| PK Stearin | 0.9 | 3379 | 24.1 | 1785 | 27.0 | 1982 | 29.0 | 2287 | 18.3 | 2116 |
| HPK Stearin | 0.03 | 4360 | 8.7 | 2566 | 8.8 | 2605 | 12.7 | 2750 | 7.2 | 2641 |

[a]RM: Malaysian Ringgit, US $ 1 ≈ RM 2.47
[b]January–June only
[c]PK: Palm kernel
[d]HPK: Hydrogenated palm kernel
*Source:* Department of Statistics, Malaysia; and *Newsletter of the Malaysian Edible Oil Manufacturer's Association* (MEOMA).

(C14:0) acids and contains lesser amounts of oleic (C18:1), capric (C10:0), and caprylic (C8:0) acids. More importantly, it has a sharp melting point of 25–27°C and a steep solid fat content (SFC) profile.

It is these unique physical characteristics that make palm kernel oil an excellent raw material for the production of cocoa butter substitutes (CBS). Coconut oil is less ideal for this purpose, as it contains a larger amount of $C_{30}$–$C_{34}$ triglycerides, whereas in palm kernel oil the major triglyceride has a carbon-number of $C_{36}$. The production technologies for palm kernel-based specialty fats, though well developed for many years, are not widely published. The know-how usually results from in-house research and development, and thus is proprietary to the companies involved.

In the past, the production of the lauric specialty fats was confined mainly to established manufacturers in developed countries. With the rapid and substantial increase in the processing of palm oil in the last two decades, enterprising Malaysian palm-oil refiners have also ventured into this area. There are now seven manufacturers performing the fractionation of palm-kernel oil and the subsequent hydrogenation of the fractions obtained to produce a variety of CBS and other specialty fats. The annual production capacity of these producers is estimated to be about 100,000–150,000 metric tons (MT). The growth of this sector of the palm-oil industry is reflected by the gradual increase in the export of various processed palm kernel fractions and related products from a meager 5,800 MT in 1985 to 104,400 MT in 1992 (Table 4). These products are exported mainly to the developed countries of the world, with the United States being the largest importer in 1992. These products are usually marketed under trade names, and the specifications of some commercial products are given in Table 5.

## Production Methods

Currently there are three different commercial methods used for the fractionation of palm kernel oil (4–6), namely, physical pressing, detergent processing, and solvent crystallization. Facilities for all these processes are available in Malaysia. The procedures involved are briefly described in the following sections.

### Physical Pressing

This is the most preferred method, as it is more economical and does not require the use of any processing aids. There are two variations, one using older vertical hydraulic presses and the other using more recently introduced horizontal presses. In this method, the dried crude palm kernel oil is first chilled in stainless steel trays for a certain length of time to cause rapid crystallization. The chilled slabs are then removed from the trays, wrapped in filter cloth, and stacked in perforated compartments in the vertical hydraulic presses. Sufficient pressure is then applied to squeeze the olein from the slabs, leaving behind the solid stearin to be unwrapped from the cloth and melted prior to storage. This process is labor inten-

**TABLE 5**
Specifications of Some Malaysian Commercial CBS from Palm Kernel Oil

| | From PKS | | From HPKS | |
|---|---|---|---|---|
| Tests | 1 | 2 | 3 | 4 |
| IV (Wijs) | 7.5 max. | 6–8 max. | 1.0 max. | 1.0 max. |
| SMP (°C) | 31–33 | 34 max. | 33–35 | 33.5–35.5 |
| Solid Fat Content by NMR (%) Temperature (°C) | | | | |
| 20 | 80 min. | 75–85 | 90 min. | 95–97 |
| 25 | 65 min. | — | 80 min. | 85–87 |
| 30 | 30 min. | 30–40 | 40 min. | 45–48 |
| 35 | 1 max. | 1 max. | 3 max. | 6 max. |
| 40 | 0 | — | 0 | 1.0 max. |

sive, and the yield is said to be about 40–45%, depending upon the quality of the feed material and the desired quality of the stearin fraction.

### *Detergent Processing*

Many palm oil refineries with detergent fractionation plants have modified them to fractionate crude palm kernel oil. The principle of operation is similar to that of detergent fractionation of palm oil, where sodium lauryl sulfate is used to trap the stearin solid in the crystallized palm oil mass, resulting in the formation of two phases. These are separated by centrifugation, and the detergent phase is then heated to melt and liberate the stearin which is recovered by centrifugation.

In the detergent fractionation of palm kernel oil, a portion of the palm kernel olein is recycled to the crystallizer prior to cooling. This is to decrease the solid:liquid ratio in the crystallized mass in order to improve pumpability, and facilitate subsequent centrifugation after addition of the detergent solution. Washing the crystallized mass with hot water is necessary to remove the detergent residue from the fractions. The yield of the palm kernel stearin in this process is comparatively low (about 25–30%).

### *Crystallizing from Solvents*

In the solvent fractionation of palm kernel oil, both hexane and acetone can be used as the solvent. However, acetone is preferred because it is found to give better results. The solvent reduces the viscosity of the system and improves the crystallization and subsequent filtration of the solid phase. Usually a high solvent:oil ratio is necessary to yield good results, and a low temperature is required for crystallization of the stearin. Filtration of the solid is accomplished in a specially designed filter, and the fractions are obtained after the removal of the solvent by distillation. The production cost of this process is much higher than the other two methods because of solvent loss, energy used to attain low temperatures for crystallization, and energy used for solvent recovery. There are also additional capital investments on solvent tanks, the solvent-recovery installation, and fire safety features.

### *Hydrogenation*

Even though palm kernel stearin is suitable for direct use as a CBS after refining, it is sometimes desirable to enhance its melting characteristics by hydrogenating it to higher melting points. Hydrogenation saturates the double bonds present in the oleic and linoleic acids, converting them into stearic acid. Some isomerization also occurs. The hardness and properties of the final product depend on its iodine value (IV) which in turn depends on the extent of hydrogenation.

Palm kernel oil often is partially hydrogenated to a suitable melting point for use as a more economical CBS in certain applications. A wider range of products can be obtained by hydrogenation of palm kernel olein to different IV and melting points for various confectionery applications (7).

### *Refining Palm Kernel Fractions and Related Products*

All the products mentioned previously are refined to remove the undesirable components before they are used in food formulations. Physical refining via the usual unit operations of degumming, bleaching earth treatment followed by filtration, and deodorization is carried out in smaller plants (relative to palm-oil refining). Deodorization temperatures, vacuum, and retention time are carefully controlled and monitored to avoid isomerization of the triglyceride molecules in the products.

## Characteristics of Cocoa Butter Substitutes

As the name implies, CBS are intended to be used to replace cocoa butter in certain chocolate or confectionery applications. Thus, it must possess physical attributes similar to those of cocoa butter, especially in terms of melting behavior and mouth-feel. Although CBS can be made from various vegetable oils, those produced from palm kernel oil are most popular. However, these palm kernel-based CBS are not compatible with cocoa butter because of differences in triglyceride species, and thus are not used as partial replacers but individually, in confectionery formulations.

The suitability of a fat product as a CBS is determined mainly by its solid fat content (SFC) profile and slip melting point (SMP). Compositional parameters such as IV, fatty acid and triglyceride compositions are useful for identification and also in production quality control.

## Survey of Malaysian Palm Kernel-Based Specialty Fats

The Palm Oil Research Institute of Malaysia, whose mission is to assist the Malaysian palm-oil industry, often carries out surveys by itself or in collaboration with other organizations, on various palm-oil products in order to obtain data on the quality, properties, and composition of palm oil. The data can then be used as a reference by industry formulating national standards where appropriate, or submitted to international organizations such as Codex Alimentarius for consideration in the preparation of Codex Standards. As exports of processed palm kernel oil products have been increasing steadily over the last few years, it was deemed necessary to perform a survey on major palm kernel specialty fats to establish their characteristics.

A total of 143 samples consisting of 49 samples of crude (CPKS) and fully refined palm kernel stearin (RBDPKS), 52 samples of crude (CPKL) and refined palm kernel olein (RBDPKL), 19 samples of fully refined, hydrogenated palm kernel stearin (RBDHPKS) and 8 samples of fully refined, hydrogenated palm kernel olein (RBDHPKL) and 5 samples of hydrogenated palm kernel oil (HPKO) were obtained from four local manufacturers using physical pressing and detergent methods. These samples were analyzed for IV, SMP, cloud point (CP), free fatty acid content (FFA), fatty acid composition (FAC), triglyceride composition by carbon number (TG by carbon-number) and SFC. The data obtained were analyzed statistically to obtain means, ranges, and standard deviations. These data were then segregated according to their production processes (i.e., physical pressing and detergent fractionation) to see if there were any differences between products from these two processes.

## Methods of Analysis

All analyses were carried out using PORIM Test Methods (9).

### Results

The data for palm kernel stearin, palm kernel olein, hydrogenated palm kernel stearin, hydrogenated palm kernel olein, and hydrogenated palm kernel oil are given in Tables 6–19.

## Discussion

### Effects of Refining

Comparison of the data on the various crude and fully refined products showed that the refining process had not altered the chemical composition and general physical properties of the crude products significantly. This indicates that refiners have taken precaution in refining, especially the deodorization step, to avoid chemical changes. However, there were differences in the SFC content of crude products, especially in crude PK olein. Their lower values could be attributed to the presence of fatty acids, which tends to lower the solids content; this is in agreement with the observation of Wong (8).

It was observed that the fractions derived from pressing and detergent processes differed slightly in terms of composition and SFC, and thus the data collected were categorized and discussed according the processes from which they were derived. In the case of SFC, comparisons were made between the crude and refined products. As the number of samples of hydrogenated products is relatively small, they were discussed together.

### General Characteristics

*Palm Kernel Stearin (PKS).* Data in Tables 6 and 8 give the results of common quality parameters and SFC data of crude and RBD palm kernel stearin from physical pressing and the detergent process. The average SMP of each of the four stearins was about 32°C and the IV averaged about 7. A narrow range was observed for the various parameters, except for the FFA. The FFA is not a characteristic, but depends on the quality of the raw material (i.e., crude palm kernel oil). These figures are in agreement with the specifications of some of the commercial products shown in Table 5. Graphically the SFC profiles were all rather sharp between 20–30°C, and at 35°C all the PKS samples analyzed were completely melted (Fig. 1).

In terms of composition (see Tables 7 and 9), there was an enrichment in the content of lauric and myristic acids and a reduction in oleic acid. This is in agreement with increases in $C_{36}$, $C_{38}$, and $C_{40}$, and reductions in the $C_{50}$–$C_{54}$ triglycerides. There was very little difference between crude and RBD products.

There were slight differences between the PK stearins derived from the physical-pressing method and that obtained from the detergent process. The RBD PKS from detergent processing appeared to have a marginally wider range in the SMP, was slightly higher in lauric and oleic acid contents, and also slightly higher in SFC (at lower temperature of 10–30°C, Figs. 1 and 2). Such minor differences, however, are not likely to have any significant effect on the suitability of the product for use as CBS. The sharp SFC profiles between 20–35°C coupled with the SMP of about 32°C make them excellent materials for CBS.

*Palm Kernel Olein (PKL).* Palm kernel olein is the less-valuable product from the fractionation process and is

**TABLE 6**
**Common Characteristics of Palm Kernel Stearin from a Physical Pressing Method**

| Parameters | CPKS | | | RBDPKS | | |
| --- | --- | --- | --- | --- | --- | --- |
| | Range | Mean | SD | Range | Mean | SD |
| SMP (°C) | 31.3–32.6 | 31.9 | 0.97 | 31.8–32.6 | 32.2 | 0.34 |
| IV (Wijs) | 5.8–8.1 | 6.9 | 0.75 | 6.0–7.7 | 6.8 | 0.62 |
| FFA (% as C12:0) | 0.35–1.12 | 0.79 | 0.31 | 0.02–0.20 | 0.08 | 0.04 |
| Solid Fat Content by Pulse NMR (%) Temperature (°C) | | | | | | |
| 5 | 91.3–93.9 | 92.9 | 0.92 | 91.6–93.9 | 92.9 | 0.94 |
| 10 | 89.3–92.8 | 91.5 | 1.19 | 90.2–93.2 | 91.9 | 1.03 |
| 15 | 88.7–91.0 | 89.6 | 0.85 | 88.1–91.6 | 89.9 | 1.32 |
| 20 | 78.0–86.5 | 83.2 | 2.84 | 79.8–84.2 | 82.6 | 1.76 |
| 25 | 60.7–73.0 | 67.2 | 4.18 | 64.5–70.9 | 67.9 | 2.56 |
| 30 | 24.7–43.2 | 33.9 | 6.14 | 30.2–38.4 | 34.9 | 2.88 |
| 35 | melted | | | melted | | |

## TABLE 7
### Composition of Palm Kernel Stearin from a Physical Pressing Method

| Parameters | CPKS | | | RBDPKS | | |
|---|---|---|---|---|---|---|
| | Range | Mean | SD | Range | Mean | SD |
| Fatty Acid Composition (%) | | | | | | |
| C6:0 | 0–0.1 | 0.1 | 0.01 | 0–0.1 | 0.1 | 0.01 |
| C8:0 | 1.5–1.9 | 1.8 | 0.15 | 1.6–2.1 | 1.8 | 0.19 |
| C10:0 | 2.5–2.9 | 2.7 | 0.16 | 2.5–2.9 | 2.7 | 0.12 |
| C12:0 | 54.8–57.2 | 56.2 | 0.78 | 55.1–57.8 | 56.6 | 0.89 |
| C14:0 | 22.2–24.1 | 23.0 | 0.71 | 21.1–23.3 | 22.3 | 0.77 |
| C16:0 | 7.8–8.3 | 8.1 | 0.20 | 7.9–8.6 | 8.2 | 0.22 |
| C18:0 | 1.6–2.1 | 1.8 | 0.19 | 1.5–2.1 | 1.8 | 0.18 |
| C18:1 | 4.6–6.4 | 5.2 | 0.63 | 4.6–6.8 | 5.4 | 0.70 |
| C18:2 | 0.8–1.1 | 0.9 | 0.13 | 0.7–1.0 | 0.8 | 0.09 |
| C20:0 | 0–0.2 | 0.1 | 0.04 | 0–0.3 | 0.1 | 0.08 |
| Triglyceride Composition by Carbon-Number (%) | | | | | | |
| $C_{26}$ | 0–0.2 | 0.1 | 0.01 | 0 | 0 | — |
| $C_{28}$ | 0.1–0.2 | 0.1 | 0.01 | 0.1–0.3 | 0.1 | 0.05 |
| $C_{30}$ | 0.4–0.7 | 0.5 | 0.05 | 0.4–0.9 | 0.5 | 0.12 |
| $C_{32}$ | 2.8–3.7 | 3.3 | 0.30 | 3.1–3.5 | 3.3 | 0.12 |
| $C_{34}$ | 6.0–6.9 | 6.5 | 0.30 | 6.1–6.5 | 6.3 | 0.14 |
| $C_{36}$ | 25.9–27.7 | 26.7 | 0.21 | 26.1–28.7 | 27.0 | 0.71 |
| $C_{38}$ | 23.4–25.7 | 24.6 | 0.73 | 24.1–25.7 | 24.6 | 0.36 |
| $C_{40}$ | 14.3–16.0 | 15.2 | 0.54 | 14.9–15.8 | 15.2 | 0.24 |
| $C_{42}$ | 9.1–9.5 | 9.3 | 0.15 | 9.2–10.1 | 9.5 | 0.28 |
| $C_{44}$ | 5.1–5.6 | 5.4 | 0.19 | 5.2–5.7 | 5.4 | 0.17 |
| $C_{46}$ | 2.6–3.6 | 3.1 | 0.16 | 2.4–3.5 | 3.0 | 0.26 |
| $C_{48}$ | 1.3–3.4 | 2.5 | 0.71 | 1.4–3.3 | 2.5 | 0.56 |
| $C_{50}$ | 0.9–1.2 | 1.0 | 0.10 | 1.1–1.5 | 1.3 | 0.16 |
| $C_{52}$ | 0.7–1.1 | 0.8 | 0.12 | 0.6–0.9 | 0.7 | 0.07 |
| $C_{54}$ | 0.7–1.0 | 0.9 | 0.10 | 0.5–0.8 | 0.6 | 0.14 |
| $C_{56}$ | 0 | 0 | — | 0 | 0 | — |

## TABLE 8
### Common Characteristics of Palm Kernel Stearin from a Detergent Method

| Parameters | CPKS | | | RBD PKS | | |
|---|---|---|---|---|---|---|
| | Range | Mean | SD | Range | Mean | SD |
| SMP (°C) | 31.3–32.9 | 32.2 | 0.35 | 31.8–33.1 | 32.3 | 0.36 |
| IV (Wijs) | 6.8–8.0 | 7.4 | 0.33 | 6.7–7.3 | 7.0 | 0.21 |
| FFA (% as C12:0) | 1.10–2.65 | 1.56 | 1.47 | 0.03–0.08 | 0.05 | 0.02 |
| Solid Fat Content by Pulse NMR (%) Temperature (°C) | | | | | | |
| 5 | 88.1–92.3 | 90.8 | 1.17 | 91.7–94.9 | 93.5 | 0.69 |
| 10 | 87.5–91.4 | 89.8 | 1.37 | 90.9–92.9 | 92.2 | 0.71 |
| 15 | 83.6–88.4 | 86.5 | 1.28 | 88.9–89.2 | 90.2 | 0.91 |
| 20 | 77.0–81.3 | 79.1 | 1.36 | 79.4–84.2 | 82.9 | 1.41 |
| 25 | 60.7–66.4 | 63.6 | 1.48 | 63.6–70.6 | 68.4 | 2.13 |
| 30 | 25.7–32.5 | 29.5 | 1.80 | 30.6–36.8 | 34.2 | 2.04 |
| 35 | melted | | | melted | | |

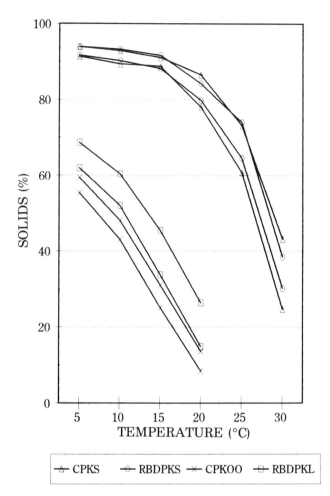

Fig. 1. SFC of the palm kernel fractions from the pressing process.

often sold to oleochemical producers at a discount, or hydrogenated into more valuable products for edible use. Palm kernel olein containing 2 ppm of silicone oil is reported to be suitable for use as a frying oil (10).

Tables 10 and 12 give some general properties and SFC profiles of palm kernel olein from pressing and detergent processes while Tables 11 and 13 give the data on their composition. The CP and SMP of crude palm kernel olein were lower than those of RBD olein. The difference is smaller in the case of olein derived from detergent processing.

The previous observations in CP, SMP, and SFC, in the case of olein from the pressing method, could be attributed to the differences in their FFA, FAC, and triglyceride composition. The crude olein generally had lower caprylic, capric, lauric, and myristic acids than those of RBD palm kernel olein and this is consistent with the triglyceride composition. In the case of olein from the detergent process, the smaller difference in the properties mentioned is likely to be the result of a lower FFA (shown by lower mean). It is interesting to note that in the samples surveyed, the crude olein had higher lauric and myristic (and likewise $C_{36}$ and $C_{38}$ triglycerides) than the RBD olein.

The data in Tables 10 and 12 also show that the SFC of the crude palm kernel olein derived from both methods were much lower than those of the RBD products. This can be seen in Figs. 1 and 2. At each temperature from 5 to 25°C, the mean SFC differed by about 8–11% being higher in the pressing process, and the corresponding differences in the palm kernel olein from detergent process were 1.9–4%. This difference between the SFC profiles of crude and refined products can be attributed to the FFA, the mean of which is higher in the crude olein from the pressing process. This is especially so because these lauric products have higher amounts of short-and medium-chain fatty acids (from caprylic to myristic).

Since the palm kernel olein is the softer fraction, it has less lauric, myristic, caprylic, and capric acids, but more oleic and linoleic acids when compared with palm kernel stearin. Similarly the $C_{36}$, $C_{38}$, and $C_{40}$ triglycerides were lower while the $C_{28}$, $C_{30}$, $C_{32}$, $C_{34}$, $C_{48}$, $C_{50}$, $C_{52}$, and $C_{54}$ triglycerides were higher compared to the stearin. This is to be expected, because the lower carbon-number triglycerides consist of the shorter chain fatty acids and the high-

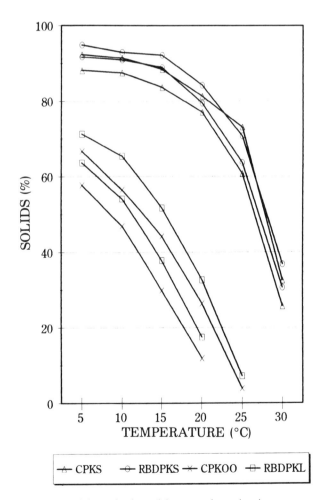

Fig. 2. SFC of the palm kernel fractions from the detergent process.

## TABLE 9
### Composition of Palm Kernel Stearin from the Detergent Method

| Parameters | CPKS | | | RBDPKS | | |
|---|---|---|---|---|---|---|
| | Range | Mean | SD | Range | Mean | SD |
| Fatty Acid Composition (%) | | | | | | |
| C6:0 | 0–0.1 | 0.1 | 0.01 | 0–0.1 | 0.1 | 0.01 |
| C8:0 | 1.6–2.3 | 1.9 | 0.19 | 1.7–2.1 | 1.9 | 0.13 |
| C10:0 | 2.5–2.9 | 2.7 | 0.14 | 2.6–2.9 | 2.7 | 0.12 |
| C12:0 | 55.1–57.8 | 56.6 | 0.75 | 56.0–58.2 | 56.9 | 0.62 |
| C14:0 | 21.1–23.3 | 22.3 | 0.77 | 21.3–22.5 | 22.1 | 0.37 |
| C16:0 | 7.2–8.5 | 7.9 | 0.29 | 7.5–8.2 | 7.9 | 0.22 |
| C18:0 | 1.5–1.8 | 1.7 | 0.13 | 1.3–2.2 | 1.8 | 0.30 |
| C18:1 | 5.1–6.6 | 5.8 | 0.41 | 5.2–6.1 | 5.7 | 0.27 |
| C18:2 | 0.6–1.0 | 0.8 | 0.12 | 0.7–0.9 | 0.8 | 0.05 |
| C20:0 | 0–0.2 | 0.1 | 0.05 | 0–0.2 | 0.1 | 0.09 |
| Triglyceride Composition by Carbon-Number (%) | | | | | | |
| $C_{26}$ | 0 | 0 | — | 0 | 0 | — |
| $C_{28}$ | 0.1–0.2 | 0.1 | 0.03 | 0–0.1 | 0.1 | 0.03 |
| $C_{30}$ | 0.4–0.6 | 0.5 | 0.05 | 0.4–0.7 | 0.5 | 0.11 |
| $C_{32}$ | 3.2–3.7 | 3.4 | 0.16 | 3.2–3.5 | 3.3 | 0.08 |
| $C_{34}$ | 6.2–6.9 | 6.5 | 0.19 | 6.3–6.7 | 6.5 | 0.13 |
| $C_{36}$ | 26.3–29.2 | 27.8 | 0.65 | 27.1–28.8 | 27.6 | 0.38 |
| $C_{38}$ | 24.1–25.6 | 24.8 | 0.39 | 24.3–25.6 | 24.9 | 0.33 |
| $C_{40}$ | 14.6–15.7 | 15.1 | 0.35 | 14.8–15.6 | 15.2 | 0.23 |
| $C_{42}$ | 8.1–9.5 | 8.9 | 0.43 | 8.4–9.3 | 8.9 | 0.28 |
| $C_{44}$ | 4.9–5.3 | 5.1 | 0.15 | 5.0–5.2 | 5.1 | 0.08 |
| $C_{46}$ | 2.3–3.2 | 3.0 | 0.20 | 2.9–3.1 | 3.0 | 0.07 |
| $C_{48}$ | 1.1–3.1 | 2.4 | 0.41 | 1.4–2.6 | 2.4 | 0.35 |
| $C_{50}$ | 0.4–1.6 | 0.9 | 0.31 | 0.4–1.1 | 0.9 | 0.16 |
| $C_{52}$ | 0.3–1.5 | 0.7 | 0.22 | 0.3–0.9 | 0.7 | 0.18 |
| $C_{54}$ | 0.4–1.0 | 0.8 | 0.15 | 0.8–1.1 | 0.9 | 0.04 |
| $C_{56}$ | 0 | 0 | — | 0 | 0 | — |

## TABLE 10
### Common Characteristics of Palm Kernel Olein from a Physical Pressing Method

| Parameters | CPKL | | | RBDPKL | | |
|---|---|---|---|---|---|---|
| | Range | Mean | SD | Range | Mean | SD |
| CP (°C) | 11.1–12.3 | 11.6 | 0.45 | 11.0–15.5 | 12.8 | 1.47 |
| SMP (°C) | 21.8–24.4 | 22.5 | 1.05 | 22.1–25.1 | 23.5 | 0.94 |
| IV (Wijs) | 23.9–25.3 | 24.7 | 0.63 | 20.6–25.0 | 23.0 | 1.54 |
| FFA (% as C12:0) | 2.19–4.43 | 3.25 | 0.86 | 0.04–0.09 | 0.07 | 0.02 |
| Solid Fat Content by Pulse NMR (%) Temperature (°C) | | | | | | |
| 5 | 55.4–59.4 | 57.4 | 1.52 | 62.0–68.7 | 65.4 | 2.30 |
| 10 | 43.1–48.1 | 46.1 | 2.10 | 52.0–60.4 | 56.6 | 3.61 |
| 15 | 25.1–31.1 | 28.1 | 2.39 | 33.8–45.5 | 39.6 | 4.97 |
| 20 | 8.5–13.7 | 11.7 | 2.18 | 14.9–26.4 | 20.3 | 4.68 |
| 25 | melted | | | 0–3.6 | 1.28 | 1.31 |
| 30 | | | | melted | | |

**TABLE 11**
**Composition of Palm Kernel Olein from a Physical Pressing Method**

| Parameters | CPKL | | | RBDPKL | | |
|---|---|---|---|---|---|---|
| | Range | Mean | SD | Range | Mean | SD |
| Fatty Acid Composition (%) | | | | | | |
| C6:0 | 0.2–0.3 | 0.3 | 0.05 | 0.2–0.3 | 0.3 | 0.05 |
| C8:0 | 3.8–4.7 | 4.3 | 0.37 | 3.6–4.7 | 4.2 | 0.39 |
| C10:0 | 3.3–3.8 | 3.6 | 0.17 | 3.3–3.9 | 3.5 | 0.17 |
| C12:0 | 42.1–44.6 | 43.6 | 0.97 | 43.3–46.3 | 45.4 | 1.15 |
| C14:0 | 13.1–14.6 | 13.7 | 0.40 | 12.9–15.5 | 14.1 | 0.67 |
| C16:0 | 8.0–9.0 | 8.4 | 0.35 | 7.4–8.5 | 8.1 | 0.31 |
| C18:0 | 2.0–2.7 | 2.3 | 0.27 | 1.9–2.8 | 2.3 | 0.32 |
| C18:1 | 18.7–21.3 | 20.1 | 0.95 | 14.6–21.0 | 18.8 | 1.71 |
| C18:2 | 3.4–3.8 | 3.6 | 0.16 | 2.6–3.6 | 3.2 | 0.30 |
| C20:0 | 0–0.2 | 0.1 | 0.09 | 0–0.2 | 0.1 | 0.08 |
| Triglyceride Composition by Carbon-Number (%) | | | | | | |
| $C_{26}$ | 0–0.1 | 0.0 | 0.04 | 0–0.1 | 0.0 | 0.03 |
| $C_{28}$ | 0.3–0.5 | 0.3 | 0.05 | 0.2–0.5 | 0.3 | 0.07 |
| $C_{30}$ | 1.6–1.8 | 1.7 | 0.07 | 1.4–1.7 | 1.5 | 0.11 |
| $C_{32}$ | 7.8–8.3 | 8.1 | 0.21 | 6.9–8.0 | 7.6 | 0.33 |
| $C_{34}$ | 9.0–10.2 | 9.5 | 0.39 | 8.7–10.1 | 9.3 | 0.38 |
| $C_{36}$ | 16.3–18.1 | 17.4 | 0.67 | 16.8–19.6 | 18.3 | 0.76 |
| $C_{38}$ | 10.1–13.2 | 11.3 | 0.73 | 10.8–14.0 | 12.5 | 1.00 |
| $C_{40}$ | 6.1–7.3 | 6.8 | 0.41 | 6.5–8.3 | 7.5 | 0.54 |
| $C_{42}$ | 8.8–9.5 | 9.1 | 0.29 | 9.1–9.9 | 9.5 | 0.27 |
| $C_{44}$ | 7.4–8.2 | 8.0 | 0.29 | 7.3–8.1 | 7.7 | 0.28 |
| $C_{46}$ | 6.6–7.1 | 6.9 | 0.23 | 6.1–6.8 | 9.5 | 0.24 |
| $C_{48}$ | 7.6–10.8 | 9.3 | 1.20 | 7.0–12.7 | 8.4 | 1.47 |
| $C_{50}$ | 3.3–3.8 | 3.5 | 0.16 | 2.8–3.7 | 3.2 | 0.29 |
| $C_{52}$ | 3.0–3.8 | 3.6 | 0.31 | 2.7–3.9 | 3.3 | 0.35 |
| $C_{54}$ | 4.1–4.7 | 4.4 | 0.23 | 3.5–4.6 | 4.1 | 0.35 |
| $C_{56}$ | 0–0.2 | 0.2 | 0.08 | 0–0.2 | 0.1 | 0.09 |

**TABLE 12**
**Common Characteristics of Palm Kernel Olein from a Detergent Method**

| Parameters | CPKL | | | RBDPKL | | |
|---|---|---|---|---|---|---|
| | Range | Mean | SD | Range | Mean | SD |
| CP (°C) | 11.7–14.7 | 12.7 | 0.99 | 11.8–15.5 | 13.2 | 1.22 |
| SMP (°C) | 22.0–24.9 | 23.6 | 0.82 | 22.0–25.4 | 23.7 | 1.00 |
| IV (Wijs) | 21.1–24.1 | 22.3 | 0.91 | 21.3–24.3 | 23.0 | 1.01 |
| FFA (% as C12:0) | 2.23–5.07 | 2.96 | 0.78 | 0.03–0.10 | 0.07 | 0.03 |
| Solid Fat Content by Pulse NMR (%) Temperature (°C) | | | | | | |
| 5 | 57.7–66.7 | 62.2 | 2.45 | 63.6–71.2 | 65.8 | 2.5 |
| 10 | 46.9–56.5 | 53.3 | 3.52 | 54.1–65.4 | 57.5 | 3.38 |
| 15 | 29.9–44.2 | 37.3 | 4.51 | 37.8–51.7 | 41.4 | 4.40 |
| 20 | 11.9–26.4 | 18.9 | 4.08 | 17.5–32.7 | 21.7 | 4.60 |
| 25 | 0–3.9 | 0.8 | 1.20 | 0–7.2 | 1.7 | 2.42 |
| 30 | melted | | | melted | | |

**TABLE 13**
Composition of Palm Kernel Olein from a Detergent Method

| Parameters | CPKL Range | Mean | SD | RBDPKL Range | Mean | SD |
|---|---|---|---|---|---|---|
| Fatty Acid Composition (%) | | | | | | |
| C6:0 | 0.3–0.4 | 0.3 | 0.05 | 0.2–0.4 | 0.3 | 0.05 |
| C8:0 | 3.8–4.9 | 4.3 | 0.33 | 3.6–5.0 | 4.4 | 0.41 |
| C10:0 | 3.3–4.5 | 3.6 | 0.28 | 3.2–3.8 | 3.6 | 0.17 |
| C12:0 | 43.3–46.3 | 44.6 | 1.00 | 43.1–45.9 | 44.2 | 0.92 |
| C14:0 | 12.3–14.7 | 14.1 | 1.02 | 12.8–14.8 | 13.8 | 0.70 |
| C16:0 | 7.8–10.6 | 8.4 | 0.64 | 7.9–9.4 | 8.4 | 0.41 |
| C18:0 | 2.1–2.4 | 2.3 | 0.11 | 1.8–2.5 | 2.3 | 0.21 |
| C18:1 | 16.8–20.7 | 19.1 | 1.22 | 17.2–21.1 | 19.7 | 1.13 |
| C18:2 | 2.8–3.7 | 3.2 | 0.24 | 2.8–3.5 | 3.2 | 0.25 |
| C20:0 | 0–0.1 | 0.1 | 0.03 | 0–0.2 | 0.1 | 0.07 |
| Triglyceride Composition by Carbon-Number (%) | | | | | | |
| $C_{26}$ | 0 | 0 | — | 0 | 0 | — |
| $C_{28}$ | 0.3–0.4 | 0.3 | 0.01 | 0.2–0.3 | 0.3 | 0.03 |
| $C_{30}$ | 1.4–1.8 | 1.6 | 0.11 | 1.3–1.8 | 1.6 | 0.17 |
| $C_{32}$ | 7.8–8.4 | 7.8 | 0.36 | 6.6–8.0 | 7.5 | 0.44 |
| $C_{34}$ | 8.8–9.9 | 9.4 | 0.26 | 8.4–9.6 | 9.2 | 0.34 |
| $C_{36}$ | 17.5–19.8 | 18.7 | 0.71 | 17.6–19.6 | 18.4 | 0.62 |
| $C_{38}$ | 11.8–13.9 | 12.9 | 0.72 | 11.9–14.0 | 12.5 | 0.71 |
| $C_{40}$ | 7.2–8.5 | 7.8 | 0.45 | 7.3–8.7 | 7.7 | 0.59 |
| $C_{42}$ | 8.1–9.6 | 9.1 | 0.33 | 9.1–9.9 | 9.4 | 0.28 |
| $C_{44}$ | 7.5–8.1 | 7.7 | 0.17 | 7.3–8.2 | 7.9 | 0.25 |
| $C_{46}$ | 6.1–6.8 | 6.4 | 0.20 | 6.1–6.8 | 6.5 | 0.25 |
| $C_{48}$ | 7.3–8.3 | 7.7 | 0.31 | 7.3–10.2 | 8.1 | 0.72 |
| $C_{50}$ | 1.7–3.3 | 2.9 | 0.99 | 2.8–3.5 | 3.2 | 0.23 |
| $C_{52}$ | 3.0–3.6 | 3.2 | 0.20 | 3.1–3.8 | 3.4 | 0.22 |
| $C_{54}$ | 3.4–4.7 | 3.8 | 0.38 | 3.5–4.5 | 4.2 | 0.32 |
| $C_{56}$ | 0–0.2 | 0.1 | 0.09 | 0–0.2 | 0.1 | 0.09 |

**TABLE 14**
Common Characteristics of Hydrogenated Palm Kernel Stearin from Physical Pressing and Detergent Methods

| Parameters | Physical Pressing RBD HPKS Range | Mean | SD | Detergent RBD HPKS Range | Mean | SD |
|---|---|---|---|---|---|---|
| SMP (°C) | 32.7–34.6 | 33.3 | 0.70 | 33.6–35.6 | 34.0 | 0.88 |
| IV (Wijs) | 0.1–0.3 | 0.23 | 0.08 | 0.4–1.2 | 0.55 | 0.21 |
| FFA (% as C12:0) | 0.03–0.15 | 0.11 | 0.15 | 0.02–0.21 | 0.09 | 0.14 |
| Solid Fat Content by Pulse NMR (%) Temperature (°C) | | | | | | |
| 5 | 95.6–95.8 | 95.7 | 0.13 | 94.1–96.7 | 95.6 | 0.53 |
| 10 | 95.5–95.7 | 95.5 | 0.10 | 93.5–95.5 | 95.2 | 0.52 |
| 15 | 95.2–95.3 | 95.2 | 0.09 | 91.8–95.1 | 94.7 | 0.86 |
| 20 | 93.1–94.6 | 94.2 | 0.58 | 85.6–94.1 | 93.1 | 2.14 |
| 25 | 85.1–89.6 | 86.4 | 1.60 | 64.9–88.4 | 73.7 | 3.80 |
| 30 | 40.3–48.3 | 43.9 | 2.77 | 30.0–48.8 | 42.6 | 4.27 |
| 35 | 0–5.4 | 1.6 | 1.97 | 2.2–5.2 | 3.3 | 1.08 |
| 40 | 0–1.2 | 0.3 | 0.40 | 0–4.0 | 1.5 | 1.86 |
| 45 | melted | | | melted | | |

## TABLE 15
### Composition of Hydrogenated Palm Kernel Stearin from Physical Pressing and Detergent Methods

| Parameters | Physical Pressing RBD HPKS | | | Detergent RBD HPKS | | |
|---|---|---|---|---|---|---|
| | Range | Mean | SD | Range | Mean | SD |
| Fatty Acid Composition (%) | | | | | | |
| C6:0 | 0–0.1 | 0.1 | 0.05 | 0–0.2 | 0.1 | 0.06 |
| C8:0 | 1.5–1.9 | 1.7 | 0.17 | 1.5–2.4 | 1.8 | 0.20 |
| C10:0 | 2.5–2.9 | 2.7 | 0.18 | 2.4–2.8 | 2.6 | 0.10 |
| C12:0 | 54.8–58.8 | 56.6 | 1.88 | 54.8–56.3 | 55.0 | 1.27 |
| C14:0 | 22.2–23.7 | 22.9 | 0.52 | 20.8–22.7 | 21.4 | 0.99 |
| C16:0 | 7.4–9.3 | 8.3 | 0.75 | 7.8–10.3 | 9.2 | 0.63 |
| C18:0 | 5.9–9.1 | 7.4 | 1.45 | 4.6–11.5 | 9.5 | 1.64 |
| C18:1 | 0–0.1 | 0 | 1.05 | 0–1.0 | 0.2 | 0.29 |
| C18:2 | 0–0.2 | 0.1 | 0.02 | 0–0.1 | 0 | 0.04 |
| C20:0 | 0–0.3 | 0.1 | 0.11 | 0–0.2 | 0.1 | 0.04 |
| Triglyceride Composition by Carbon–Number (%) | | | | | | |
| $C_{26}$ | 0–0.1 | 0.0 | 0.04 | 0–0.1 | 0.0 | 0.02 |
| $C_{28}$ | 0.1–0.2 | 0.1 | 0.04 | 0.1–1.3 | 0.1 | 0.14 |
| $C_{30}$ | 0.4–0.5 | 0.4 | 0.05 | 0.4–0.8 | 0.6 | 0.09 |
| $C_{32}$ | 3.1–3.3 | 3.2 | 0.10 | 3.1–3.6 | 3.3 | 0.16 |
| $C_{34}$ | 6.3–6.7 | 6.5 | 0.16 | 5.8–6.5 | 6.0 | 0.23 |
| $C_{36}$ | 24.9–27.9 | 26.8 | 1.03 | 24.9–28.1 | 27.0 | 2.00 |
| $C_{38}$ | 20.1–25.9 | 25.4 | 1.82 | 20.0–25.7 | 24.3 | 1.56 |
| $C_{40}$ | 15.3–16.0 | 15.5 | 0.26 | 13.2–15.8 | 15.0 | 0.61 |
| $C_{42}$ | 8.8–9.4 | 9.1 | 0.25 | 8.3–12.6 | 9.3 | 0.63 |
| $C_{44}$ | 4.7–5.4 | 5.1 | 0.26 | 5.0–6.0 | 5.2 | 0.76 |
| $C_{46}$ | 2.3–3.6 | 2.9 | 0.26 | 2.9–4.9 | 3.2 | 0.51 |
| $C_{48}$ | 1.9–3.9 | 2.5 | 0.72 | 2.4–4.2 | 2.7 | 0.83 |
| $C_{50}$ | 0.7–1.2 | 0.9 | 0.19 | 0.9–2.3 | 1.2 | 0.34 |
| $C_{52}$ | 0.5–0.9 | 0.8 | 0.15 | 0.8–1.4 | 1.0 | 0.19 |
| $C_{54}$ | 0.5–0.9 | 0.7 | 0.13 | 1.0–1.3 | 1.1 | 0.13 |
| $C_{56}$ | 0–0.1 | 0.0 | 0.05 | 0 | 0 | — |

## TABLE 16
### Common Characteristics of Hydrogenated Palm Kernel Olein from Physical Pressing and Detergent Methods

| Parameters | Physical Pressing RBD HPKL | | | Detergent RBD HPKL | | |
|---|---|---|---|---|---|---|
| | Range | Mean | SD | Range | Mean | SD |
| SMP (°C) | 32.0–33.9 | 33.0 | 0.90 | 31.5–39.9 | 35.4 | 2.79 |
| IV (Wijs) | 6.1–8.4 | 7.3 | 0.88 | 0.5–10.7 | 5.5 | 3.34 |
| FFA (% as C12:0) | 0.02–0.04 | 0.03 | 0.01 | 0.02–0.14 | 0.1 | 0.19 |
| Solid Fat Content by Pulse NMR (%) Temperature (°C) | | | | | | |
| 5 | 92.3–93.8 | 93.2 | 0.56 | 89.0–94.8 | 93.2 | 1.37 |
| 10 | 90.9–92.3 | 91.5 | 0.64 | 86.2–93.2 | 91.1 | 1.93 |
| 15 | 85.6–88.5 | 87.4 | 1.11 | 71.9–91.0 | 86.6 | 5.44 |
| 20 | 68.3–75.1 | 71.0 | 3.87 | 45.4–84.6 | 71.7 | 10.20 |
| 25 | 40.9–52.2 | 45.3 | 4.50 | 19.7–65.5 | 46.7 | 12.60 |
| 30 | 15.9–28.8 | 21.7 | 4.76 | 9.4–42.9 | 32.0 | 13.10 |
| 35 | 6.0–11.5 | 9.1 | 2.00 | 5.0–21.9 | 11.0 | 6.69 |
| 40 | 0–2.0 | 0.8 | 1.79 | 0–9.5 | 3.2 | 3.43 |
| 45 | melted | | | melted | | |

er carbon-number triglycerides contain one or more of the longer chain fatty acids, such as oleic or linoleic acid.

*Hydrogenated Palm Kernel Stearin (HPKS).* In order to enhance the SFC profile further, palm kernel stearin is often partially or completely hydrogenated to produce hydrogenated palm kernel stearin, which is a premium CBS. The hydrogenated products are usually refined before they are marketed. Thus, our data were all derived from fully refined, hydrogenated palm kernel stearin, and they are shown in Tables 14 and 15.

The hydrogenated stearin from the pressing and detergent processes has narrow ranges for SMP and IV. The IV of the product derived from the detergent process had a wider range, although this could be due to intentional partial hydrogenation to suit the needs of different buyers.

The SFC profiles of the two products were generally in agreement with the commercial specifications given in Table 5. At lower temperatures (5–30°C) the two products have very similar SFC profiles (Figs. 3 and 4), but at higher temperatures (35–40°C), the product from detergent fractionation had about 1–2% higher solids. The fatty acid and triglyceride compositions of the two products however were very similar to one another. The product from the detergent process had slightly higher oleic (0–1.0%) and stearic acid contents (4.6–11.5%). This is consistent with the presence of higher triglycerides ($C_{50}$–$C_{54}$).

*Hydrogenated Palm Kernel Olein (HPKL).* By hydrogenating the palm kernel olein to different IV, products of different hardness and composition can be obtained. This explains the occurrence of wide ranges in the IV, SMP, and SFC of the samples of the hydrogenated palm kernel oleins surveyed (see Tables 17 and 18). The different ranges observed for the products from the two different processes simply indicate that they had been hydrogenated to different levels, probably to suit the requirements of different applications. It is interesting to note that though RBD HPKL may have similar ranges of IV and SMP compared to RBD HPKS, they can, however be differentiated by their fatty acid and triglyceride compositions and SFC profiles.

The higher fat SFC at 35–40°C (Figs. 3 and 4) indicate the presence of higher melting triglycerides resulting from

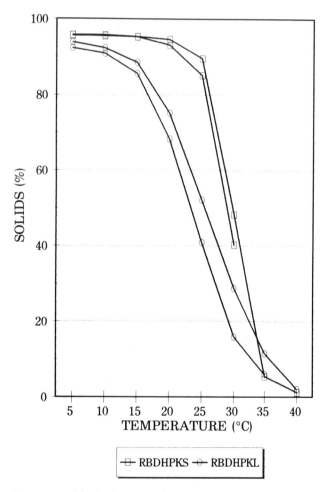

**Fig. 3.** SFC of the hydrogenated palm kernel fraction from the pressing process.

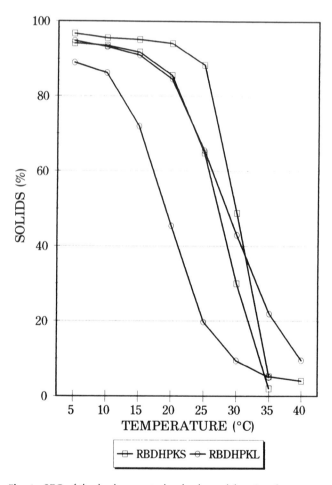

**Fig. 4.** SFC of the hydrogenated palm kernel fraction from the detergent process.

## TABLE 17
### Composition of Hydrogenated Palm Kernel Olein from Physical Pressing and Detergent Methods

| Parameters | Physical Pressing RBD HPKL | | | Detergent RBD HPKL | | |
|---|---|---|---|---|---|---|
| | Range | Mean | SD | Range | Mean | SD |
| Fatty Acid Composition (%) | | | | | | |
| C6:0 | 0.1–0.3 | 0.2 | 0.07 | 0.3–0.4 | 0.3 | 0.03 |
| C8:0 | 3.4–4.1 | 3.7 | 0.29 | 3.4–4.7 | 4.2 | 0.32 |
| C10:0 | 3.0–3.4 | 3.2 | 0.15 | 3.1–3.6 | 3.4 | 0.14 |
| C12:0 | 42.0–46.0 | 43.4 | 1.70 | 40.8–46.0 | 43.5 | 1.44 |
| C14:0 | 13.1–15.7 | 14.3 | 0.99 | 12.1–15.1 | 13.8 | 0.66 |
| C16:0 | 8.0–8.6 | 8.4 | 0.30 | 8.1–9.9 | 8.9 | 0.53 |
| C18:0 | 14.6–22.1 | 19.3 | 3.09 | 13.9–26.5 | 19.5 | 4.03 |
| C18:1 | 6.1–8.5 | 7.4 | 0.90 | 0.5–12.0 | 5.9 | 3.96 |
| C18:2 | 0–0.3 | 0.1 | 0.13 | 0–0.2 | 0.1 | 0.06 |
| C20:0 | 0 | 0 | — | 0–0.3 | 0.1 | 0.15 |
| Triglyceride Composition by Carbon-Number (%) | | | | | | |
| $C_{26}$ | 0 | 0 | — | 0 | 0 | — |
| $C_{28}$ | 0.2–0.3 | 0.3 | 0.05 | 0.2–0.3 | 0.3 | 0.05 |
| $C_{30}$ | 1.3–1.6 | 1.4 | 0.13 | 1.2–1.7 | 1.5 | 0.12 |
| $C_{32}$ | 6.7–8.0 | 7.4 | 0.53 | 6.1–7.7 | 7.0 | 0.41 |
| $C_{34}$ | 7.4–10.0 | 8.8 | 0.95 | 8.0–9.3 | 8.8 | 0.32 |
| $C_{36}$ | 16.2–18.8 | 17.7 | 1.06 | 17.1–19.1 | 18.1 | 0.65 |
| $C_{38}$ | 11.9–13.9 | 12.6 | 0.79 | 11.9–13.7 | 12.8 | 0.60 |
| $C_{40}$ | 7.3–8.7 | 7.8 | 0.58 | 7.3–8.7 | 8.0 | 0.45 |
| $C_{42}$ | 9.1–10.0 | 9.6 | 0.32 | 8.6–10.3 | 9.2 | 0.45 |
| $C_{44}$ | 7.9–8.9 | 8.3 | 0.40 | 7.6–8.1 | 7.9 | 0.13 |
| $C_{46}$ | 6.4–8.0 | 6.9 | 0.65 | 6.3–6.7 | 6.5 | 0.13 |
| $C_{48}$ | 7.7–9.4 | 8.5 | 0.65 | 7.5–9.8 | 8.0 | 0.50 |
| $C_{50}$ | 3.1–3.8 | 3.5 | 0.27 | 2.9–4.6 | 3.5 | 0.53 |
| $C_{52}$ | 2.5–3.5 | 3.2 | 0.40 | 3.1–5.2 | 3.8 | 0.57 |
| $C_{54}$ | 3.4–4.1 | 3.9 | 0.29 | 3.8–5.0 | 4.3 | 0.38 |
| $C_{56}$ | 0–0.2 | 0.1 | 0.02 | 0–0.2 | 0.2 | 0.08 |

## TABLE 18
### Common Characteristics of Hydrogenated Palm Kernel Oil

| Parameters | Range | Mean | SD |
|---|---|---|---|
| SMP (°C) | 31.9–35.3 | 33.9 | 1.24 |
| IV (Wijs) | 1.7–4.4 | 4.0 | 0.41 |
| FFA (% as C12:0) | 0.02–0.12 | 0.07 | 0.09 |
| Solid Fat Content by Pulse NMR (%) Temperature (°C) | | | |
| 5 | 94.3–95.2 | 94.6 | 0.24 |
| 10 | 93.7–94.2 | 94.0 | 0.33 |
| 15 | 91.2–93.5 | 92.5 | 0.85 |
| 20 | 80.8–88.9 | 83.9 | 2.00 |
| 25 | 54.8–68.8 | 58.8 | 1.73 |
| 30 | 23.6–34.0 | 27.0 | 1.45 |
| 35 | 6.7–8.1 | 7.3 | 0.59 |
| 40 | 0–2.0 | 0.5 | 1.00 |
| 45 | melted | | |

the hardening of the unsaturated acids. This is consistent with the fatty acid composition data where there were higher percentages of stearic acid (average about 19–19.5%) and higher contents of $C_{50}$–$C_{56}$ triglycerides.

*Hydrogenated Palm Kernel Oil (HPKO).* Hydrogenated palm kernel oil is another possible choice for CBS manufacture. Its properties and composition depend mainly on the extent of hydrogenation; normally the trading specification allows a maximum IV of 4. The data in Table 19 indicate that the Malaysian HPO had IV ranging from 1.7 to 4.4 and SMP of 31.9–35.3°C. The SFC of this product is fairly sharp (Fig. 5), but as expected, it is not as steep as the stearin and the hydrogenated stearin. The presence of 6.7–8.1% solids at 35°C is likely to impart poorer mouthfeel when compared to the more expensive HPKS or PKS products.

As expected, the fatty acid composition of HPKO was high in lauric, myristic, palmitic, and stearic acids. The

**TABLE 19**
**Composition of Hydrogenated Palm Kernel Oil**

| Parameters | Range | Mean | SD |
|---|---|---|---|
| Fatty Acid Composition (%) | | | |
| C6:0 | 0–0.2 | 0.1 | 0.06 |
| C8:0 | 2.9–3.6 | 3.3 | 0.28 |
| C10:0 | 2.8–3.4 | 3.1 | 0.17 |
| C12:0 | 46.2–48.8 | 47.7 | 1.38 |
| C14:0 | 16.5–17.1 | 16.8 | 0.38 |
| C16:0 | 7.9–9.2 | 8.3 | 0.42 |
| C18:0 | 15.2–19.0 | 16.6 | 1.40 |
| C18:1 | 1.4–4.3 | 3.7 | 1.03 |
| C18:2 | 0–0.2 | 0.1 | 0.06 |
| C20:0 | 0–0.2 | 0.2 | 0.08 |
| Triglyceride Composition by Carbon-Number (%) | | | |
| $C_{26}$ | 0 | 0 | — |
| $C_{28}$ | 0.2–0.3 | 0.2 | 0.05 |
| $C_{30}$ | 1.0–1.3 | 1.2 | 0.09 |
| $C_{32}$ | 5.9–6.2 | 6.1 | 0.39 |
| $C_{34}$ | 7.9–8.8 | 8.4 | 0.39 |
| $C_{36}$ | 19.2–22.1 | 20.9 | 1.02 |
| $C_{38}$ | 16.0–18.3 | 16.5 | 0.84 |
| $C_{40}$ | 10.0–11.2 | 10.6 | 0.47 |
| $C_{42}$ | 9.0–9.7 | 9.4 | 0.25 |
| $C_{44}$ | 6.6–7.4 | 7.1 | 0.29 |
| $C_{46}$ | 4.9–5.9 | 5.4 | 0.36 |
| $C_{48}$ | 5.0–6.6 | 6.0 | 0.46 |
| $C_{50}$ | 2.5–2.8 | 2.7 | 0.12 |
| $C_{52}$ | 2.4–2.8 | 2.6 | 0.18 |
| $C_{54}$ | 2.2–3.6 | 3.1 | 0.53 |
| $C_{56}$ | 0 | 0 | — |

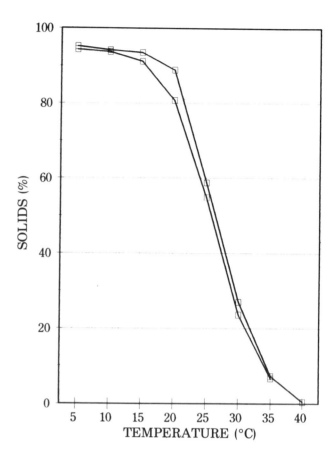

**Fig. 5.** SFC of hydrogenated palm kernel oil (HPKO).

oleic and linoleic acids were low, as most had been converted to stearic acid.

The triglyceride composition of HPKO was the same as that of normal palm kernel oil because hydrogenation does not affect the carbon-number of triglycerides.

*Relationship between IV and SMP of Palm Kernel Products.* Figure 6 gives the plot of SMP and IV of the various products surveyed. The results for the palm kernel oil and palm kernel olein can be distinguished clearly from the other products as they are clustered within ranges characteristic of their respective compositions. However, the hydrogenated products have a broad spread, with two distinct clusters embedded within it. The smaller cluster in the range of IV 0–1.2, and SMP of 32–35°C is that of hydrogenated palm kernel stearin. However, the results of the palm kernel stearin samples are concentrated in the "region" formed by the IV range of 5.8–8.1 and SMP range of 31.3–33.1°C. As expected, because of the different extent of hydrogenation, the results for the hydrogenated palm kernel olein and hydrogenated palm kernel oil are widely scattered over a larger region.

## Conclusion

Based on the results of the survey of 143 samples of palm kernel stearin, olein, their hydrogenated products, and hydrogenated palm kernel oil, a number of conclusions can be made: The palm kernel stearins from both pressing and detergent processes differ only slightly, and both products

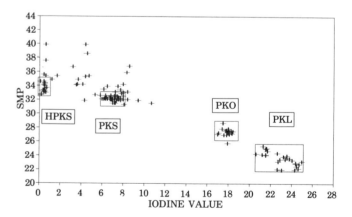

**Fig. 6.** Slip melting points and iodine values of pk products.

generally have consistent properties and compositions. Both have sharp SFC profiles. Based on the combined data, the ranges of IV and SMP for RBD palm kernel stearin are 5.1–8.1 and 31.3–33.1°C, respectively. These figures conform to the normal requirement in the trading specifications. The FFA of the fractions affects their physical properties such as CP (in the case of olein), SMP, and SFC. Removal of FFA (especially the shorter chain acids) by refining changes the physical properties, thus enhancing the SFC profile. Hydrogenating the fractions enhances their SFC profiles and also makes available a wider range of palm kernel specialty fats. This is especially important in PK olein where hydrogenation certainly will provide more lucrative alternatives to its use as a raw material for oleochemical manufacture. The IV–SMP plot of all the palm kernel products surveyed indicate that palm kernel oil, palm kernel olein, palm kernel stearin, and hydrogenated palm kernel stearin cluster, whereas those of hydrogenated palm kernel olein and hydrogenated palm kernel oil are spread over a wider area.

## Acknowledgments

The author acknowledges the technical assistance provided by staff of the Chromatography/Analytical and Physics laboratories, the statistical analysis by Chow Chee Sing, and the cooperation of the four manufacturers of palm kernel specialty fats in supplying the samples for this survey. The author would like to thank the Director-General of PORIM for permission to present this paper.

## References

1. Siew, W.L., *PORIM Bulletin 19*:19 (1989).
2. Siew, W.L., C.L. Chong, Y.A. Tan, T.S. Tang, and F.C.H. Oh, *Elaeis 4*:79 (1992).
3. Siew, W.L., T.S. Tang, F.C.H. Oh, C.L. Chong, and Y.A. Tan, *Elaeis 4*:38 (1993).
4. Rossell, J.B., *J. Am. Oil Chem. Soc. 62*:385 (1985).
5. Wong, S., in *Specialty Fats versus Cocoa Butter*, Atlanto Sdn. Bhd., Malaysia, 1991, pp. 145–148.
6. Willner, T., W. Sitzmann, and E.W. Münch, in *Proceedings of the World Conference on Edible Fats and Oils*, D.R. Erickson, ed., the American Oil Chemists' Society, Champaign, IL, 1989, pp. 239–245.
7. Berger, K.G., *Leatherhead Fd. R.A. Symposium Proceedings 19*:89 (1984).
8. Wong S., in *Specialty Fats versus Cocoa Butter*, Atlanto Sdn. Bhd., Malaysia, 1991, p. 167.
9. PORIM, PORIM Test Methods, 1988.
10. Lee, L.S., and R.E. Timms, *Asean Fd. J. 3*:11 (1987).

# Formulation of Lauric Oil-Containing Food Products and Their Performance

E.M. Goh

Guan Soon Heng Edible Oil Sdn. Berhad, P.O. Box 163, 34008 Taiping, Perak, Malaysia.

## Abstract

Presently, the two major lauric oils that are of commercial importance are coconut and palm kernel oil. Due to similarity of properties, coconut and palm kernel oil are interchangeable in many food applications. With modification, these two lauric oils are ideal raw materials for the production of a variety of confectionery fats that can be used with good performance in the formulation of various food products.

While palm kernel oil and olein are suitable feedstocks for margarines, they are more suitable and perform better than cocoa butter as fast-setting chocolate-coating fats for ice cream and other deep-frozen confections. Hydrogenated (hardened) coconut and palm kernel fats can be used for coating wafers, in filler creams, and in toffees. Palm kernel fats, interesterified with palm oil, can also be used for the preparation of coatings with firm textures and good melting properties. Coffee whiteners can be formulated based on hardened coconut oil, palm kernel oil or olein with melting points of 35–40°C and high solid contents at room temperature. Fractionated and hardened palm kernel stearin is used as a cocoa butter substitute in substitute chocolates (compound coatings) with snap and eating qualities similar to real chocolate.

Due to the low unsaturation in lauric oils, and especially in the hardened fats, lauric oil containing food products are highly stable against oxidative rancidity. The only serious disadvantage is that in the presence of moisture and lipase, hydrolysis can occur liberating short-chain fatty acids, which may give rise to a soapy off-flavor. Nevertheless, because of their unique properties and good performance in various food products, coconut and palm kernel oils will remain attractive in the confectionery industry at least through the 1990s.

---

Edible oils and fats are essential nutrients of the human diet. They play a vital role in an individual's daily diet by supplying essential fatty acids and providing energy for bodily requirements. In addition to nutritional qualities, edible oils and fats offer consistency and melting characteristics to fat-containing products, act as heat transfer media during frying, and act as carriers for fat-soluble vitamins and flavors. Of the total worldwide production of major oils and fats, over 80% are for food use. Table 1 provides figures for world vegetable-oil production in 1990, 1991, and 1992. The production and demand of these oils for food-based applications is increasing and is expected to continue to rise.

**TABLE 1**
World Production of Vegetable Oils (1990, 1991, and 1992)

| Oil | Million Tons | | |
|---|---|---|---|
| | 1990 | 1991 | 1992 |
| Soybean oil | 16.08 | 15.96 | 16.95 |
| Palm oil | 10.95 | 11.43 | 12.02 |
| Rapeseed oil | 8.16 | 8.95 | 9.41 |
| Sunflower oil | 7.85 | 8.01 | 8.08 |
| Groundnut oil | 3.96 | 3.96 | 3.96 |
| Cotton oil | 3.75 | 4.18 | 4.27 |
| Coconut oil | 3.37 | 3.03 | 2.89 |
| Olive oil | 1.86 | 1.76 | 2.17 |
| Corn oil | 1.46 | 1.51 | 1.57 |
| Palm kernel oil | 1.45 | 1.46 | 1.53 |
| Linseed oil | 0.66 | 0.68 | 0.64 |
| Sesame oil | 0.61 | 0.65 | 0.63 |
| Castor oil | 0.43 | 0.48 | 0.46 |
| Total | 60.59 | 62.06 | 64.58 |

*Source*: Oil World Annual 1993 (April 1993).

## Classification of Oils and Fats

According to their fatty acid compositions, the oils and fats can be classified into two groups, the lauric acid group and the nonlauric acid group. Table 2 shows the fatty acid compositions of some nonlauric vegetable oils in commercial use. Although there are several varieties of lauric oils and fats, such as coconut oil, palm kernel oil, babassu, and

**TABLE 2**
Fatty Acid Composition of Some Nonlauric Vegetable Oils

| Saturated acids | Medium $C_{10}$–$C_{15}$ | | |
| | Long $C_{16}$–$C_{24}$ | | |
| Unsaturated acids | Mono | | |
| | Poly | | |

| | Medium | Long | Mono | Poly |
|---|---|---|---|---|
| Palm oil | 1 | 49 | 39 | 11 |
| Groundnut oil | | 19 | 56 | 25 |
| Soybean oil | | 18 | 21 | 61 |
| Cottonseed oil | 1 | 29 | 20 | 50 |
| Maize oil (corn oil) | | 14 | 33 | 53 |
| Olive oil | | 14 | 76 | 10 |

tucum, the two lauric oils which are of commercial importance are coconut oil and palm kernel oil. Therefore in this paper only the formulations of food products containing these two lauric oils are considered.

## Origins of Coconut Oil and Palm Kernel Oil

Coconut and palm kernel oils originate from the tree crops of coconut and oil palm, respectively. Coconut oil is obtained by pressing copra, the dried coconut meat, while palm kernel oil is obtained from the kernel of the fruit of the palm species *Elaeis guineensis*.

## Characteristics of Coconut Oil and Palm Kernel Oil

Table 3 shows the main chemical and physical characteristics of coconut oil and palm kernel oil. Coconut oil is softer than palm kernel oil. This is indicated by the slightly lower slip melting point (SMP) and lower solid fat content (SFC) at 20°C of coconut oil. The fatty acid composition of these two oils is shown in Table 4. Both oils have almost equal amounts of the main fatty acid, lauric acid (C12:0). Because of their high content of lauric acid, these two oils are called lauric oils. However coconut oil has a higher content of the short-chain fatty acids and is more stable than palm kernel oil because of its low content of unsaturated fatty acids. It is more appropriate in certain food products due to the mild, pleasant coconut flavor. The higher amount of oleic acid (C18:1) in palm kernel oil makes it a suitable oil for hydrogenation (hardening) to produce confectionery fats with different degrees of hydrogenation, providing different end-use melting points. Due to the similarities in properties, coconut and palm kernel oils are interchangeable in many food applications.

## Commercial Utilization of Lauric Oils in Food Products

Because of the availability of various sources of oils and fats and the flexibility of oil-modification processes, formulation of fat blends has now become increasingly complex. A considerable variety of lauric fat blends and end-

**TABLE 3**
Main Characteristics of Coconut Oil and Palm Kernel Oil

|  | Coconut Oil | Palm Kernel Oil |
|---|---|---|
| Slip Melting Point (°C) | 24–26 | 27–29 |
| Iodine Value (Wijs) | 7–11 | 16–20 |
| Temperature (°C) | SFC (% by NMR) | |
| 20 | 35–42 | 40–50 |
| 30 | 1 | 1 |
| 35 | 0 | 0 |

**TABLE 4**
Fatty Acid Composition of Coconut Oil and Palm Kernel Oil

| Fatty Acid | Coconut Oil | Palm Kernel Oil |
|---|---|---|
| Short-chain saturated acids | 15% | 8% |
| Lauric acid (C12:0) | 47% | 48% |
| Myristic acid (C14:0) | 18% | 16% |
| Palmitic acid (C16:0) | 9% | 8% |
| Stearic acid (C18:0) | 2% | 2% |
| Oleic acid (C18:1) | 6% | 15% |
| Linoleic acid (C18:2) | 3% | 3% |

products with different rheological properties can be produced. However the ultimate aims are to formulate the least expensive products from the available raw materials without compromising product-quality requirements.

Coconut oil and palm kernel oil are two of the world's most desirable oils for confectionery formulations because of their high degree of saturation and stability. They are suitable raw materials for specific food applications, and together with palm oil, they are significant components in the production of specialty fats (e.g., lauric hard butters) for the formulation of confectionery products which include nondairy/imitation dairy products, coffee whiteners, and biscuit-filler creams. Although coconut oil and palm kernel oil can be used interchangeably in many food products, palm kernel oil performs better than coconut oil in certain food applications.

## Fat Modification and Product Variation

Coconut and palm kernel oils, like most of other vegetable oils in their original state, have limited application when utilized as such. Their properties have to be modified in order to extend their degree of usefulness. The modification processes commonly used are hydrogenation, fractionation, interesterification, and blending. Through these processes, various confectionery fats and end-use food products can be formulated. However, only some of the products containing lauric oil are considered here.

## Manufacture of Lauric Oil-Containing Products

### Nondairy/Imitation Dairy Products

Unhardened (or unhydrogenated) palm kernel oil and olein are ideal fats for the manufacture of nondairy products. They can be used as fast-setting chocolate-coating fats for ice cream and other deep-frozen confections. The coatings set quickly when applied to ice cream. These palm kernel fats are more suitable and perform better than cocoa butter for this purpose because the coatings formed are hard but not brittle, whereas pure chocolate coating (with cocoa butter) sets too hard, is brittle, and flakes off easily. A typical ice cream-coating formulation is given in Table 5.

**TABLE 5**
**Typical Ice Cream-Coating Formulation**

| Ingredient | % |
|---|---|
| Cocoa powder (10–12%) | 14.0 |
| Icing (fine) sugar | 25.5 |
| Fat | 60.0 |
| Lecithin | 0.5 |

## Coatings and Filler Creams

The simplest lauric fat (or hard butter) to be used as a coating fat can be made by hardening coconut oil and palm kernel oil to SMP between 32 and 41°C. These lauric hard butters have high solids at room temperature and melt rapidly enough to produce fairly good quality coating systems. Hardened palm kernel oil with a melting point of 39-41°C is used for coating wafers. However, those with high melting points melt above body temperature and therefore produce a waxy mouth-feel. Figure 1 shows the SFC/melting profile of hardened palm kernel products, which can be interesterified and blended with palm oil products to modify or improve the melting profile and the SFC curve, therefore making the fats more suitable for certain applications. The improvement in the melting profile of hardened palm kernel oil after interesterification is clearly demonstrated in Fig. 2. Coatings made from hardened palm kernel oil and interesterified palm kernel oil have moderate resistance to

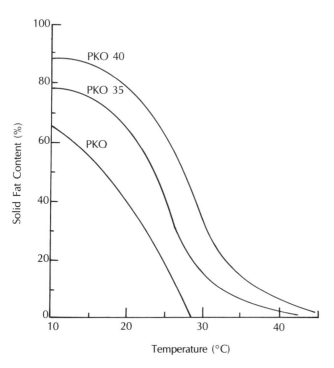

**Fig. 1.** Solid fat contents of hardened palm kernel products. *Abbreviations:* PKO, palm kernel oil; PKO 35, hydrogenated palm kernel oil with a SMP of 35°C; PKO 40, hydrogenated palm kernel oil with a SMP of 40°C.

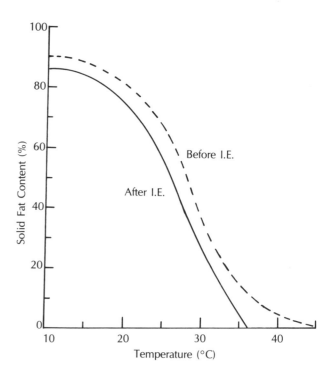

**Fig. 2.** Solid fat contents of fully hydrogenated palm kernel oil before and after interesterification (I.E.).

fat bloom and a gloss inferior to those based on fractionated palm kernel stearin. Palm kernel fats interesterified with or without a small amount of palm oil can also be used for the preparation of cheap and intermediate grade coatings with a firm but not brittle texture at ambient temperature and good melting properties.

Hydrogenated palm kernel olein can be used in the formulation of biscuit creams, and bakery coatings and glazes for cakes. Cream fillings between wafers require fats which set quickly but melt cleanly in the mouth, producing a cool sensation on the palate. Hydrogenated palm kernel oleins and coconut stearin are ideal for this formulation.

## Toffee Formulation

Hydrogenated coconut oil and palm kernel oil or olein are now largely used as a less expensive toffee-fat alternate, either partially or completely replacing the more expensive dairy butter. The physical characteristics of some of these fats are shown in Table 6. The inclusion of these fats in toffees retains good texture by giving body to the products and offering resistance to moisture penetration. They also provide lubrication and chewiness to the products. Typical recipes for toffees are given in Table 7.

## Coffee Whiteners

Since coffee whiteners are used at higher temperatures, they can be formulated based on hydrogenated coconut oil, hydrogenated palm kernel oil, or hydrogenated palm kernel olein with melting points of 35–42°C and a high solids

**TABLE 6**
Physical Characteristics of Toffee Fats

|  | HCNO | HPKO | HPKO | HPKOL | HPKOL | Butter Fat |
|---|---|---|---|---|---|---|
| Melting Point (°C) | 34 | 37 | 41 | 37 | 41 | 34 |
| Temperature (°C) | | | SFC (% by NMR) | | | |
| 20 | 57 | 64 | 77 | 63 | 73 | 22 |
| 25 | 20 | 39 | 58 | 37 | 53 | 13 |
| 30 | 5 | 14 | 31 | 12 | 28 | 7 |
| 35 | 2 | 7 | 15 | 4 | 14 | 4 |
| 40 | 0 | 2 | 7 | 0 | 6 | 0 |

*Abbreviations:* HCNO, Hydrogenated coconut oil; HPKO, hydrogenated palm kernel oil; HPKOL, hydrogenated palm kernel olein.

content at room temperature. These hydrogenated fats have little or no unsaturation and, thus are more resistant to the development of off-flavors due to oxidation. A typical coffee-whitener formulation is given in Table 8.

## Whipped Toppings

These products are regularly formulated using a mixture of hydrogenated coconut or palm kernel oils, or hydrogenated palm kernel stearin, or blends of the two (for economic reasons) which impart easy air incorporation properties during whipping and give the high solids content required for foam stiffness.

**TABLE 7**
Recipes for the Manufacture of Toffee

| Ingredient | "European" Recipe | "Tropical" Recipe |
|---|---|---|
| Granulated sugar | 23.0% | 15.0% |
| Glucose syrup | 34.0% | 43.0% |
| Skimmed sweetened condensed milk | 28.0% | 28.0% |
| Vegetable fat | 11.0% | 11.0% |
| Water | 4.0% | 3.0% |
| Salt | 0.4% | 0.4% |

**TABLE 8**
Typical Powder Coffee-Whitener Formulation

| Ingredient | % |
|---|---|
| Corn syrup solids | 55–60 |
| Fat[a] | 35–40 |
| Sodium caseinate | 4.5–5.5 |
| Emulsifiers | 0.2–0.5 |
| Phosphate salt | 1.2–1.8 |

[a]Hardened palm kernel oil or olein, SMP of 40°C.

## Substitute Chocolates/Compound Coatings

Coconut and palm kernel oils can be fractionated by dry, detergent, or solvent fractionation to give stearins with much better melting properties than the hardened coconut and palm kernel fats. However, coconut stearins are generally softer than palm kernel stearins (Table 9). Therefore, in many confectionery applications palm kernel stearin is superior to coconut stearin.

Palm kernel stearin with physical properties, such as the high SFC and steep melting profile, resembling the more expensive cocoa butter is usually described as a cocoa butter substitute (CBS). It is often hydrogenated to improve its melting profile further and used for a wide variety of confectionery applications requiring improved eating qualities. A typical melting curve of hardened palm kernel stearin is shown in Fig. 3. Such a CBS is particularly suitable for the manufacture of solid or hollow-molded products with excellent mold release, good snap, steep melting characteristics, good flavor release, and resistance to fat bloom.

Although the substitute chocolate formulated from rapidly cooled lauric fats forms stable crystals without tempering, simplifying the production plant and reducing

**TABLE 9**
Characteristics of Coconut Stearin and Palm Kernel Stearin

|  | Coconut Stearin | | | Palm Kernel Stearin |
|---|---|---|---|---|
|  | 1 | 2 | 3 |  |
| Iodine value (Wijs) | 5.6 | 6.4 | 7.2 | 6–8 |
| Melting point (°C) | 28.0 | 27.4 | 27.2 | 32–34 |
| Temperature (°C) | | SFC (% by NMR) | | |
| 10 | 89.3 | 87.9 | 87.0 | 80–90 |
| 20 | 64.9 | 59.0 | 56.1 | 75–85 |
| 25 | 24.8 | 23.9 | 17.4 | 58–68 |
| 30 | 0 | 0 | 0 | 24–33 |
| 35 | — | — | — | 2 max |

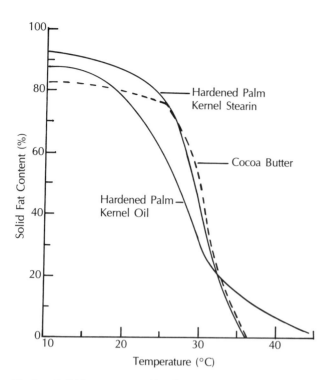

**Fig. 3.** Solid fat contents of hardened palm kernel stearin, cocoa butter, and hardened palm kernel oil.

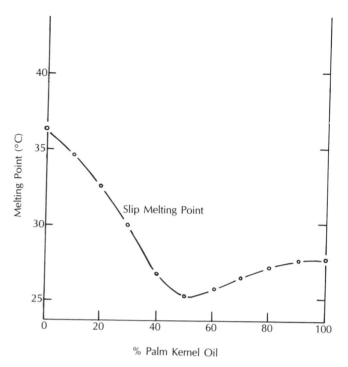

**Fig. 4.** Slip melting point curve of mixtures of palm and palm kernel oils.

cost, palm kernel stearin-based CBS forms eutectic mixtures with cocoa butter. In other words, the fat is incompatible with cocoa butter. It can only tolerate a small amount of cocoa butter (1,2), and therefore must substitute for all the cocoa butter in the recipe if a good chocolate is to be made. Therefore, in substitute chocolate or other coating formulations, the CBS-based coating must be formulated with low-fat cocoa powder in order to avoid incompatibility with cocoa butter. Coatings formulated with these fractionated lauric hard butters are very similar to real chocolate (made using cocoa butter) in set-up, shrink, snap, and eating qualities. A typical substitute chocolate made with CBS is shown in Table 10.

## Margarine and Shortening

Although both palm and palm kernel oils come from the same palm fruit, they are incompatible with each other and form eutectic mixtures. The eutectic interaction is clearly shown by the minima in the SMP curve in Fig. 4 and in the isosolids diagram of the mixture of these two oils in Fig. 5. The eutectic interaction of palm and palm kernel oils can be advantageously used in the blending of by-products from specialty fat production to produce fats for margarines and shortenings.

In palm oil-based margarines, the inclusion of a certain amount of palm kernel oil improves the slow "get-away" on the palate due to palm oil. Palm kernel oil (or coconut oil) with high lauric acid content is also utilized in cake margarine due to its fast crystallization and good creaming properties. Interesterified blends of palm and palm kernel

**TABLE 10**
**Typical Substitute Chocolate Formulation**

| Ingredient | % |
|---|---|
| Cocoa powder | 8 |
| Icing sugar | 44 |
| Skimmed milk powder | 20 |
| CBS | 28 |
| Lecithin | 0.4 |
| Vanillin | 0.1 |

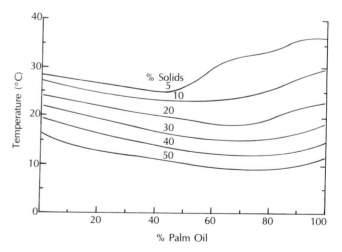

**Fig. 5.** Isosolid diagram for simple blends of palm and palm kernel oils.

**TABLE 11**
Interesterified Fat Blends for Margarines

|  | PO:PKOL (80:20) | PO:PKO (80:20) | PST:PKOL (60:40) |
|---|---|---|---|
| Melting point (°C) | 37.0 | 36.7 | 36.5 |
| Temperature (°C) | | SFC (% by NMR) | |
| 10 | 51.5 | 53.0 | 60.0 |
| 20 | 31.0 | 31.0 | 42.0 |
| 25 | 20.0 | 23.0 | 28.5 |
| 30 | 13.5 | 13.5 | 19.5 |
| 35 | 8.0 | 7.0 | 9.0 |
| 40 | 3.0 | 3.5 | 0 |

*Abbreviations:* PO, Palm oil; PKO, palm kernel oil; PST, palm stearin; PKOL, palm kernel olein.

oils and their fractions are suitable fat blends for table margarines (Table 11). Such blends when formulated with liquid vegetable oils, such as rapeseed oil, into margarines (minimum 80% fat content) give products with good spreadability and fairly good mouth-feel. Palm oil interesterified with lauric oils can also be incorporated in shortening which, when used for making creams, improves the creaming volume and mouth-feel of the cream.

## Stability of Products

Stability against oxidative rancidity is attributable to the low unsaturation in lauric oils, especially after they are hardened. The only serious disadvantage with using lauric fats which contain short-chain fatty acids is that hydrolytic fat splitting can occur in the presence of moisture and lipase enzymes, liberating short-chain caproic, caprylic, capric, and lauric acids (C6:0–C12:0) which, due to their low flavor thresholds (Table 12), may give rise to unpleasant soapy off-flavors. This makes the product unpalatable. When lipase is produced by molds, yeasts, and bacteria from infected cocoa powder and milk powder or through postinfection, the fat becomes the victim of this enzymatic attack. In view of this sensitivity of lauric fats to hydrolysis, good manufacturing practices, such as using fresh and good quality raw materials, and precautions to prevent post infection, are needed when using lauric fats in food products.

**TABLE 12**
Flavor Threshold Values of Some Fatty Acids

| Butyric acid (C4:0) | 0.6 ppm |
|---|---|
| Caproic acid (C6:0) | 2.5 ppm |
| Capric acid (C10:0) | 200 ppm |
| Lauric acid (C12:0) | 700 ppm |
| Stearic acid (C18:0) | 15000 ppm |

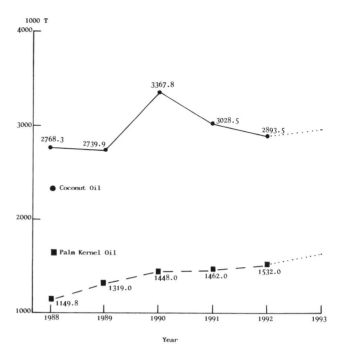

**Fig. 6.** World production of coconut and palm kernel oils, 1988–1992. *Source:* Oil World Annual 1993 (April 1993).

## Conclusion

Despite the undesirable possibility of lipolytic activity, the unique properties of coconut and palm kernel oils have made these two lauric oils particularly attractive to the confectionery industry. With modification, coconut and palm kernel oils are ideal raw materials for the production of various confectionery fats, which perform well in many food products. Today, it appears that the utilization of palm kernel-based confectionery fats is increasing at the expense of coconut oil-based fats. With the continuous increase in production of palm kernel oil (Fig. 6), these two major lauric oils are expected to remain available in quantities sufficient to meet the demands by industrial, institutional, and household consumers in market countries at least through the 1990s.

## Acknowledgment

The author is grateful to Guan Soon Heng Edible Oil Sdn. Bhd. for the permission to present this paper.

## References

1. Paulicka, F.R., *Chem. Ind.* (London) 835 (1973).
2. Gordon, M.H., F.B. Padley, and R.E. Timms, *Fette Seifen Anstrichm. 81*:116 (1979).

# Nutritional Aspects of Lauric Oils

M.I. Gurr

Maypole Scientific Services, Vale View Cottage, Maypole, St. Mary's, Isles of Scilly TR21 0NU, U.K.

## Abstract

The lauric oils, palm kernel and coconut oils have been termed "cholesterol raising," a term used pejoratively to imply that their high content of saturated fatty acids promote hypercholesterolemia, increasing the risk of coronary heart disease (CHD). Coconut oil is widely used in some eastern countries, although some epidemiological studies have indicated that plasma total cholesterol (TC) tends to reflect the level of CNO in the diet, the absolute concentrations of TC are lower than in western industrialized countries. In a study of two Polynesian communities with significantly different intakes of CNO and significantly different TC, CHD was virtually unknown, and the health characteristics of the populations were not obviously different. One study found that CNO had a moderate effect on blood-clotting time, being only slightly more thrombotic than olive oil. The adverse health effects of the lauric oils have been exaggerated. In western countries, diets contain insufficient amounts to have significant effects, whereas in communities in which high levels are consumed, there are no adverse health problems attributable to lauric fats. Products derived from fractionation of CNO are valuable sources of fat for those with impaired ability to adsorb fat and may also find use in weight management programs.

## Lauric Oils: Subjects of Nutritional Controversy

For centuries, the lauric oils, palm kernel and coconut oils, have contributed to the diet of people in countries where these palm trees are grown. During the 1980s, advertisements appeared in newspapers and journals in the United States warning consumers that the inclusion of these oils in their diet should be avoided since it greatly increased their chance of developing heart disease and succumbing to a heart attack. Insofar as there was any scientific basis for these statements, the arguments were based mainly on research done several decades earlier, demonstrating that edible oils containing a high proportion of saturated fatty acids (SFA) had the effect of increasing the circulating concentration of total cholesterol (TC) in the blood. Fatty acids with chain lengths of 12, 14, and 16 carbon atoms were found to be particularly potent in raising TC. Epidemiological studies have associated a high concentration of TC with an increased risk for coronary heart disease (CHD). Since both palm kernel and coconut oils contain a high proportion of total SFA and are particularly rich in the "cholesterol-raising" SFA (Table 1), many have as-

**TABLE 1**
**Fatty Acid Composition of Coconut and Palm Kernel Oils**

| Fatty Acid | | Coconut | Palm Kernel |
|---|---|---|---|
| Caprylic | 8:0[a] | 8 | 4 |
| Capric | 10:0 | 7 | 4 |
| Lauric | 12:0 | 48 | 45 |
| Myristic | 14:0 | 16 | 18 |
| Palmitic | 16:0 | 9 | 9 |
| Stearic | 18:0 | 2 | 3 |
| Oleic | 18:1 | 7 | 15 |
| Linoleic | 18:2 | 2 | 2 |
| Others | | 1 | 0 |

[a]The number before the colon indicates the number of carbon atoms in the fatty acid hydrocarbon chain and the number after the colon, the number of double bonds.

sumed that inclusion of lauric oils in a diet would automatically lead to raised TC, increased risk of CHD, and that these oils should be avoided altogether.

It is not the intention of this paper to present a balanced review of the evidence for and against the view that dietary SFA are major contributors to the risk of CHD; the author has reviewed such evidence elsewhere (1). The purpose of this paper is to argue, with supporting data, that the contribution of SFA, including those present in the lauric oils, to CHD risk has been grossly overemphasized.

## Saturated Fatty Acids and Blood Lipids

Scientific evidence relating dietary intake of SFA with blood-lipid concentration (and with disease patterns) is of three main types: epidemiological, experimental studies in man, and animal experiments. While the latter have yielded important information about biochemical pathways and potential mechanisms, differences between species in biology, as well as in the diets normally consumed and mode of living, require the utmost caution in relating experimental results from animals to human health. In this paper, the results of animal experiments will be used sparingly, and most emphasis will be placed on human studies.

### Epidemiological Evidence

*International and Within-Country Studies.* In seeking confirmation from cross-cultural studies that high intake of SFA is directly associated with a high concentration of TC, most commentators have relied almost entirely on the Seven Countries Study of Keys et al. (1). These authors found a positive correlation (r = 0.87) between the percentage of total dietary energy from SFA and the concentration of TC. Although other international studies have tended to show that countries with a low average intake of dietary fat also tend to have populations with low average TC, most have been cross-sectional in design and their usefulness is limited in this respect. Other aspects of diet and lifestyle (protein, fiber, alcohol, and energy intakes; exercise; stress patterns; and smoking habits) are also quite different, and it is impossible to isolate specific effects of diet.

Comparisons of individuals or groups of people within countries have not demonstrated significant associations between type or level of dietary fat and blood lipids using a number of study techniques. One of the most famous and long-running prospective studies, the Framingham Study, has not been able to show differences in energy, proteins, carbohydrates, fat, or cholesterol in subgroups representing tertiles of the cholesterol distribution (1). In the Caerphilly Heart Disease Study in the United Kingdom, diet accounted for only 1.9% of the variance in TC (1). It has been argued that the poor correlation between dietary fat and TC within countries as compared to correlations between countries is due to the limited range of intakes within one country. However, examination of the intakes reported in many within-country studies suggests that differences between highest and lowest intakes can be as large as those, which in carefully controlled experimental studies, have demonstrated a significant difference in TC.

*Effects of Coconut Consumption in Asia.* In many Asian communities, particularly the Philippines, Polynesia, and parts of India, coconut has traditionally made a substantial contribution to food and energy intake. However, there are surprisingly few published studies providing information on possible associations between consumption of coconut oil and cholesterolemia. If lauric oils contribute importantly to hypercholesterolemia, then it would be expected to be manifested in the TC of these communities.

Prior et al. have published studies of diet, TC, and CHD of the inhabitants of various Polynesian islands (2,3). They compared the inhabitants of the islands of Pukapuka and Tokelau (2). The percentage of energy derived from fat in the communities was about 35 and 50%, respectively, and in each community was derived mainly from coconut. Saturated fatty acid provided about 28 and 48% of total energy, respectively, and PUFA about 2% in each community. These data, obtained by 7-day weighing of household food items, were reflected in the adipose tissue fatty acid composition, which is regarded as being a good reflection of the dietary fat composition over a long period of time. The proportions of 12:0, 14:0, and 18:2 in the adipose tissue were about 11, 17, and 3%, respectively, in both communities compared with 0.3, 4, and 3% in New Zealanders of European origin. There is no doubt that coconut oil had dominated the diets of these communities over a very long period and that, as a result, intakes of the "cholesterol-raising" fatty acids were far higher in some of these communities than in the West.

Total cholesterol concentrations of Pukapukan men and women were 170 mg/dL and 176 mg/dL, while those of Tokelauans were 208 and 216 mg/dL. The higher levels in Tokelauans were judged to result from their higher intakes of SFA. Nevertheless, these TC are considerably lower than those found in New Zealand (on average about 225 mg/dL), where SFA intakes are lower, and much lower than would be predicted from application of the Keys/Hegsted formulas.

When Tokelauans migrated to New Zealand, their TC increased (3), despite decreases in their consumption of saturated fatty acids (from 45% to 21% of energy). However, their consumption of cholesterol increased (from 85 mg/d to 340 mg/d). It is clear that with most other lifestyle attributes being equal, differences in SFA intake may affect TC to some extent, but overall the influence of SFA is not dominant.

## Experimental Evidence

*Predictive Equations.* The first very extensive systematic investigation of the effects of different dietary fats on human plasma cholesterol was that of Keys et al. published in 1957 (4). They studied men in a metabolic ward setting, giving them diets with or without fat in amounts up to 35% of energy, provided by maize, soybean, sunflower seed, rapeseed, safflower, cottonseed, hydrogenated coconut, olive, pilchard, menhaden, and butterfat. This study gave rise to the well-known "Keys equation" which is used to predict the change in cholesterol that would occur in response to changes in the consumption of SFA and PUFA. Hegsted et al., as a result of similar painstaking and thorough nutritional experiments (5), derived a similar equation. Briefly, the equations summarized the finding that SFA were about twice as effective in raising TC as PUFA were in lowering it and that there was a small contribution of dietary cholesterol to plasma cholesterol.

There are several limitations to these equations:

- Saturated fatty acids and PUFA are broad terms covering a wide variety of chemical structures. Saturated fatty acids with carbon chain lengths below 12 or of 18 and above do not influence plasma cholesterol. This fact was recognized by Keys et al. (4) and Hegsted et al. (5), but the information was not incorporated into the equations. It was revealed later that the cholesterol-lowering effect of PUFA was due mainly to n-6 PUFA. More recent research has shown that n-3 PUFA have little effect on TC, but effectively lower plasma TAG (6).
- The equations did not include a term for MUFA, even though oleic acid is quantitatively the most important fatty acid in the diet and in the body. Keys and Hegsted (4,5) regarded oleic acid as "neutral" in its effects on TC. More recently, the effects of diets rich in MUFA on plasma lipids have been reevaluated and, in general, MUFA are now regarded as having effects quite similar to n-6 PUFA (6).

- Although Hegsted et al. (5) made brief reference in their paper to "β-lipoprotein cholesterol" (now referred to as low-density lipoprotein (LDL) cholesterol), only results on TC were presented. More recent research has pointed up the fact that changes in TC are of limited significance and that changes in individual lipoproteins are more closely related to vascular health. It is, for example, quite possible for rises in TC to be the result of rises in both LDL, regarded as "atherogenic," and HDL, regarded as "protective."
- The equations do not take into account interactions between different fatty acids at different levels of intake.

In carefully controlled experiments with different species of monkeys chosen to simulate human "high," "medium," and "low" responders to dietary fatty acids, Hayes et al. (7) found that 12:0 and 14:0 were considerably more potent than 16:0 in raising TC and LDL-C. Similar results were obtained (8) in human subjects. The relative potencies of 12:0 and 14:0 remain in question, but results suggest that 14:0 is the major cholesterol-raising fatty acid. Hayes et al. (9) have developed a hypothesis to explain the differential effects of SFA and the apparent discrepancies between previous publications. They propose that the different hypercholesterolemic SFA have different thresholds at which they exert an effect on plasma cholesterol, which are dependent on: the level of 18:2 in the diet, the level of dietary cholesterol, and the initial concentration of LDL in the plasma.

The interaction of these various factors operates at the level of the apoB/E receptors on cell surfaces which are critically important for the removal of LDL particles from plasma. When 18:2 intake exceeds about 6.5% of dietary energy, the LDL apoB/E receptors are "upregulated." That is, they readily interact with LDL particles, removing them from plasma, thus maintaining relatively low plasma concentrations of LDL. As the dietary level of 18:2 is gradually reduced below 6.5%, 14:0 and 12:0 begin to override the effects of 18:2 and "downregulate" the receptors. Low-density lipoprotein particles are thus less efficiently removed, and plasma concentrations of LDL begin to rise. Because of its lower potency, more 16:0 is required to observe an effect, and then only when 18:2 levels in the diet are at the low end of the intake distribution.

This hypothesis begins to explain alternate conclusions in the literature, coming from different experiments employing widely differing levels of dietary 18:2, different cholesterol intakes, and widely different plasma TC and LDL-C concentrations. Thus, discrepancies between the original studies of Keys and Hegsted (4,5), which observed no cholesterolemic effect of 18:1 and more recent studies (10,11) which observed a hypocholesterolemic effect, can be explained. Keys and Hegsted (4,5) exchanged 18:1 for 18:2 at 1 and 6% of energy from 18:2 and in the presence of relatively large amounts of 14:0. Mattson and Grundy (10) and Mensink and Katan (11) worked at levels of 18:2 above the 6.5% threshold and in the presence of relatively small amounts of 14:0. At low levels of 14:0, minimal 18:2 is required to ensure maximum apoB/E receptor activity, so that 18:1 appears to be as efficient as 18:2 in maintaining a low plasma LDL when exchanged for what is effectively an "excess" of 18:2.

*Specific Effects of Coconut Oil.* Contrary to the impression given by many commentators, Keys et al. (4) did not use coconut oil, but *hydrogenated* coconut oil. Their results are thus not strictly relevant to communities that use the unprocessed oil in their diets. However, this and several other studies found that providing most of the dietary fat from hydrogenated coconut oil (having a higher content of SFA than the natural oil) did not result in a significantly higher TC than when the subjects consumed their normal (usually American) diets. Several studies reported minimal effects on plasma cholesterol after giving unprocessed coconut oil. They are difficult to compare, since some used liquid formula diets, others normal dietary patterns; some gave predominantly coconut, others gave mixtures; the amount of fat as a percentage of energy, the amount of dietary cholesterol, and the initial TC were also markedly different. In contrast, Hegsted et al. (5) found that natural coconut oil was among the most hypercholesterolemic of the oils tested. Its effects were smallest when its was mixed with other oils, as it normally would be, and when dietary cholesterol intakes were relatively low. These results are consistent with the hypothesis of Hayes et al. (9), outlined previously. Table 2 presents results from a few studies that are reasonably comparable. In general, those that have been omitted found little effect of giving coconut oil on TC but did not present statistical analyses.

A more recent study (12), while confirming a hypercholesterolemic effect of coconut oil in human subjects, found only a modest rise (12%) in TC when compared with the subjects' habitual diets. These authors (12) also measured plasma high-density lipoproteins (HDL), a fraction not analyzed in the earlier studies of Keys and Hegsted (4,5). High-density lipoprotein cholesterol rose by 22%. Thus, half the rise of 0.54 mmol/L in TC due to coconut oil consumption was accounted for by a rise in HDL cholesterol. Since current views are that HDL exerts a protective effect in relation to CHD, it is doubtful whether rises in TC associated with inclusion of coconut oil in diets under experimental conditions can be regarded as conferring significantly increased CHD risk.

*Dietary Strategies to Lower Plasma Cholesterol.* Whereas replacement of SFA with unsaturated fatty acids (UFA) can result, on average, in a reduction in TC in an experimental setting as in the studies of Keys and Hegsted (4,5), it has proved much more difficult to achieve substantial and lasting reductions in larger population-based studies. Ramsay et al. (13) reviewed 16 intervention studies designed to reduce TC and CHD. Diets equivalent to the American Heart Association "Step 1" diet (total fat < 30% of energy; P/S = 1.0; cholesterol < 300 mg/d) resulted in reductions in cholesterol between about 0.2% and 4% in nine studies, no change in one and an increase of 1% in another. Diets equivalent to, or more intensive than, the "Step 2" diet (P/S = 1.4) resulted in drops of TC ranging

**TABLE 2**
**Effect of Coconut Oil on Plasma Total Cholesterol in Dietary Studies with Human Subjects**

| Reference | Total Fat (% Energy) | CNO (% Fat) | Dietary Cholesterol (mg/d) | Plasma TC Normal Diet (mg/dL) | Plasma TC CNO Diet (mg/dL) |
|---|---|---|---|---|---|
| 34 | — | 90 | — | 170–370 | 140–240 |
| 5 | 38 | 85 | 306 | 220 | 343 |
| 12 | 32 | 75 | 200 | 167 | 187 |
| 35 | 42 | 50 | — | — | 256 |
| 5 | 38 | 42.5 | 306 | 220 | 283 |

Abbreviations: CNO, coconut oil; TC, total cholesterol.

from 6.5–15.5%. These results contrast very sharply with assertions in many guidelines and reviews that TC will fall by 10–25% in response to a Step 1 diet.

## Saturated Fatty Acids and Coronary Heart Disease

### Epidemiological Studies

*Cross-Cultural Comparisons.* The Seven Countries Study found a strong association (r = +0.84) between CHD mortality and intake of SFA but not total fat. The data were selected, however, and it has been pointed out that comparisons of other countries for which similar data were available would have produced correlation coefficients ranging from −0.9 to +0.9, and that only those selections containing Japan and Finland produced coefficients greater than zero (1).

*Fatty Acid Consumption by Cases and Controls.* If there were a strong connection between the type of fatty acids that people eat and their risk of dying from CHD, then one would expect to be able to observe significant differences in the amount or type of fat, or both, consumed by people who experienced a CHD event and those who did not. The results of many studies indicate that this is not the case (1). In at least two studies, those who experienced CHD events ate more PUFA and less SFA than those who did not.

*Fatty Acid Consumption in Communities.* There is little evidence that communities that have relied for generations on coconut oil as their main source of fat, experience greater CHD incidence and mortality than those consuming less saturated fats. Indeed, studies of the inhabitants of several Polynesian (2) and Melanesian (14) islands with very high intakes of saturated fatty acids from coconut oil and, in one case (2) very high total fat consumption as well, had extremely low rates of CHD. In Africa, CHD is rare among the Masai (15) despite their high consumption of SFA and Nigerians, with a high consumption of saturated fatty acids from palm oil, albeit with relatively low total fat intakes, also suffer little CHD (16).

*Trends in Consumption and Disease Over Time.* Mortality rates in many western countries reached peaks at different times this century and then began to decline. There is no evidence that the decline is a result of a reduction in consumption of SFA (1,17). Indeed, declines in mortality have often predated declines in fat consumption, and in several countries the PUFA/SFA ratio has continued to rise gradually during a period when CHD mortality first increased, peaked, and then declined. Fat consumption in Japan has been increasing steadily while CHD, already at a low level compared with the West has continued to decline. Such trends are more likely to be explained by environmental influences on growth and development in utero and in infancy coupled with failure to adapt to altered environmental conditions later in life (18).

### Intervention Studies

The ultimate test of the view that SFA are prime causes of CHD is to demonstrate experimentally that reducing intakes of SFA, sufficient to cause significant reductions in the circulating concentrations of LDL cholesterol, will reduce morbidity and mortality from CHD. A recent report (19) on the diets of Scots, who have one of the highest rates of CHD mortality in the world, states that: "evidence linking specific SFA to increases in blood cholesterol is overwhelming, as is the causal role of the elevated levels of the cholesterol-containing low-density lipoproteins in promoting heart disease." The results of many intervention trials, designed to modify CHD risk factors and reduce CHD, dispute this assertion. Many have been "multiple risk factor trials" and, even when demonstrating overall improvement, have found difficulty in ascribing the benefits specifically to reducing consumption of SFA (1,17).

A trial most often cited as providing evidence for the benefits of dietary modification (20), involved interventions on diet and smoking. The authors confessed that "if this had been a dietary trial only, the difference in myocardial infarction would probably not have reached statistical significance." Moreover, the nutritional information provided in the paper was extremely poor and consisted of inadequate data obtained 4 years into a 12-year study. The quality of the nutritional data in this and in many other similar studies would not have warranted publication in a specialized nutrition journal. This study demonstrated an improvement in CHD mortality as well as morbidity. However, few studies have provided evidence for improvements in total mortality due to intervention. Indeed, there is growing evidence that noncardiovascular mortality may actually be increased, sometimes from other illness, sometimes from accidental, often violent, causes (1,21–24). These have been brushed aside by those convinced that dietary change is imperative to reduce CHD (19,25) but there is no rational reason to ignore such evidence and it should not be dismissed lightly.

*Dietary Fatty Acids and Atherosclerosis.* Few studies have examined the intake of fatty acids in relation to atherosclerosis in man. Moore et al. (26) found that the degree of atherosclerosis found during autopsy was unrelated to previous intakes of animal fats or SFA. Numerous angiographic and autopsy studies have been cited in support of the contention that high TC and LDL-C is associated with a high degree of atherosclerosis, but the correlations were weak (no greater than $r = 0.35$ [27]). Such a correlation may be biologically, if not statistically, insignificant especially as most studies were flawed by the inclusion of large numbers of people with familial hypercholesterolemia and by a failure to adjust for age. Several follow-up studies observing atherosclerosis by angiography found no association between progression or regression of atherosclerosis and secular changes in TC (17,27).

*Dietary Fatty Acids and Thrombosis.* There are few studies on the influence of individual dietary fatty acids or even dietary oils on the process of thrombosis, mainly due to severe methodological problems. Using a rat model for studying arterial thrombosis in vivo, Hornstra et al. (28,29) found that PUFA (especially marine oils containing PUFA of the n-3 family) reduced thrombotic tendency, while SFA increased it. Chain length had little influence, and coconut oil was only slightly less thrombogenic than olive oil, currently regarded by many as an ideal dietary oil. In man, different components of the hemostatic system have been studied in vitro. Most published studies have focused on the n-3 fatty acids (1). Reduction of total fat intake reduced the concentration of Factor VIIc in the coagulation cascade, but the composition of the fatty acids had no influence (30).

## Other Nutritional Aspects of Lauric Oils

One consequence of the unusual fatty acid composition of lauric oils is that they have relatively low melting points for their degree of saturation. Short- and medium-chain fatty acids are released readily from the triacylglycerols by lipases. A proportion of short- and medium-chain fatty acids from milk is released by lingual and gastric lipases before the fat reaches the duodenum where it normally is digested by pancreatic lipase. It is probable that this also occurs with lauric oils. In digestibility experiments with rats, coconut oil had the highest digestibility of all oils tested (99.7%), and even fully hydrogenated coconut oil was 98.5% digested (31).

Fatty acids with chain lengths of 12 or less are absorbed by a mechanism different from those with longer chain lengths. Instead of being absorbed as components of mixed micelles into the intestinal epithelial cells, they are transferred directly into the portal vein, carried to the liver and rapidly oxidized there. This has two important consequences. First, they are not incorporated to any large extent into chylomicrons and so do not contribute to hyperlipidemia. This may in part account for why coconut oil does not raise plasma cholesterol as much as would be predicted on the basis of its total saturation. Second, they are not deposited in adipose tissue. This may be an important point in their favor in relation to maintenance of energy balance, but definitive results on effects of diets rich in short- and medium-chain fatty acids on long-term weight maintenance have yet to be established.

People who are unable to digest or absorb long-chain fatty acids are usually able to assimilate short- and medium-chain fatty acids easily (32). This has led to the important therapeutic use of medium-chain triacylglycerol (MCT) preparations by fractionation of coconut oil for use in diets of such patients. Such preparations are expensive and only available in small quantities. Processes for the more efficient and cheaper production of MCT from lauric oils would be nutritionally and clinically advantageous.

## Conclusions

Concerns about the harmful effects of lauric oils with respect to raised TC and CHD have been largely unfounded. In practice, populations in which coconut oil contributes a considerable proportion of energy and fat intake do not have notably high TC or CHD mortality. In Western countries, factors other than dietary fats are more important for contributing to CHD and the intake of lauric oils is very low in relation to other sources of SFA. In the United States, lauric oils contribute less than 2% of daily energy intake and less than 6% of SFA intake (33). There is a case for developing cheaper sources of MCT to test their potential role in weight-maintenance programs.

## References

1. Gurr, M.I., *Progr. Lipid Res. 31*:195 (1992).
2. Prior, I.A., F. Davidson, C.E. Salmond, and Z. Czochanska, *Am. J. Clin. Nutr. 34*:1552 (1981).
3. Stanhope, J.M., V. Sampson, and I.A.M. Prior, *J. Chron. Dis. 34*:45 (1981).
4. Keys, A., J.T. Anderson, and F. Grande, *Lancet II:*959 (1957).
5. Hegsted, D.M., R.B. McGandy, M.L. Myers, and F.J. Stare, *Am. J. Clin. Nutr. 17*:281 (1965).
6. Grundy, S.M., and M.A. Denke, *J. Lipid Res. 31*:1149 (1990).
7. Hayes, K.C., A. Pronczuk, S. Lindsey, and D. Diersen-Schade, *Am. J. Clin. Nutr. 53*:491 (1991).

8. Ng, T., K.C. Hayes, G. DeWitt, M. Jegathesan, N. Satganasingam, A. Ong, and D. Tan, *J. Am. Coll. Nutr. 11*:383 (1994).
9. Hayes, K.C., and P. Khosla, *Fed. Am. Soc. Exp. Biol. J. 6*:2600 (1992).
10. Mattson, F.H., and S.M. Grundy, *J. Lipid Res. 26*:194 (1985).
11. Mensink, R.P., and M.B. Katan, *N. Engl. J. Med. 321*:436 (1989).
12. Ng, T.K.W., K. Hassan, J.B. Lim, M.S. Lye, and R. Ishak, *Am. J. Clin. Nutr. 53*:1015S (1991).
13. Ramsay, L.E., W.W. Yeo, and P.R. Jackson, *Br. Med. J. 303*:953 (1991).
14. Lindeberg, S., and B. Lundh, *J. Int. Med. 233*:269 (1993).
15. Mann, G.V., R.D. Shaffer, R.S. Anderson, and H.H. Sandstead, *J. Atheroscl. Res. 4*:289 (1964).
16. Kesteloot, H., V.O. Oviasu, A.O. Obasohan, A. Olomu, C. Cobbaert, and W. Lissens, *Atherosclerosis 78*:33 (1989).
17. Rosenman, R.H., *Homeostasis 34*:1 (1993).
18. Barker, D.J.P., P.D. Gluckman, K.M. Godfrey, J.E. Harding, J.A. Owens, and J.A. Robinson, *Lancet 341*:938 (1993).
19. The Scottish Office Home and Health Department. *The Scottish Diet*. Report of a Working Party to the Chief Medical Officer for Scotland (1993).
20. Hjermann, I., K. Velve-Byre, I. Holme, and P. Leren, *Lancet II*:1303 (1981).
21. Muldoon, M.F., S.B. Manuck, and K.A. Matthews, *Br. Med. J. 301*:309 (1990).
22. Jacobs, D. *Circulation 86*:1046 (1992).
23. Oliver, M.F., *Eur. J. Clin. Invest. 22*:441 (1992).
24. Davey Smith, G., F. Song, and T.A. Sheldon, *Br. Med. J. 306*:1367 (1993).
25. American Heart Association, *Circulation 81*:1721 (1990).
26. Moore, M.C., M.A. Guzman, P.E. Shilling, and J.P. Strong, *J. Am. Diet. Assoc. 68*:216 (1976).
27. Ravnskov, U., *Med. Hypotheses 36*:238 (1991).
28. Hornstra, G., and R.B. Lussenburg, *Atherosclerosis 22*:499 (1975).
29. Hornstra, G., A.A.H.M. Hennissen, and J.W. Wierts, in *Essential Fatty Acids and Eicosanoids,* edited by A. Sinclair and R. Gibson. The American Oil Chemists' Society, Champaign, IL 1992, pp. 287–289.
30. Marckmann, P., B. Sandstrom, and J. Jespersen, *Atherosclerosis 80*:227 (1990).
31. Calloway, D.H., G.W. Kurtz, J.J. McMullen, and L.V. Thomas, *Food Res. 21*:621 (1956).
32. Sickinger, K., in *The Role of Fats in Human Nutrition,* edited by A.J. Vergroesen, Academic Press, New York, 1975, pp. 115–209.
33. Park, Y.K., and E.A. Yetley, *Am. J. Clin. Nutr. 51*:738 (1990).
34. Halden, V.W., and H. Lieb, *Nutr. Dieta 3*:75 (1961).
35. Hashim, S.A., R.E. Clancy, D.M. Hegsted, and F.J. Stare, *Am. J. Clin. Nutr. 7*:30 (1959).

# Health Effects of Lauric Oils Compared to Unsaturated Vegetable Oils

E.A. Emken

U.S. Department of Agriculture, Agricultural Research, ARS, National Center for Agricultural Utilization Research, Food Quality and Safety Research, Peoria, IL, USA.

## Abstract

Recent reports have suggested that the nutritional value of oleic acid may be equivalent to or better than linoleic acid (18:2n-6) because oleic acid does not lower HDL levels and/or lower apolipoprotein A-I. There are also recent studies, which agree with many older studies, reporting that linoleic acid does not lower HDL and that linoleic acid is more effective than oleic acid for lowering serum cholesterol levels. Clearly, the question of whether the hypocholesterolemic effect of oleic acid is equivalent to or better than linoleic acid is a controversy that remains to be resolved. Linolenic acid (18:3n-3) has beneficial health effects, but its health and nutritional role is separate from 18:2n-6. Lauric and myristic acids raise cholesterol levels relative to unsaturated fats and stearic acid has a lesser effect on serum cholesterol levels than other saturated fatty acids. Whether 16:0 is as hypercholesterolemic as 12:0 or 14:0 should be regarded as an unanswered question.

---

Almost three decades ago, Keys and Hegsted developed their now well-known equations for predicting the effect of saturated and polyunsaturated fatty acids on human serum cholesterol concentrations (1,2). Since then, the health effects associated with dietary fats have become increasingly complex and controversial. The effects of saturated fatty acids with different chainlength on serum cholesterol levels are not equivalent. The physiological effects of n-6 and n-3 polyunsaturated fatty acids are different, but the n-6 to n-3 ratio is believed to be important. High oleic acid oils are reported to lower serum cholesterol. High levels of linoleic acid may not be desirable. *Trans* fatty acid isomers in partially hydrogenated vegetable oils raise serum cholesterol levels relative to their *cis* counterparts. The beneficial effects of serum cholesterol reduction on coronary heart disease are being challenged, and the importance of dietary cholesterol intake on serum cholesterol concentrations is less important than was believed originally.

As the health effects of each fatty acid structure are identified, nutritional advice regarding dietary fats is modified to accommodate the new information. This continual revision of health advice has caused confusion among consumers and health professionals. The popular press has contributed to the controversy and confusion by sensationalizing results that have not been confirmed independently. In retrospect, the recognition that each individual fatty acid structure has a different physiological effect should not come as a surprise because each fatty acid structure has different biochemical properties.

## Composition of Common Fats and Oils

The fatty acid composition for a few common "saturated" and "unsaturated" fats and oils are shown in Table 1 (3). Vegetable oils that contain over 70% unsaturated fatty acids (soybean, canola, corn, sunflower, safflower, and olive) are classified as "unsaturated" fats. The saturated fatty acids in these oils are mainly palmitic and stearic acid. Fats and oils that contain about 50% or more saturated fatty acids and less than 10% polyunsaturated fatty acids (coconut, palm kernel, cocoa butter, palm, and butter) are classified as saturated fats. A distinguishing characteristic of the saturated fatty acids in coconut and palm kernel oils is their high content of 12 and 14 carbon chain-length saturated fatty acids (*ca.* 48% 12:0 and 9–15% 14:0).

## Predicted Effect on Serum Cholesterol Concentrations

The classical empirical equations derived by the Keys and Hegsted groups were based on results from controlled diet studies that used many different fats and oils (1,2). Both groups reported that saturated fats were correlated with increased serum cholesterol, and polyunsaturated fats were correlated with decreased serum cholesterol concentrations. Monounsaturated fats did not correlate with changes in serum cholesterol concentrations.

A modified version of the Keys equation does not include stearic acid as part of the sum for saturated fats and has an added term for dietary cholesterol (4). A revised equation has been recently derived by Hegsted, based on data published for 248 metabolic studies and 96 field studies (5). This equation accounts for about 84% of the observed change in total serum cholesterol. The modified Keys equation and the revised Hegsted equation still include terms for saturated and polyunsaturated fatty acids (PUFA), but the addition of a coefficient for monounsaturated fatty acids (MUFA) did not improve the predictability of the equations.

The Keys' equation predicts that 1% of total energy from saturated fatty acids will raise serum cholesterol by 2.4 mg/dL, and the revised Hegsted equation predicts a 2.1 mg/dL increase. Polyunsaturated fatty acids are predicted to reduce serum cholesterol by 1.2 mg/dL (Keys) and 1.16 mg/dL (Hegsted) per 1% of total energy.

The fatty acid compositions presented in Table 1 for saturated and unsaturated fats and oils were used to estimate the effect of these dietary fats on serum cholesterol concentrations (Table 2). The calculated values listed in Table 2 are relative to a reference diet that contained equal amounts of saturated, monounsaturated, and polyunsaturated fatty acids and are consistent with suggested guidelines for dietary fat intake (6). The modified Keys and revised Hegsted equations were used to estimate the changes in serum cholesterol concentrations shown in Table 2. The estimates from both equations were similar except for the prediction for cocoa butter. The predictions for cocoa butter were different because it is high in both palmitic and stearic acids, and the Keys equation does not include stearic (18:0) in the sum for saturated fatty acids. The large increases in serum cholesterol levels predicted for both coconut and palm kernel oils are due to the high saturated fat and low polyunsaturated fat contents of these oils.

## Controlled Diet Studies

Generally, the predicted effect on serum cholesterol concentrations (Table 2) were consistent with results from controlled diet studies. Results from both recent and older human diet studies are summarized in Tables 3–5. Important differences between the design of the studies cited in Tables 3–5 were likely responsible for the variation in the results from these studies. Differences included the amount of total dietary fat; the percent of test fat added to the base diet; the amount of cholesterol in the diets; the length of the dietary periods; and the age, sex, and initial serum cholesterol concentration of the subjects. All of the "lauric–myristic acid" diets in the cited studies contained sufficient amounts of linoleic acid to meet essential fatty acid requirements, but the amounts of linoleic acid included in the diets varied widely.

Overall, the data in Tables 3–5 show that a change in total cholesterol was paralleled by a change in low-density lipoprotein cholesterol (LDL) concentration and that both the monoene and polyene diets had only a small effect on high-density lipoprotein cholesterol (HDL). These results do not support the suggestion that the cholesterol-lowering effect of high polyunsaturated fat diets is less beneficial than high monounsaturated fat diets, because the polyunsaturated fat diets significantly reduced HDL levels (7,8).

## "Lauric–Myristic Acid" vs. Monoene Diets

The effect of diets containing "lauric–myristic acid" fats and oils compared to diets containing monounsaturated oils (monoene diet) are summarized in Table 3 (2, 9–15). These data for monoenoic diets illustrate the effect of reducing saturated fat intake by substituting monounsaturated fats for saturated fats. For all studies, the average serum cholesterol concentrations were significantly higher for the subjects fed the coconut oil- or dairy fat-enriched diets compared to the subjects fed the monoenoic acid diets. The source of the monounsaturated fat did not have an influence.

The results in Table 3 appear to suggest that monounsaturated fats are as effective as polyunsaturated fats for lowering serum cholesterol levels. The data for the last study (15) listed in Table 3 illustrate that polyunsaturated fats are more effective than monounsaturated fats for lowering cholesterol levels. In this study, serum cholesterol

**TABLE 1**
**Approximate Fatty Acid Composition of Selected Oils**[a]

| Fatty Acid | Saturated Fats and Oils | | | | Unsaturated Oils | | |
|---|---|---|---|---|---|---|---|
| | Coconut Oil | Palm Kernel[b] | Butter Fat | Cocoa Butter | Soybean Oil | Canola Oil | Corn Oil |
| 6:0 | 1.3 | 0.2 | 2.2 | — | — | — | — |
| 8:0 | 12.2 | 10.1 | 1.2 | — | — | — | — |
| 10:0 | 8.0 | 8.1 | 2.8 | — | — | — | — |
| 12:0 | 48.8 | 47.0 | 2.8 | — | — | — | — |
| 14:0 | 14.8 | 9.1 | 10.1 | — | — | — | 0.2 |
| 16:0 | 6.9 | 8.1 | 25.1 | 26.5 | 11.5 | 3.8 | 13.1 |
| 18:0 | 2.0 | 2.0 | 12.1 | 34.7 | 4.0 | 0.5 | 2.5 |
| 18:1n-9 | 4.5 | 13.3 | 27.1 | 35.6 | 22.5 | 55.5 | 30.5 |
| 18:2n-6 | 1.4 | 2.0 | 2.4 | 3.0 | 53.0 | 24.9 | 52.3 |
| 18:3n-3 | | | 1.0 | 0.2 | 7.5 | 12.0 | 1.0 |
| Other[c] | 0.1 | 0.1 | 13.3[d] | — | 1.5 | 3.3 | 0.4 |

[a]Adapted from Ref. 3.
[b]Ouricuri variety. For commercial varieties the 6:0–10:0 content is about 8%.
[c]Branched, odd chain, and $C_{20}$ fatty acids.
[d]Also includes trans fatty acids.

**TABLE 2**
Effect of Substitution of a Dietary Fat for Reference Fat on Serum Cholesterol as Estimated by Keys and Hegsted Equations[a]

| Fat or Oil | Composition, %[b] | | Predicted Change Cholesterol, mg/dL[c] | |
|---|---|---|---|---|
| | Sat | Poly | Keys[c] | Hegsted[d] |
| Reference fat[e] | 33 | 33 | — | — |
| Soybean oil | 16 | 61 | −20.0 | −23.0 |
| Corn oil | 16 | 53 | −18.1 | −22.5 |
| Canola oil | 4 | 37 | −14.5 | −19.7 |
| Cocoa butter | 61 | 3 | +14.6 | +32.4 |
| Butter fat | 50 | 3 | +24.7 | +24.2 |
| Palm oil | 49 | 10 | +27.5 | +20.6 |
| Palm kernel | 66 | 2 | +47.6 | +36.6 |
| Coconut oil | 73 | 1 | +53.3 | +41.6 |

[a]Assumes a 2500 KCal diet containing 100 g (36 en%) of fat and 350 mg of dietary cholesterol.

[b]Sat and Poly = saturated and polyunsaturated fatty acids (includes 18:2n-6 and 18:3n-3). Percentage of total dietary fat.

[c]Modified Keys equation: $\Delta$Chol = 2.4 $\Delta$Sat −1.2 $\Delta$18:2n-6 + (1.5 $\Delta$Chol$^{0.5}$). Sat does not include 18:0. Ref. 4.

[d]Revised Hegsted equation: $\Delta$Chol = 2.10 $\Delta$Sat −1.16 $\Delta$18:2n-6 + 0.067 $\Delta$Chol. Sat includes 18:0. Ref. 5.

[e]Composition of hypothetical reference fat is consistent with suggested nutritional guidelines and contains 23% (12:0, 14:0, 16:0), 10% 18:0, 33% 18:1n-9, 30% 18:2n-6, and 3% 18:3n-3.

Note: Saturated fatty acids with chain-lengths of less than 12 carbons, 18:1n-9, and 18:3n-3 are not included in the equations used to calculate the values listed.

levels were slightly lower when a diet containing a small amount of coconut oil in a high polyunsaturated fat diet was compared to a high monounsaturated fat diet.

Comparison of soybean oil diets (high 18:2n-6) to canola oil diets (high 18:1n-9) indicated that both oils have a similar effect on serum cholesterol (16,17). Compared to canola oil, the cholesterol-increasing effect of the higher saturated content of soybean oil apparently counterbalanced much of the cholesterol-decreasing effect of the higher 18:2n-6 content. These results, plus results from other studies that suggest oleic acid has a cholesterol-lowering effect (7,8,12,16,17), are not consistent with results of earlier studies (1,2). The reason for the inconsistency of results from studies with monounsaturated fats is unknown. Clearly, it is still premature to draw conclusions regarding the cholesterolemic effect of oleic acid versus linoleic acid.

## "Lauric–Myristic Acid" vs. Polyunsaturated Fatty Acid Diets

The effect on serum cholesterol levels of diets containing "lauric–myristic acid" fats and oils compared to diets containing polyunsaturated oils are summarized in Table 4 (2,10,12–14,18–21). Serum cholesterol concentrations were significantly higher for subjects fed the coconut oil- or dairy fat-enriched diets compared to those fed the polyunsaturated diets.

The differences in cholesterol levels for lauric–myristic acid diets versus polyunsaturated control diets were larger than the differences obtained when lauric–myristic diets were compared to monoenoic acid control diets (Tables 3 and 4). The larger effect of polyunsaturated fatty acid diets on serum cholesterol is consistent with results of

**TABLE 3**
Effect of Lauric–Myristic Acid Diets Compared to Monoenoic Acid Diets on Human Serum Cholesterol Concentrations

| Diets Compared | | | Change in Serum Cholesterol Concentrations (mg/dL)[a] | | | | |
|---|---|---|---|---|---|---|---|
| Experimental | | Monoene | Total | LDL | HDL | Subjects[b] | Ref. |
| Coconut | vs. | Olive | −37 | −31 | −3 | 20 M | 9 |
| Coconut | vs. | Olive | −93 | nd[c] | nd | 10 M | 2 |
| Coconut | vs. | Olive | −39 | −31 | −5 | 13 F | 9 |
| Coconut | vs. | Palm | −36 | −25 | −10 | 21 M / 7F | 10 |
| Coconut | vs. | Palm | −39 | −32 | −3 | 20 M | 9 |
| Coconut | vs. | Palm | −39 | −33 | −5 | 13 F | 9 |
| Coconut | vs. | Oleic | −34 | −32 | −3 | 7 M | 11 |
| Dairy | vs. | MUFA | −28 | −23 | −4 | 27 M / 31F | 12 |
| Dairy | vs. | HiOl Sun | −41 | −37 | 0 | 20 M | 13 |
| Butter | vs. | Olive | −24 | −21 | +3 | 14 M | 14 |
| 8% Coconut/ 92% Soybean | vs. | 67% Olive/ 33% Soybean | +4 | +6 | −4 | 14 M | 15 |

[a]Cholesterol data for monoene diet minus Experimental (lauric–myristic acid) diet cholesterol data.

[b]Number of male (M) and female (F) subjects.

[c]nd = no data reported.

**TABLE 4**
**Effect of Lauric–Myristic Acid Diets Compared to Polyunsaturated Fatty Acid Diets on Human Serum Cholesterol Concentrations**

| Diets Compared | | | Change in Serum Cholesterol Concentrations (mg/dL)[a] | | | Subjects[b] | Ref. |
|---|---|---|---|---|---|---|---|
| Experimental | | Polyene | Total | LDL | HDL | | |
| Coconut | vs. | Corn | −68 | −51 | −14 | 20 M/7 F | 10 |
| Coconut | vs. | Safflower | −31 | −21 | −6 | 19 M | 18 |
| Coconut | vs. | Safflower | −37 | −31 | −8 | 5 M/7 F | 19 |
| 12:0 + Safflower[c] | vs. | Safflower | −15 | nd[d] | nd | 10 M | 20 |
| 14:0 + Safflower[c] | vs. | Safflower | −22[c] | nd | nd | 9 M | 20 |
| Coconut | vs. | Safflower | −95 | nd | nd | 10 M | 2 |
| Dairy | vs. | PUFA | −20 | −18 | −2 | 27 M/31 F | 12 |
| Butter | vs. | Soybean | −37 | −31 | +1 | 14 M | 14 |
| Dairy | vs. | Corn | −53 | −45 | −1 | 20 M | 13 |
| P/S 1.3[e] | vs. | P/S 2.0 | −8 | −6 | −1 | 11 M/9 F[f] | 21 |

[a]Cholesterol data for Polyene diet minus Experimental (lauric–myristic acid) diet data.
[b]Number of male (M) and female (F) subjects.
[c]8.4% 12:0 or 7.8% 14:0 interesterified with safflower oil.
[d]nd = no data reported.
[e]8% more 12:0 + 14:0 in diet with a polyunsaturate to saturate (P/S) ratio of 1.3.
[f]Hyperlipidemic subjects; cholesterol increased only for subjects with Type IIb hypercholesterolemia.

studies that report monounsaturated fatty acids have a neutral effect on cholesterol. The last entry (21) in Table 4 indicates that at high 18:2n-6 levels the effect of 18:2n-6 on serum cholesterol concentrations may be reduced.

Thus, substitution of 18:1n-9 for a portion of the 18:2n-6 in high 18:2n-6 diets may be as effective as 18:2n-6 for lowering serum cholesterol levels. This concept is controversial, because it is not consistent with the regression

**TABLE 5**
**Effect of Lauric vs. Myristic Acid Diets on Human Serum Cholesterol Concentrations**

| Diets Compared | | | Change in Serum Cholesterol Concentrations (mg/dL)[a] | | | Number of Male Subjects | Ref. |
|---|---|---|---|---|---|---|---|
| Experimental | | Reference | Total | LDL | HDL | | |
| Coconut | vs. | Beef | −18 | −11 | −6 | 19 | 18 |
| 14:0 + Olive[b] | vs. | Butter | −20 | nd[c] | nd | 10 | 20 |
| Butter | vs. | Cocoa butter | −11 | −10 | −1 | 14 | 14 |
| Coconut | vs. | Butter | +4 | nd | nd | 10 | 2 |
| Coconut + cholesterol[d] | vs. | Butter | −2 | nd | nd | 10 | 2 |
| 12:0 (44%)[e] | vs. | Palm | +9 | +9 | +2 | 14 | 22 |
| 14:0 + Olive[b] | vs. | Olive | −10 | nd | nd | 11 | 20 |
| 12:0 + Olive[b] | vs. | Olive | −19 | nd | nd | 10 | 20 |
| 12:0 (44%)[d] | vs. | HiOl sun[e] | −20 | −15 | −5 | 14 | 22 |
| 12:0 + 14:0[f] | vs. | Monoene | −18 | −14 | −1 | 36 | 23 |
| 12:0 + Safflower[g] | vs. | Safflower | −15 | nd | nd | 10 | 20 |
| 14:0 + Safflower[g] | vs. | Safflower | −22 | nd | nd | 11 | 20 |

[a]Reference diet cholesterol data minus experimental diet cholesterol data.
[b]14:0 + Olive and 12:0 + Olive = 8% 14:0 or 9% 12:0 interesterified with olive oil.
[c]nd = no data reported.
[d]Coconut + Chol = coconut oil plus added cholesterol to equal butter.
[e]44% 12:0 interesterified with high oleic sunflower oil (HiOl sun).
[f]Average American diet containing 9% more 12:0 and 14:0 than monoene diet.
[g]8.4%12:0 or 7.8% 14:0 interesterified with safflower oil.

equations recently developed from the analysis of over 340 studies (5). The equations predict that the effect of 18:2n-6 on serum cholesterol should not have a maximum.

## Lauric Acid vs. Myristic Acid Diets

The effect of lauric acid-, myristic acid-containing diets and diets containing fat from a variety of sources are compared in Table 5 (2,14,18,20,22,23). The relative cholesterol-increasing effects of lauric, myristic, palmitic, and stearic acids can be evaluated from these data. Although the information is limited and not always consistent, myristic acid appears to be the most hypercholesterolemic, followed by lauric acid and then palmitic acid with stearic acid being the least hypercholesterolemic of these saturated fatty acids.

The results in Table 5 indicate that coconut oil is at least as hypercholesterolemic as animal fats containing cholesterol. However, dietary cholesterol is now considered to be less important than was previously believed, and it is thought to have less influence on serum cholesterol than saturated fatty acids. A 100 mg increase in dietary cholesterol is estimated to raise serum cholesterol by 2.2 mg/dL which means that a 50% reduction of dietary cholesterol would reduce serum cholesterol levels by about 4 mg/dL (24).

## Thrombotic Effect

In addition to the cholesterol-raising or atherogenic properties of dietary fats, the thrombotic effect of specific fatty acids needs to be considered. Serum cholesterol levels are associated with development of arterial lesions and obstruction of the arteries, but the occlusion of the coronary artery by a blood clot is the actual event that causes a coronary thrombosis responsible for myocardial infarction. Saturated fats, including stearic acid, promote thrombotic tendencies and unsaturated fats suppress thrombosis. A recent report indicates that 18:0 does not increase thromboxane $A_2$ (25) and raises the question of whether all saturated fatty acids are equally thrombotic. Data for polyunsaturated fatty acids indicate that the n-3 fatty acids are more antithrombotic than the n-6 fatty acids (26,27), demonstrating that the atherogenic and thrombogenic properties for each specific fatty acid, whether saturated or unsaturated, are different.

An index of atherogenicity and thrombogenicity for various fats has been calculated based on fatty acid composition and reported clinical data (28). Data for some selected fats are plotted in Figs. 1 and 2. When these indexes are considered together, unsaturated fats are preferred over saturated fats from a health viewpoint. Comparison of polyunsaturated fats to monounsaturated fats indicate relatively small differences.

## Animal Studies and Mechanism

A large number of animal studies have reported that coconut oil increases serum cholesterol concentrations relative to diets containing unsaturated vegetable oils. Data from a few recent animal studies are summarized in Table 6 (29–34). These animal data are consistent with the human data in Tables 3–5. An exception is the data for

TABLE 6
Effect of Lauric–Myristic Acid Diets Compared to Monoenoic Acid Diets on Serum Cholesterol Concentrations in Animals

| Diets Compared | | | Change in Serum Cholesterol Concentrations (mg/dL)[a] | | | Animal | |
|---|---|---|---|---|---|---|---|
| Experimental | | Reference | Total | LDL | HDL | Model | Ref. |
| Coconut | vs. | Palm | −49 | −31 | −16 | Monkey[b] | 29 |
| Coconut | vs. | AHA step-1[c] | −59 | −42 | −13 | Monkey[b] | 29 |
| Coconut | vs. | HiOl Safflower | −91 | nd[d] | nd | Monkey[e] | 30 |
| Coconut | vs. | Safflower | −115 | nd | nd | Monkey[e] | 30 |
| Coconut[f] | vs. | Safflower | −399 | −106 | nd | Hamster | 31 |
| Coconut | vs. | Olive | −10 | +10 | −25 | Baboon | 32 |
| Coconut | vs. | Peanut | −1 | +25 | −35 | Baboon | 32 |
| Coconut | vs. | Safflower | −16 | nd | nd | Rat | 33 |
| Coconut | vs. | Cocoa butter | −20 | nd | −1 | Rat | 34 |
| Coconut | vs. | Corn | −33 | nd | +10 | Rat | 34 |
| Coconut | vs. | Palm kernel | +13 | nd | +8 | Rat | 34 |

[a]Reference diet data minus Experimental diet data. A source of 18:2n-6 was added to the coconut oil diets.
[b]Combined data from 8 rhesus, 8 cebus, and 5 squirrel monkeys.
[c]AHA step-1 = American Heart Association step-1 diet.
[d]nd = no data reported.
[e]8–10 Cebus monkeys; cholesterol-free diets.
[f]Hydrogenated coconut.

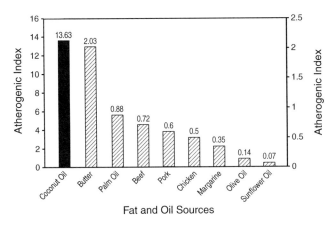

**Fig. 1.** Atherogenicity index for selected fats and oils. The scale on the left-hand side of the figure is used for coconut oil value, the scale on the right-hand side of the figure is used for values for other fats. The equation used to calculate values: Index = [12:0 + (4 × 14:0) + 16:0]/[18:1 + n-6 + n-3]. Adapted from Ref. 28.

baboons. In this study, the coconut oil diet was reported to lower LDL cholesterol relative to an olive oil or peanut oil diet, and HDL cholesterol was dramatically higher for the coconut diet. The last entry in Table 6 shows a difference in the effect of coconut oil versus palm kernel oil. This result is interesting, since the fatty acid compositions of these oils are similar. A possible explanation may be the difference in the triglyceride structure of these oils.

The primary mechanism for regulation of LDL levels involves a change in activity of the LDL-dependent receptors. The specific details of this mechanism have been reviewed recently (35). Briefly, the reason 12:0, 14:0, and 16:0 have hypercholesterolemic properties is because these fatty acids decrease removal of LDL from plasma by reduc-

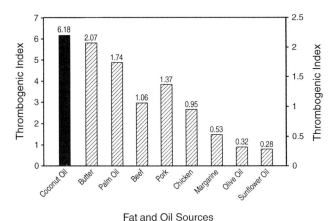

**Fig. 2.** Thrombogenicity index for selected fats and oils. The scale on the left-hand side of the figure is used for coconut oil value, the scale on the right-hand side of the figure is used for values for other fats. The equation used to calculate values: Index = [14:0 + 16:0 + 18:0]/[0.5 × (18:1 + n-6) + (3 × n-3) + n-3/n-6]. Adapted from Ref. 28.

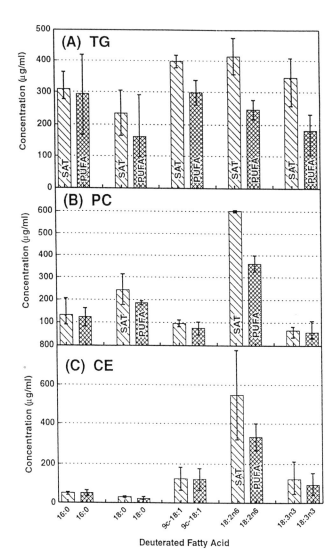

**Fig. 3.** Comparison of deuterated fatty acids incorporated into plasma triglyceride (A), phospholipid (B), and cholesteryl ester (C) from 7 young adult male subjects prefed diets with different fatty acid compositions. Fatty acid composition of prefed diets: SAT = 48.4 g saturates, 48.5 g 18:1n-9, and 15.1 g 18:2n-6; and PUFA = 36.3 g saturates, 45.7 g 18:1n-9, and 29.8 g 18:2n-6. Mixtures containing approximately equal amounts of each deuterated fatty acid were fed as their triglycerides. Adapted from Ref. 37 and 38.

ing the LDL-receptor-dependent activity of liver cells. Because liver LDL receptors are also responsible for the removal of VLDL remnants, the precursors of LDL particles, LDL production is increased when LDL-receptor-dependent activity is decreased. In contrast, 18:1n-9 and 18:2n-6 increased LDL-receptor-dependent activity, which increases LDL removal and decreases LDL production. The medium-chain (6:0, 8:0, 10:0) fatty acids, 18:0, and *trans* 18:1n-9 did not influence LDL-receptor-dependent activity, and these fatty acids were classified as biologically neutral.

Addition of dietary cholesterol to the diets enhanced the effect of saturated fatty acids on LDL-receptor-dependent activity. The combination of coconut oil plus cholesterol resulted in an eightfold increase in LDL concentrations compared to diets containing short-chain triglycerides and no added cholesterol (36). In baboons fed coconut oil, a reduction in hepatic LDL-mRNA levels occurred and correlated negatively with total serum cholesterol (32). These result indicate that 12:0, 14:0, and 16:0 influence cholesterol-regulation mechanisms by altering the production and removal of cholesterol and by reducing the synthesis of LDL-mRNA.

Analysis of data from monkeys fed diets containing various sources of saturated fatty acids indicates that 16:0 is less hypercholesterolemic than 12:0 or 14:0, and that 14:0 was the fatty acid mainly responsible for the hypercholesterolemic effect of saturated fatty acids (30). Reanalysis of data from a number of human studies indicated that the effect of 16:0 was variable, and dietary 16:0 does not raise cholesterol levels in normal (<220 mg/dL) cholesterolemic subjects when diets contained less than 300 mg cholesterol/d (30).

Based on tissue-lipid composition, it is unclear why dietary fats effect serum-cholesterol concentrations. If the composition of a diet contains a reasonable distribution of the common fatty acids, metabolic studies with stable isotope-labeled fatty acids indicate that the fatty acid compositions of plasma and tissue lipids are controlled by differences in the selectivity of the acyltransferases for various fatty acid structures. The result is that the fatty acid composition of dietary fats normally have little influence on tissue-lipid composition because tissue-lipid fatty acid composition is well regulated. For example, the isotope-tracer data shown in Fig. 3 for human-plasma triglyceride, phospholipid, and cholesterol ester compare the effect of saturated and polyunsaturated fatty acid-enriched diets on stable-isotope-labeled fatty acid incorporation. The concentrations of each deuterated fatty acid in plasma phosphatidylcholine were significantly different ($p < 0.01$). For plasma cholesterol ester, the concentrations of the deuter-

**TABLE 7**
Effect of Dietary Fat on Serum Total Phospholipid Fatty Acid Composition of Men ($n = 20$)[a]

| Fatty Acid % S:M:P[b] | Diets | | |
|---|---|---|---|
| | Butter Fat (52:35:12) | Hi Oleic Sunflower Seed Oil (17:68:15) | Corn Oil (20:34:46) |
| Phospholipid Fatty Acid Composition (%)[c] | | | |
| 14:0 | 0.5 | 0.4 | 0.4 |
| 16:0 | 31.4 | 27.6 | 28.1 |
| 18:0 | 15.4 | 15.6 | 16.2 |
| 18:1n-9 | 12.9 | 16.9 | 11.2 |
| 18:2n-6 | 16.7 | 17.4 | 21.3 |
| 18:3n-3 | 0.2 | 0.3 | 0.2 |
| 20:4n-6 | 6.1 | 6.6 | 6.9 |
| 20:5n-3 | 0.4 | 0.4 | 0.3 |
| Other | 16.4 | 14.8 | 15.4 |

[a]Adapted from Ref. 13.
[b]Saturated (S), monounsaturated (M), polyunsaturated (P) fatty acid composition of diets.
[c]Fatty acid composition of plasma phospholipid after diet fed for 5 weeks. Source indicated provided 85% of dietary fat.

**TABLE 8**
Effect of Dietary Fat on Plasma Total Lipid Fatty Acid Composition (%) of Male Rats ($n = 8$)[a]

| Fatty Acid | Lipid Composition[b] | | | Dietary Fat Composition (%) | | |
|---|---|---|---|---|---|---|
| | Coconut Oil | Palm Kernel | Corn Oil | Coconut Oil | Palm Kernel | Corn Oil |
| 10:0 | — | — | — | — | 3.1 | — |
| 12:0 | 1.2 | 0.6 | 0.3 | 47.3 | 51.2 | — |
| 14:0 | 3.3 | 1.4 | 1.6 | 18.6 | 17.5 | — |
| 16:0 | 22.3 | 22.4 | 25.4 | 9.4 | 9.6 | 12.3 |
| 16:1 | 7.6 | 4.8 | 2.6 | — | — | — |
| 18:0 | 14.3 | 14.3 | 18.8 | 11.1 | 2.0 | 1.8 |
| 18:1n-9 | 29.5 | 32.1 | 23.0 | 6.8 | 13.9 | 24.6 |
| 18:2n-6 | 12.7 | 15.4 | 19.0 | 0.8 | 2.6 | 60.1 |
| 18:3n-3 | — | — | — | — | — | 0.7 |
| 20:4n-6 | 9.1 | 9.0 | 9.3 | — | — | — |
| SUM SAT | 41.1 | 38.7 | 46.1 | — | — | — |
| SUM MONO | 37.1 | 36.9 | 25.6 | — | — | — |
| SUM POLY | 21.8 | 24.4 | 28.3 | — | — | — |

[a]Adapted from Ref. 34.
[b]Composition of plasma lipid fatty acids after diets fed for 3 weeks. Source indicated provided 100% of dietary fat. Original plasma lipid data normalized to 100%. Note abnormally high percentage of 18:0 for coconut oil diet.

**TABLE 9**
Estimated Intake of Lauric Oils in U.S. Food Supply[a]

| Dietary Fat | Amount | Men | Women | Children |
|---|---|---|---|---|
| Total fat | g/d | 105 | 69 | 56 |
| Saturated fat | g/d | 38 | 25 | 23 |
|  | % total fat | 36 | 36 | 41 |
| Palm kernel | g/d | 1.3 | 0.9 | 0.7 |
|  | % total fat | 1.2 | 1.3 | 1.3 |
| Coconut oil | g/d | 1.3 | 0.9 | 0.7 |
|  | % total fat | 1.2 | 1.3 | 1.3 |
| Total lauric oils | g/d | 2.6 | 1.8 | 1.4 |
|  | % total fat | 2.5 | 2.6 | 2.5 |
|  | % total calories | 1 | 1 | 1 |
|  | % total sat. fat | 5.8 | 6.0 | 5.2 |

[a]Adapted from Ref. 42.

ated saturated fatty acids were different ($p < 0.01$) from *cis* 18:1n-9, 18:2n-6, and 18:3n-3 and the concentrations of *cis* 18:1n-9 and 18:3n-3 were different from 18:2n-6. However, the composition of the prefed diets had little influence on the incorporation of fatty acids into plasma-lipid classes. A dietary influence ($p < 0.01$) was observed only for the concentration of deuterated 18:2n-6 in the plasma phosphatidylcholine and cholesterol ester fractions (37,38).

Thus, one of the problems with understanding why dietary 12:0 and 14:0 raise cholesterol levels relative to unsaturated fatty acids is that the levels of 12:0 and 14:0 in animal and human tissues are low even when diets are high in 12:0 and 14:0. Examples of fatty acid composition data are shown in Tables 7 (13) and 8 (34). The data in these tables are from studies that used diets with extreme differences in fatty acid composition. The modest differences in plasma-lipid fatty acid composition indicate that tight metabolic control of 12:0 and 14:0 results in low incorporation of both fatty acids. The mechanisms that control plasma-lipid composition should have an effect on the regulation of LDL-receptor activity since they are responsible for the rapid modification of the fatty acid composition of dietary fats. A possible explanation is that the effect of dietary 12:0 and 14:0 on LDL-receptor-dependent activity occurs as these fatty acids are removed from plasma by the liver.

A similar problem is associated with understanding why the results reported for the influence of oleic and linoleic acids on HDL cholesterol have been inconsistent. The oleic and linoleic acid content of human tissues are very similar and are not easily altered by a change in dietary fat composition. However, a number of studies have reported that linoleic acid, but not oleic acid, lowers HDL cholesterol (7,8,12,16,39–41). Others have reported that linoleic acid does not lower HDL cholesterol concentrations (1,2,5,20,42).

## The Cholesterol Hypothesis

Because high serum cholesterol correlates strongly with incidence of coronary heart disease, it has been assumed that lowering serum cholesterol concentration will reduce the risk of coronary heart disease and mortality. This assumption has been questioned (43,44). A disturbing and interesting observation is that results from dietary intervention trials designed to test the cholesterol hypothesis have not been very conclusive. Also of concern is the observation that when deaths due to CHD were lower in the experimental groups, the experimental groups did not experience a significant overall lower mortality rate than the subjects in the control groups (43–45). Whether these concerns are valid is still controversial.

## Importance of Lauric Oils in U.S. Diets

The estimated contribution of lauric oils to the average U.S. diet is small (Table 9) (46). According to the Keys and Hegsted equations, the level of lauric oils in the typical U.S. diet (2.5% of the total fat intake, or about 1% of total energy) would be predicted to raise serum cholesterol concentration by less than 2.0 mg/dL. If the 12:0, 14:0, 16:0, and 18:0 fatty acids from lauric oils are replaced with PUFA, the predicted change in serum cholesterol is less than 3.0 mg/dL (< 2.0 mg/dL decrease due to removal of saturated fat and about 1.0 mg/dL decrease due to replacement by PUFA).

Given the lack of a clear benefit associated with lowering serum cholesterol concentration by dietary intervention and the modest predicted effect on serum cholesterol of the small amount of lauric oils in the U.S. diet, it is not reasonable that the people with serum cholesterol levels below 200 mg/dL should be overly concerned about low levels of lauric oils in their diets. A caveat to this statement is that for some individuals, a relatively small increase in dietary saturated fatty acid content causes a much larger than predicted increase in serum cholesterol levels (47). For these hyperresponders, a small amount of lauric oils in their diet may be undesirable.

## References

1. Keys, A., J.T. Anderson, and F. Grande, *Metabolism 14*:747 (1965).
2. Hegsted, D.M., R.B. McGandy, M.L. Myers, and F.J. Stare, *Am. J. Clin. Nutr. 17*:281 (1965).
3. Gunstone, F.D., J.L. Harwood, and F.B. Padley, *The Lipid Handbook*, Chapman and Hall, New York. Ch. 3, pp. 49–129 (1986).
4. Keys, A., J.T. Anderson, and F. Grande, *Metabolism 14*:776 (1965).
5. Hegsted, D.M., L.M. Ausman, J.A. Johnson,, and G.E. Dallal, *Am. J. Clin. Nutr. 57*:875 (1993).
6. National Research Council, *Designing Foods: Animal Product Options in the Marketplace*. National Academy Press, Washington, D.C., 1988.
7. Grundy, S.M., D. Nix, M.F. Whelan,, and L. Franklin, *J. Am. Med. Assoc. 256*:2351 (1986).
8. Mattson, F.H., and S.M. Grundy, *J. Lipid Res. 26*:194 (1985).

9. Ng, T.K.W., K.C. Hayes, G.F. DeWitt, M. Jegathesan, N. Satgunasingam, A.S.H. Ong, and D. Tan, *J. Am. Coll. Nutr. 11*:383 (1992).
10. Ng, T.K.W., K. Hassan, J.B. Lim, M.S. Lye, and R. Ishak, *Am. J. Clin. Nutr. 53*:1015S (1991).
11. Grundy, S.M. *N. Engl. J. Med. 314*:745 (1986).
12. Mensink, R.P., and M.B. Katan, *N. Engl. J. Med. 321*:436 (1989).
13. Wardlaw, G.M., and J.T. Snook, *Am. J. Clin. Nutr. 51*:815 (1990).
14. Kris-Etherton, P.M., J. Derr, D.C. Mitchell, V.A. Mustad, M.E. Russell, E.T. McDonnell, D. Salabsky, and T.A. Pearson, *Metabolism 42*:121 (1993).
15. Chang, N.W., and P.C. Huang, *J. Lipid. Res. 31*:2141 (1990).
16. Chan, J.K., V.M. Bruce, and B.E. McDonald, *Am. J. Clin. Nutr. 53*:1230 (1991).
17. Shane, J., P. Walker, and E.A. Emken, *Fed. Am. Soc. Exp. Biol. 7*:A84 (1993).
18. Reiser, R., J.L. Probstfield, L.W. Scott, M.L. Shorney, R.D. Wood, B. O'Brien, A.M. Gotto, P. Insull, and W. Insull, *Am. J. Clin. Nutr. 42*:190 (1985).
19. Rassias, G., M. Kestin, and P.J. Nestel, *Euro. J. Clin. Nutr. 45*:315 (1991).
20. McGandy, R.B., D.M. Hegsted, and M.L. Myers, *Am. J. Clin. Nutr. 23*:1288 (1970).
21. Gustafsson, I.-B., J. Boberg, B. Karlstrom, H. Lithell, and B. Vessby, *Brit. J. Nutr. 50*:531 (1983).
22. Denke, M.A., and S.M. Grundy, *Am. J. Clin. Nutr. 56*:895 (1992).
23. Ginsberg, H.N., S.L. Barr, A. Gilbert, W. Karmally, R. Deckelbaum, K. Kaplan, R. Ramakrishnan, S. Holleran, and R.B. Dell, *N. Engl. J. Med. 322*:574 (1990).
24. McNamara, D.J., *Adv. Meat Sci. 6*:63 (1990).
25. Mustad, V.A., P.M. Kris-Etheron, J. Derr, C.C. Reddy, and T.A. Pearson, *Metabolism 42*:463 (1993).
26. Vericel, E., M. Lagarde, F. Mendy, P. Courpron, and M. Dechavanne, *Thrombosis Res. 42*:499 (1986).
27. Renaud, S., in *Dietary Omega-3 and Omega-6 Fatty Acids: Biological Effects and Nutritional Essentiality*, edited by C. Galli and A. Simopoulos, Plenum Press, New York, 1989, pp. 263–271.
28. Ulbricht, T.L.V., and D.A.T. Southgate, *Lancet 338*:985 (1991).
29. Hayes, K.C., A. Pronczuk, S. Lindsey, and D. Diersen-Schade, *Am. J. Clin. Nutr. 53*:491 (1991).
30. Hayes, K.C., and P. Khosla, *Fed. Am. Soc. Exp. Biol. 6*:2600 (1992).
31. Spady, D.K., and J.M. Dietschy, *Proc. Natl. Acad. Sci. USA 82*:4526 (1985).
32. Fox, J.C., H.C. Mcgill Jr., K.D. Carey, and G.S. Getz, *J. Biol. Chem. 262*:7014 (1987).
33. Huang, Y.S., M.S. Manku, and D.F. Horrobin, *Lipids 19*:664 (1984).
34. Kritchevsky, D., S. Tepper, L. Lloyd, L. Davidson, and D. Klurfeld, *Nutr. Res. 8*:287 (1988).
35. Dietschy, J.M., S.D. Turley, and D.K. Spady, *J. Lipid Res. 34*:1637 (1993).
36. Woollett, L.A., D.K. Spady, and J.M. Dietschy, *J. Clin. Invest. 84*:119 (1989).
37. Emken, E.A., R.O. Adlof, W.K. Rohwedder, and R.M. Gulley, *Biochim. Biophys. Acta 1170*:173 (1993).
38. Emken, E.A., R.O. Adlof, and R.M. Gulley, *Biochim. Biophys. Acta 1213*:277 (1994).
39. Gustafsson, I.B., B. Vessby, and M. Nydahl, *Arteriosclerosis 96*:109 (1992).
40. Foley, M., M. Ball, A. Chisholm, A. Duncan, G. Spears, and J. Mann, *Eur. J. Clin. Nutr. 46*:429 (1992).
41. Wahrbung U., H. Martin, M. Sandkamp, H. Schulte, and G. Assmann, *Am. J. Clin. Nutr. 56*:678 (1992).
42. Berry, E.M., S. Eisenberg, D. Haratz, Y. Friedlander, Y. Norman, N.A. Kaufmann, and Y. Stein, *Am. J. Clin. Nutr. 53*:899 (1991).
43. Windeler, J., and O. Gelfeller, in *Medinfo,* edited by K.C. Lun, Elsevier Science Publishers, Holland, 1992, pp. 1190–1194.
44. Rossouw, J.E., and B.M. Rifkind, *Endocrinol. Metab. Clin. Nutr. America 19*:279 (1990).
45. British Nutrition Foundation, *Unsaturated Fatty Acids: Nutritional and Physiological Significance*, Chapman and Hall, London, 1992.
46. Parks, Y.K., and E.A. Yetley, *Am. J. Clin. Nutr. 51*:738 (1990).
47. Jacobs, D.R., Jr., J.T. Anderson, P. Hannan, A. Keys, and H. Blackburn, *Arteriosclerosis 3*:349 (1983).

# Health Aspects of Coconut Oil

Conrado S. Dayrit

United Laboratories, Inc., P.O. Box 3594, United St., Mandaluyong, Metro Manila 3119, Philippines.

## Abstract

Coconut oil, a saturated fat, and the saturated, animal-derived fats have been labeled unhealthy. The saturated fatty acids of coconut oil are predominantly medium-chain length ($C_6$–$C_{12}$), with absorption, transport, metabolism, and distribution characteristics radically different from those of the long-chain fats.

Coconut oil-feeding experiments in various animal species, particularly rabbits, have been flawed by the nonphysiologic amounts of coconut oil given to these animals and the failure to supplement the diet with essential oils, leading to essential fatty acid deficiency. To date, no evidence has been presented to indicate that coconut oil causes coronary heart disease in human subjects or in populations that consume large amounts in their diets. On the contrary, the Polynesian study showed that the people of Puka Puka and Tokelau had normal serum cholesterol and very low incidence of coronary heart disease even though they received 35–55% of their calories from fat, mostly from coconut and fish.

A controlled, randomized clinical trial on high fat-consuming (37% of total calories) Caucasian subjects given coconut oil as high as 60% of their daily fat intake showed no significant increase in total cholesterol and some increase in HDL while lowering the cholesterol:HDL ratio towards a more favorable ratio. Recent findings on atherogenesis explain why the lipid–cholesterol–atherogenesis theory is too simplistic and fails to explain many observations on coronary heart disease and cerebrovascular disease occurrence.

---

Diets rich in cholesterol and saturated fats are believed to contribute to the elevation of blood lipids and the progression of atherosclerosis. Coconut oil and its mostly saturated fats have been labeled "unhealthy." Demonstration that the saturated fatty acids of coconut oil are predominantly of medium-chain length ($C_6$–$C_{12}$), whose absorption, transport, metabolism, and distribution differ radically from those of the long-chain fats (1) has done little to change this unfavorable impression. Coconut oil-feeding experiments in various animal species, particularly rabbits, have been flawed by the nonphysiologic amounts of coconut oil given to these animals, and the failure to supplement their diets with essential oils, leads to essential fatty acid deficiency (2). On the other hand, there are experiments with pigs (3) and rats (4) where coconut oil was found to be neutral regarding atherogenecity or cholesterolemia. No evidence has been presented that coconut oil causes coronary heart disease in human subjects or populations who take it liberally in their diets. This is contrary to the well-known findings of increased prevalence of coronary heart disease in populations with diets high in long-chain animal fats (5–10). Since the long-chain and the medium-chain triglycerides are both saturated, the term "saturated fats" has come to denote fats that produce a high risk of hypercholesterolemia and atherogenecity. This paper seeks to correct this erroneous classification and to emphasize that while long-chain saturated fats, when taken in large amounts, may indeed be associated with coronary heart disease, particularly in persons genetically predisposed, that as a whole the medium-chain triglycerides and coconut oil have not been shown to have any such association and are safe and useful.

## Is Coconut Oil Cholesterogenic?

### Theoretical Considerations

According to present concepts, hypercholesterolemia may be induced by intake and adequate absorption of a diet rich in cholesterol; by synthesis of cholesterol in the body and its transport by lipoproteins, low-density lipoproteins (LDL) in particular; and by inhibition or down-regulation of LDL receptors. Coconut oil contains almost no cholesterol—only 0–14 ppm (Table 1). This concentration translates to 1.2 mg of cholesterol in 83 g of coconut oil, the entire 30% fat intake for a 2500 calorie diet. Other vegetable oils (i.e., palm, soybean, and corn) contain slightly more cholesterol, but vegetable oils have far less preformed cholesterol than animal and dairy fats.

The triglycerides in coconut oil, like those in palm kernel oil and unlike all other oils, are predominantly (over 60%) of the $C_6$–$C_{12}$ variety, known as medium-chain triglycerides (MCT [Table 2]). The properties of

**TABLE 1**
**Cholesterol Content of Various Fats and Oils**

|  | ppm |
|---:|:---:|
| Coconut oil | 0–14 |
| Palm oil | 18 |
| Soybean oil | 28 |
| Corn oil | 50 |
| Butter | 3150 |
| Lard | 3500 |

MCT are well known as discussed by Bach and Babayan (1). Fatty acids having a length of $C_{12}$ or less are easily digested by intestinal lipases and do not require pancreatic lipase, which is essential for digestion of the long-chain triglycerides (LCT). Absorption of MCT is consequently easier and faster with no need for their incorporation into chylomicra by the intestinal mucosal cell. Medium-chain fatty acids (MCFA) comprise only 8–13% of the total chylomicron triglyceride fatty acids (11). Medium-chain triglycerides ($C_8$–$C_{10}$) are transported by portal circulation directly to the liver for metabolic degradation, while LCT ($C_{14}$ and longer) are absorbed by the lymphatics and are circulated systemically before finally reaching the liver (Fig. 1). Lauric acid ($C_{12}$) can go by either the portal or lymphatic route with a slight preference for the former. The MCT enter the mitochondria of liver cells with no assistance by carnitine, and are rapidly oxidized producing energy and carbon dioxide (1). Medium-chain triglycerides therefore behave as energy providers, like simple carbohydrates, but require no insulin, and for this reason have been used as food for newborn babies and premature infants. Because of their rapid degradation, MCFA do not provide much material for incorporation into very low-density lipoproteins (VLDL). Medium-chain triglycerides therefore contribute little, if any, to the lipoprotein transport of lipids, the normal route for the LCT, and their deposition in fat stores of the body is low.

About 30% of the fatty acids of coconut oil are saturated $C_{14}$–$C_{18}$ LCFA. Consumption of large amounts (35–45% of total calories) as animal fat, with a high content of $C_{14}$–$C_{18}$ LCFA, has been blamed for the high incidence of coronary heart disease in many populations (5–10). Since then, much work has been done on LCFA, and several studies suggest that stearic acid ($C_{18}$) is one

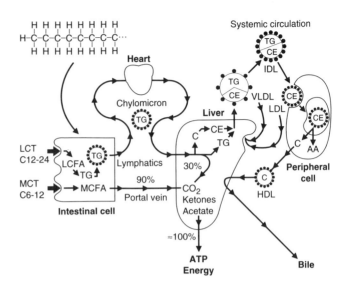

**Fig. 1.** MCT is metabolized differently from LCT. Abbreviations: LCT, Long-chain triglycerides; LCFA, long-chain fatty acids; MCT, medium-chain triglycerides; MCFA, medium-chain fatty acids; TG, triglycerides; C, cholesterol; CE, cholesterol esters; AA, amino acids.

LCFA that does not raise serum cholesterol significantly (12–15)—at least not as much as palmitic acid (15). In 1991, Hayes et al. (16) reported that palmitic acid ($C_{16}$) was neutral also, and that it had no effect on LDL-receptor activity in hamsters and monkeys, and that this effect was present when serum cholesterol levels were less than 200 mg/dL. Furthermore, they noted that the cholesterolemic impact of any saturated fatty acid appears to be countered (up to a saturable "threshold" level) by dietary linoleic

**TABLE 2**
**Fatty Acid Composition of Various Fats and Oils[a]**

| Fatty Acid | Coconut | Palm | Soybean | Corn | Butter | Lard | Beef[b] |
|---|---|---|---|---|---|---|---|
| $C_4$ | — | — | — | — | 3 | — | — |
| $C_6$ | 0.5 | — | — | — | 1 | — | — |
| $C_8$ | 7.8 | — | — | — | 1 | — | — |
| $C_{10}$ | 6.7 | — | — | — | 3 | — | — |
| $C_{12}$ | 47.5 | 0.2 | — | — | 4 | — | — |
| $C_{14}$ | 18.1 | 1.1 | — | — | 12 | 3 | 3 |
| $C_{16}$ | 8.8 | 44.0 | 11 | 11.5 | 29 | 24 | 29 |
| $C_{18}$ | 2.6 | 4.5 | 4 | 2.2 | 11 | 18 | 22 |
| $C_{20}$ | 0.1 | — | — | — | 5 | 1 | — |
| $C_{16:1}$ | — | 0.1 | — | — | 4 | — | — |
| $C_{18:1}$ | 6.2 | 39.2 | 25 | 26.6 | 25 | 42 | 43 |
| $C_{18:2}$ | 1.6 | 10.1 | 51 | 58.7 | 2 | 9 | 1.4 |
| $C_{18:3}$ | — | 0.4 | 9 | 0.8 | — | — | — |
| % saturated | 92.1 | 45.2 | 15 | 13.7 | 69 | 46 | 54 |
| % unsaturated | 7.9 | 44.8 | 85 | 86.1 | 31 | 54 | 46 |

[a]Modified from: Padolina, W.C., L.Z. Lucas and L.G. Torres, *Phil J. Coconut Studies XII:2*:4 (1987).
[b]Banzon, J.A., O.N. Gonzalez, S.Y. de Leon, and P.C. Sanchez, *Coconut as Food*, 1990, p. 10.

acid, which up-regulates the LDL receptor, and that once above this threshold of $C_{18:2}$, the major fatty acids appear to exert an equal influence on the circulating cholesterol concentration.

These findings bring up three important considerations. First, linoleic acid should indeed be an essential part of the diet, particularly when saturates make up a large portion of the fat. Second, inasmuch as the protective action of $C_{18:2}$ has an optimal limit, the amount of saturated LCT consumed should be kept within this limit. The recommendation of the National Cholesterol Education Program of the National Institutes of Health to reduce dietary fats to 30% or less of the dietary intake is a step in the right direction, but may need to be reduced more. Third, the LCFA are not equally cholesterogenic, as was once believed, and stearic acid ($C_{18}$) and perhaps palmitic acid ($C_{16}$) are neutral when taken in amounts within the "threshold" protection of linoleic acid. Myristic acid ($C_{14}$) has been reported to down-regulate LDL receptors and appears to be the most cholesterogenic of the LCFA. Unfortunately, as yet there are no studies on how protective $C_{18:2}$ is against $C_{14}$.

Coconut oil has very little linoleic acid (Table 2) and for this reason, requires dietary supplementation with this fatty acid. The omission of linoleic acid in coconut oil-feeding experiments with development of essential fatty acid deficiency (EFAD) might explain the atherogenic findings in some studies (17).

Coconut oil also has about 18% myristic acid ($C_{14}$) and much lower amounts of $C_{16}$ and $C_{18}$ (Table 2). If $C_{14}$ is indeed the lone fatty acid that has the highest potential for elevating serum cholesterol, the question is: How much dietary coconut oil must be taken to raise serum cholesterol when adequate linoleic acid is taken? In addition to linoleic acid, a protective effect of fish oil taken with coconut oil was proposed by Blackburn (17). Thromboxane ($TxA_2$) is known to initiate thrombosis by platelet aggregation, and thrombosis plays an important part in atherogenesis. Thromboxane $A_2$ is derived from arachidonic acid ($C_{20:4n-6}$) that is formed by the desaturation and elongation of dietary linoleic acid. Eicosapentaenoic acid (EPA, or $C_{20:5n-3}$) in fish oil can displace arachidonic acid in membrane phospholipids, and when released by phospholipase $A_2$, is converted to an inactive thromboxane ($TxA_3$) that has little platelet aggregational and thrombotic activity. Since coconut oil has very little linoleic acid, it may suppress arachidonic acid formation, and favor arachidonate substitution by EPA thus reducing thrombogenicity and atherogenicity.

The studies that can supply answers to the questions posed in this paper must be done on humans, provided a high dietary intake of coconut oil (at least 35% of daily calories) can be taken for a long enough period (at least 4 weeks), and include foods with linoleic acid or fish oil.

## Human Studies

Population studies on coconut oil are rare because few countries consume significant amounts of coconut oil (18) and those that do, such as the Philippines, Sri Lanka, and Indonesia, have a low total fat consumption. In these countries, both serum cholesterol levels and coronary heart disease incidence are low. To produce adequate data, at least 35% of dietary calories should be from fat, and coconut oil must form a large portion of this fat because the type of fat may not be significant at low levels. There are two studies with these characteristics.

*The Polynesian Study.* The much-cited study by Prior (19) on the Puka Pukans and Tokelauans fulfills the conditions for a high-fat and high-coconut oil intake. Their dietary intakes are listed in Table 3. Their source of protein is predominantly fish, which should provide dietary fish oil. The intake of unsaturated fat provides an adequate amount of essential fatty acids. In addition to fish oil, this intake should satisfy the essential fatty acid requirements. The serum cholesterol of the Puka Pukans is low. The Tokelauans, with their very high fat intake, have slightly higher serum cholesterol levels. From this study, it is concluded that coconut oil is not hypercholesterolemic.

*The Deaconess Hospital (Harvard) Study.* This study by Blackburn et al. (20) was designed to determine whether lipid metabolism is affected differently by fats of varying chainlengths. Coconut oil (CNO) was used as a source of medium-chain triglycerides, soybean oil (SBO) as a source of polyunsaturated fatty acid, and hydrogenated

TABLE 3
Coconut Diet—Polynesian Atolls (19)

|  | Males | | Females | | |
| --- | --- | --- | --- | --- | --- |
|  | Puka Puka | Tokelau | Puka Puka | Tokelau | Remarks |
| kCAL | 2120 | 2520 | 1810 | 2100 |  |
| Protein (g) | 31 | 34 | 53 | 63 | Mostly fish |
| Fat (total g) | 83 | 156 | 80 | 131 | Mostly coconut |
| % of total calories | 35.2% | 55.7% | 39.8% | 56.1% |  |
| Fat, saturated (g) | 63 | 137 | 64 | 120 | Mostly coconut |
| Fat, unsaturated (g) | 7 | 6 | 4 | 4 |  |
| Cholesterol (mg) | 73 | 51 | 70 | 48 |  |
| Carbohydrate (g) | 283 | 229 | 230 | 189 |  |
| Serum cholesterol (mg%) | 170 | 208 | 176 | 216 |  |

soybean oil (HSBO) as a source of hydrogenated polyunsaturates and *trans*-fatty acids. Dietary fat was maintained at 37% of daily caloric intake. Twenty-eight free-living males completed the 6-month study which consisted of 6-week dietary oil intervention periods with a minimum of a 4-week washout phase between each oil intervention phase. The oil intervention periods were high in either coconut oil (92% saturated), soybean oil (60% polyunsaturated), or hydrogenated soybean oil (19% *trans*-fatty acids). During each intervention phase, one-half of the fat in each subject's diet was replaced with one of the three test oils in a randomized crossover design. The authors concluded:

1. Coconut oil intake levels of up to 50% of total fat in a 37% fat diet have no effect on serum total cholesterol when compared to baseline values.
2. There was a reduction in both the total cholesterol/HDL ratio and the apoB levels with the CNO diet.
3. Compared to baseline values and to the HSBO phase, there was an increase in HDL cholesterol during the CNO phase, which may convey beneficial effects from CNO use.
4. The study contradicts the Key's Equation prediction that a 50% increase of CNO as a dietary fat will increase serum cholesterol concentration by 0.73 mmol/L. The experimental results from this study produced only a 0.13 mmol/L increase.
5. There were no adverse effects observed in lipid profiles during any of the three dietary intervention phases, suggesting that it may be the overall fat content of the typical American diet, rather than specific fatty acids that contribute to hypercholesterolemia and coronary heart disease.

*Sri Lanka.* Mendis et al. (21) replaced the dietary coconut oil with corn oil in the diets of 16 free-living healthy young adult Sri Lankan males. Like Filipinos, typical Sri Lankan meals are low in fat. The observation periods for the two diets was 6 weeks. Phase I used the regular Sri Lankan diet with coconut oil, coconut milk, and coconut kernel; phase II used powdered cow's milk and corn oil as substitute for the coconut items. The blood-lipid values taken at the end of each phase are shown in Table 4. Two notable findings are:

1. While corn oil does lower the total serum cholesterol to 146 ± 13.4 mg%, it also lowers HDL from 43.43 to 25.43 mg% and raises the LDL:HDL ratio from 3:1 to a less-favorable ratio of almost 4:1.
2. When the subjects were taking their regular Sri Lankan diet of coconut oil, their serum cholesterol was only 179.6 mg/dL. The National Cholesterol Education Program of the U.S. NHLBI-NIH recommends serum cholesterol levels of 200 mg/dL or less. These Sri Lankan males therefore are at a low level of coronary risk when on coconut oil.

## Atherogenicity, Coconut Oil and Other Factors

Hypercholesterolemia is multifactorial, and atherogenesis is even more complex. While coronary artery disease and cerebrovascular disease are both atherosclerotic processes, they have significant differences in epidemiology and risk factors. In both cases a genetic predisposition (Fig. 2) appears to be the most important factor. Genes exert influence over every mechanism of the body.

Thus, the body is regulated by enzymes, hormones, cytokines, and innumerable regulatory factors that are still being discovered. The risk factors of atherosclerosis tend to swing reactions in one direction, but are probably met by antagonistic forces. In situ variations also occur. Thus, while coronary artery disease and cerebrovascular disease are both atherosclerotic processes, the influence of risk factors on these two conditions differ. Hypercholesterolemia has been found to have a positive influence on the development of coronary artery disease, but only in young to middle-aged adults who happen to be responders. No relation between these two conditions has been established in elderly people or in young adult females. Also, there is not a good relationship between hypercholesterolemia and cerebrovascular disease. The regulatory factors for these conditions have yet to be found.

Fig. 3 is an attempt to display some of the atherogenic and antiatherogenic mechanisms as they are known at the present. Platelets (P) plug endothelial lesions and release various aggregating factors to initiate clotting. Platelet-derived growth factor (PDGF), a strong chemotactic factor, induces monoclonal smooth muscle and fibroblast migration. Vasoconstrictor factors also are released from platelets, endothelin from endothelial cells and

**TABLE 4**
**Blood Lipids Before and After Replacement of Coconut Oil in Sri Lankan Diet (21)**

|  | Total Cholesterol, Mean ± SE (mg/dL) | LDL-Cholesterol, Mean ± SE (mg/dL) | HDL-Cholesterol, Mean ± SE (mg/dL) | LDL:HDL |
|---|---|---|---|---|
| Phase 1 | 179.6 ± 9.1 | 131.6 ± 8.9 | 43.43 ± 5.01 | 3.0:1 |
| Phase 2 | 146.0 ± 13.4 | 100.3 ± 8.8 | 25.43 ± 3.95 | 3.9:1 |
| *t*-test | $p < 0.05$ | $p < 0.05$ | $p < 0.025$ | |

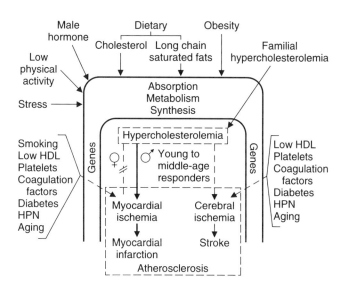

Fig. 2. Risk factors for hypercholesterolemia coronary and cerebrovascular diseases.

angiotensin II (AT II) by plasma and tissue angiotensin converting enzyme (ACE). Antithrombotic factors from the endothelium, such as heparin, antithrombin III (AT-III) and thrombomodulin, as well as, antiplatelet factors [prostacyclin ($PGI_2$)], and tissue plasminogen activator (TPA), counteract the thrombotic cascade. Lipoprotein (a), a gene-controlled lipoprotein, is prothrombotic by inhibiting plasmin formation and promotes plaque formation by means of its highly adhesive apolipoprotein (a). Vasoconstriction is counterbalanced by vasodilators, such as $PGI_2$ and the endothelial relaxing factor (EDRF-NO). Low-density lipoproteins penetrate the endothelial barrier and are oxidized by free radicals to oxidized LDL (LDLox). Macrophages engulf the toxic LDLox and are transformed into foam cells and contribute necrotic debris. The macrophages release various cytokines, such as interleukin I (Il-I) and tumor necrosis factor alpha (TNFα) to attract more leukocytes and monocytes. The free radicals come from various metabolic sources, particularly peroxidation of long-chain unsaturated fatty acids (PUFA). High-density lipoproteins are thought to prevent atherosclerosis by reverse transport of the cholesterol to the liver for excretion. It is not know which initiates the

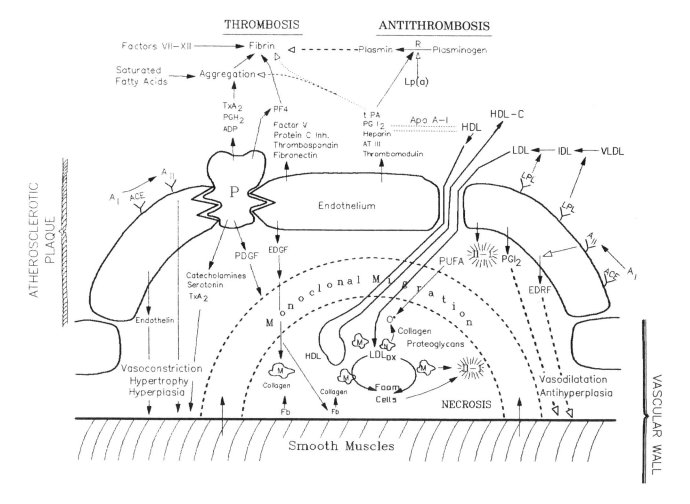

Fig. 3. Atherogenesis 1992.

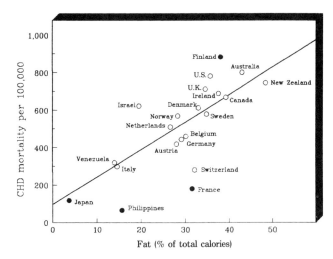

Fig. 4. Dietary fat and coronary heart disease in various countries.

atherosclerotic process—whether endothial injury and platelet aggregation occur first, or LDL infiltration. Arteries are blocked by both the growing plaque and the increasing thrombotic mass. Thus, formation of oxidized LDL by free radicals is suggested to be an important step in atherosclerotic plaque formation. Also lipoprotein (a) is a genetically dependent, highly atherogenic factor (22).

The prevention, control, and regression of atherosclerosis, therefore, will need measures not only to lower LDL and raise HDL for a lower LDL:HDL ratio, but also measures to decrease platelet activity and to enhance antithrombotic mechanisms, measures to decrease lipoprotein (a), and the use of antioxidants (such as β-carotene, vitamin C, vitamin E, selenium, tocotrienol, and flavanoids), as well as, measures to offset vasoconstriction in favor of vasodilation, and agents to normalize the various growth, inflammatory, and regulatory cytokines.

France, which consumes as much animal fat as Finland has a much lower coronary mortality rate (Fig. 4) and is a striking exception to the high-fat, high-coronary heart disease theory and provides strong evidence for the oxidation theory. The "French Paradox" has been attributed to their liberal intake of wine. Recently, Frankel et al. (23) showed in vitro inhibition of oxidation of human LDL bodies by flavanoid substances in red wine. This new role in atherogenesis prevention by antioxidants may be another important explanation for the unpredictability of atherogenesis and why the high-fat, high-cholesterol theory fails when applied to general populations.

The incidence of coronary heart disease among Polynesians was low in 1981 (19) and has remained low (24). In Filipinos, heart disease is reported to have a mortality rate of 74.6 per 100,000 population (24). These figures include deaths from coronary, rheumatic, and other heart diseases. The heart disease group ranks ninth in morbidity rates, outranked by the infectious diseases (respiratory, gastrointestinal, tuberculosis, malaria, measles, and dengue). The coronary heart disease mortality rate in the Philippines, is as low as that of Japan and France, which have the lowest coronary deaths among developed countries (Fig. 4).

The low-serum cholesterol and low-coronary heart disease mortality of coconut-consuming populations, like Filipinos, Sri Lankans, Polynesians, and most probably the Indonesians, suggest that coconut oil is neither cholesterogenic nor atherogenic for these general populations. The atherogenic process is multifactorial and highly complex, with a strong genetic influence and control. Individuals who are prone to or already afflicted with coronary or cerebrovascular atherosclerosis should lower their high-serum cholesterol by strict diet and lifestyle changes (26) and/or by drugs (27), and should prevent atherosclerosis progression by antiplatelet-aggregation and antithrombotic agents, antioxidants, and agents such as ω3 fatty acids, having regulatory functions over the chemotactic and growth-regulating inflammatory cytokines.

## References

1. Bach, A.C., and V.K. Babayan, *Am. J. Clin. Nutr.* 36:950 (1982).
2. Wigand, G., *Acta. Med. Scand.* 351:1 (1959).
3. Hill, E., V. Lundberg, and J.L. Titus, *Mayo Clin. Proc.* 46:613 (1971).
4. Hostmark, A.T., O. Spydevod, and E. Eilertson, *Artery* 7:367 (1980).
5. Keys, A., O. Mickelsen, E. Miller, E.R. Hayes, and R.L. Tood, *J. Clin. Invest.* 29:1347 (1950).
6. Keys, A., F. Vivanco, J.E. Rodriguez Minon, M.H. Keys, and H. Castro Mendoza, *Metabolism* 3:195 (1954).
7. Keys, A., F. Fidanza, V. Scardi, G. Bergami, M.H. Keys, and F. di Lorenzo, *Arch. Int. Med.* 99:328 (1954).
8. Bronte-Steward, B., A. Keys, and J.F. Brock, *Lancet* 2:1103 (1955).
9. Keys, A., J.T. Anderson, M. Aresu, G. Biorck, J.F. Brock, F. Bronte-Stewart, M.H. Fidanza, M.H. Keys, H. Malmros, A. Poppi, T. Postili, B. Swahn, and A. Del Vecchio, *J. Clin. Invest.* 35:1173 (1956).
10. Keys, A., *J. Chronic. Dis.* 6:552 (1957).
11. Swift, L.L., J.O. Hill, J.C. Peters, and H.L. Greene, *Am. J. Clin. Nutr.* 52:834 (1990).
12. Ahrens, B.N. Jr., J. Hirsch, W. Insull, Jr., T.T. Tsaltas, R. Blomstrand, and M.I. Peterson, *Lancet* 1:943 (1957).
13. Hegsted, D.M., R.B. McGandy, M.I. Myers, and F.J. Stare, *Am. J. Clin. Nutr.* 17:281 (1965).
14. Keys, A., J.T. Anderson, and J. Grande, *Metabolism* 14:776 (1965).
15. Bonanome, A., and S.M. Grundy, *New Eng. J. Med.* 318:19:1244 (1988).
16. Hayes, K.C., P. Khosia, A. Pronczuk, and S. Lindsey, *Proceedings of the 6th Asian Congress on Nutrition*, Nutrition Society of Malaysia and Federation of Asian Nutrition Societies, 1991, p. 33.
17. Blackburn, G.L., G. Kater, E.A. Mascioli, M. Kowalchuk, V.K. Babayan, and B.R. Bistrian, "A Reevaluation of Coconut Oil's Effect on Serum Cholesterol and Atherogenesis, *Coconuts Today, Special Issue for the XIth World Congress of Cardiology*, United Coconut Association of the Philippines, Manila, 1990.

18. Kaunitz, H., and Dayrit, C.S., *Phil. J. Int. Med. 30*:165 (1992).
19. Prior, I.A., F. Davidson, E.C. Salmono, and Z. Czochanska, *Am. J. Clin. Nutr. 84*:1552 (1986).
20. Blackburn, G.L., S.P. Angerman, D. Norton, N. Istfan, S.M. Lopes, V.K. Babyan, M.C. Putz, and S.N. Steen, *A Comparative Study of Coconut Oil, Soybean Oil, and Hydrogenated Soybean Oil.* Report to the Philippine Coconut Research and Development Foundation, May 1992.
21. Mendis, S., R.W. Wissler, R.T. Bridensten, and F.J. Prodbielski, *Nutr. Rpt. Internat. 40*:4 (1989).
22. Lawn, R.M., *Sci. Am.*:June (1992).
23. Frankel, E.N., J. Kannes, J.B. German, E. Parks, and J.E. Kinsella, *Lancet 341*:454 (1993).
24. Prior, I.A., Personal Communication (1990).
25. Philippine Health Statistics, 1987.
26. Ornish, D., S.E. Brown, L.W. Scherwitz, J.H. Billings, W.I. Armstrong, T.A. Ports, S.M. McLanahan, R.L. Kirkeside, R.J. Brand, and K.L. Gould, *Lancet 336*:129 (1990).
27. Blankenhorn, D.H., S.A. Nessim, R.L. Johnson, M.E. Sanmarco, S.P. Azen, and L. Cashin-Hemphill, *J. Am. Med. Assoc. 257*:3233 (1987).

# Discussion

The chairperson made some comments about the French paradox and the possible reasons for their low rate of coronary heart disease. They also discussed the edible applications of lauric oils as specialty fats, leaving frying applications to other fats and oils. It was stressed that medium-chain triglycerides (MCT) have wide application in infant formulas and intravenous feeding.

As for heart disease, questions were raised about the multifactorial nature of coronary heart disease (CHD), the differences noted with various oils compared to their component fatty acids, and the fact that palm olein is similar to olive oil in its effects on low-density lipoprotein (LDL) and high-density lipoprotein (HDL) cholesterol and thrombosis. The involvement of LDL-cholesterol oxidation also was mentioned and tied to the fact that in the six country study Vitamin E levels correlated better with CHD than did serum cholesterol. It was stressed that there is a possibility of an imbalance of the immune system with an excessive intake of polyunsaturated fatty acids (PUFA). A balanced diet with adequate levels of PUFA and monounsaturates was suggested.

In a discussion of MCT, it was noted that it is metabolized much as carbohydrates are, and is not deposited as fat. On the other hand, it was pointed out that irrespective of the calorie source, excess calories, even from MCT, are stored as fat. In questioning the studies on the effects of coconut oil on serum cholesterol levels and CHD, it was stressed that many of these were done with essential fatty acids (EFA) deficient diets. At this conference, however, all of the studies reviewed were those where adequate EFA had been provided.

Also, there were extensive comments on the research that has been done around the world with palm and coconut oils, with the conclusion that these do not pose a threat to human health. Instead, the use of mega-analyses, the Key's equation, and studies on fatty acids were questioned as possibly being misleading. Last, the point was made that the use of high-fat diets (>30% of calories) in human studies may be misleading and even unethical.

# Uses and Applications of Alkyl Polyglycoside in Personal Care Products

John Fallon

Henkel Corporation, Emery Group, 300 Brookside Ave., Ambler, PA 19002, USA.

## Abstract

A wide variety of surfactant types can be found in personal care products. For cleaning purposes, the most common are alkyl and alkyl ether sulfates, due to their wide availability and good economy.

Alkyl polyglycosides, however, represent a new generation of surfactants that are mild, biodegradable, and offer excellent value and performance when compared to conventional surfactants. Mildness plays a vital role in products formulated for bodily use. Testing, to demonstrate this characteristic, is therefore increasingly important. Other important factors in cleaning and washing products include foam, foam characteristics, and viscosity building in the final formulated product.

Alkyl polyglycosides, in various test methods, perform very well relative to other surfactants. Its nonionic nature offers a wide degree of flexibility compared to anionic or cationic surfactants. This flexibility also means performance beyond basic cleansing. Alkyl polyglycosides can be made from natural renewable resources, such as corn sugar or glucose and long-chain fatty alcohols made from many sources, for example, coconut oil. Increasing environmental awareness is beginning to play a role in the development of personal care products. This emerging class of surfactants is of significant importance to the personal care industry.

---

For centuries, the use of natural renewable fats and oils has provided the basic ingredient in personal hygiene in the form of soap. Today's personal care market, by definition, includes many product types but, for the purposes of this discussion, will focus on cleansing agents for skin and hair.

Soap remains the oldest and certainly the most heavily consumed product in this market. Its use continues, however, to undergo a slow, gradual decline with changing consumer needs. Excluding soap, it has been estimated that 175,000 tons of surfactants are consumed annually in the U.S., Japanese and West European markets (1). Products within this segment of the market include shampoos, foam baths, shower gels, facial cleansers, so-called "liquid soap," and a variety of specialty cleansers. Of this product grouping, shampoo has the highest consumption.

Consumer trends, while influenced by advertising, are based on real needs. In today's market, key words include mild, nonirritating, natural, nonanimal, safe, biodegradable and, of course, performance and value. As each of these basic needs is examined, their development can be better understood. The growing demand for milder products can be attributed to the more frequent consumer use brought about by a health-conscious, sports-oriented society. Showering and shampooing once a day is more commonplace than it was just a few years ago. Products now, more than ever before, must offer excellent skin compatibilities to accommodate increased usage without negative results. Furthermore, today's consumers pay more attention to their personal appearance leading to increased use of permanent waves and hair-coloring products. Ingredients in these products tend to be harsh and, as a result, increase damage to the hair itself. Aggressive surfactant systems only further this damage.

Although a number of surfactant types are used in the formulation of these products, fatty alcohol sulfates (AS) and fatty alcohol ether sulfates (AES) dominate this group. Alcohol sulfates and AES have become the workhorse surfactants in personal care products since they provide good detergency, excellent foaming and, for the most part, are economically suitable. Despite these positive attributes, AS and AES have some limitations. For example, lauryl sulfate is often considered to be an aggressive surfactant to skin and mucous membranes. This can be offset somewhat by ethoxylation of the alcohol prior to sulfation. Skin and mucous membrane irritation is reduced, but at the expense of foam and detergency. New entries in the surfactant market in the past decade have been limited. Alkyl polyglycosides represent a new generation of surfactants which are well suited to the preparation of modern, mild concepts in cosmetics and toiletries.

Like AS and AES, alkyl polyglycosides possess good foaming and detergency properties and at the same time exhibit excellent skin and mucous membrane compatibility. They can be based completely on a renewable vegetable-derived raw material source. With low ecotoxicity and high biological degradability, these surfactants fit within current marketing concepts of "earth friendly" products. Since alkyl polyglycosides are not based on nitrogen or ethylene oxide chemistry, the formation of troublesome by-products, such as nitrosamines and 1,4-dioxane, is not a factor.

Commercially this class of surfactants has been known for many years. Only in the recent decade have technological advances made large-scale commercial availability possible. Growing acceptance of this class of compounds is evidenced by the increasing number of branded products in which they are found.

## Technology

Alkyl polyglycosides were first described by Emil Fischer in 1893 (3) when he identified ethyl glucoside as the reaction product of ethanol and glucose. Today, the "Fischer synthesis" reacts a carbohydrate, such as glucose, with fatty alcohol in the presence of an acid catalyst.

During the reaction, alcohol condenses with the glucose molecule forming an alkyl monoglycoside, the acetal of alcohol and glucose. Although the monoglycoside is kinetically favored, each hydroxyl is a potential glucose acceptor and concurrent reactions readily form polyglycosides. The reaction is driven by removal of water. Monoglycosides further react to form diglycosides, triglycosides, and so on (Fig. 1).

As the polymerization continues, a novel oligomer distribution is created, containing decreasing amounts of higher oligomers (Fig. 2). That is, there are more monoglycosides than diglycosides, more diglycosides than triglycosides, and so on. The final average degree of polymerization is controlled by carefully run reaction conditions.

The resulting glycoside structures have both hydrophobic and hydrophilic characteristics. The alkyl groups display the hydrophobic characteristics, and the carbohydrate contributes the hydrophilic characteristics (Fig. 3). As expected, variation in either group affects final product characteristics. For example, the higher the degree of polymerization, the more water soluble the molecule becomes. Physical behavior, for example solidification point and clear point, differs by the alkyl chain length and the degree of glycosidation. (4,5)

Alkyl polyglycosides are nonionic in nature, because the hydrophilic portion bears no charge when solubilized in water. However, they differ structurally from other nonionics, such as fatty alcohol ethoxylates. Their nonionic nature is particularly beneficial in formulations since it is compatible in both anionic and cationic systems.

Although synthesis of an alkyl polyglycosides can be from a synthetically derived alcohol, there is the opportunity to produce this unique surfactant completely from renewable resources for both the hydrophilic and hydrophobic moieties. That is, fatty alcohol can be derived from coconut and the glucose from corn. Reaction of the fatty alcohol and glucose can be direct or through transacetalization via short-chain alkyl glycosides, such as butyl glycoside.

Significant commercial use of alkyl polyglycosides in personal care cleaning products is only a recent occurrence. However, the growth in the number of marketed products containing alkyl polyglycosides is truly notable.

Although a growing number of commercial alkyl polyglycosides product types is available, this discussion details two of the most widely used alkyl polyglycosides, lauryl polyglucose (LPG) and decyl polyglucose (DPG). All test data referred to are based on specific product types; LPG is based on a typical lauryl alcohol chain length of $C_{12}$–$C_{16}$, and DPG, based on an alkyl chain length ranging from $C_8$–$C_{16}$, with an average of 10.5 carbons. In both cases, the degree of polymerization is an average of 1.4 molecules of glucose per molecule of fatty alcohol.

## Foaming and Viscosity Behavior of Alkyl Polyglycosides

In comparison to other nonionics, alkyl polyglycosides have outstanding foaming properties. Decyl glucose has

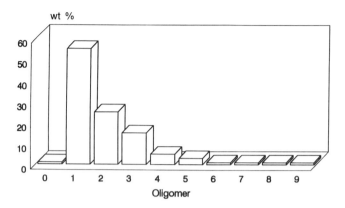

Fig. 2. Alkyl polyglycoside typical oligomer distributions.

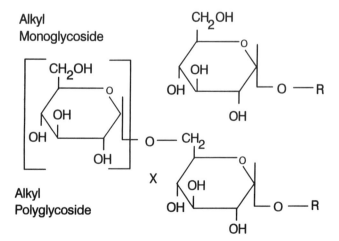

Fig. 1. Production of alkyl polyglycosides.

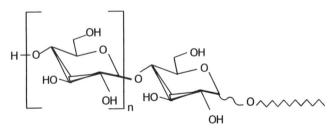

Fig. 3. Hydrophobic and hydrophilic characteristics of alkyl polyglycoside.

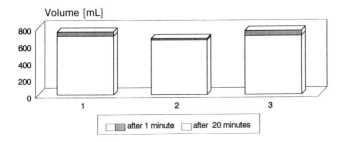

**Fig. 4.** Foam behavior of alkyl polyglycosides. Method, Beating Method (DIN 53902); water hardness, 0 mmol/L $Ca^{2+}$; concentration, 1 g AS/L. 1. SLES. 2. SLES/LPG 3:1. 3. SLES/DPG 3:1.

similar foaming properties to anionics and can be considered as a primary surfactant in cleansing preparations.

Combinations of polyglycosides were studied in mixtures with anionic surfactants. By combining sodium laureth sulfate (SLES) with LPG and with DPG, little effect was seen on the good foaming characteristic of SLES alone (Fig. 4). Foam stability of some surfactants can be improved by the addition of alkyl polyglycosides. A marked improvement in stability can be seen when LPG is combined with disodium laureth sulfosuccinate (Fig. 5).

The foam structure of alkyl polyglycosides differ from that of fatty alcohol ether sulfates. Sodium laureth sulfate forms large irregular bubbles with low foam lamellae thickness which yields a drier, less stable structure. Alkyl glycoside, however, generates smaller round bubbles of various sizes with thicker foam lamellae. This structural difference gives alkyl polyglycoside a more stable foam with a wetter feel. In some shampoo formulations, alkyl polyglycoside has been said to impart a rough, squeaky feel to the hair (6). However, it is known to be remedied by addition of anionic surfactants or cationic additives.

## Viscosity Behavior of Alkyl Polyglycoside

Viscosity serves real and perceived needs in personal cleansing products. Consumers frequently associate higher product viscosity with being richer or more concentrated. But viscosity actually serves a functional need. In a shampoo, a high viscosity is needed to prevent the product from being rinsed off of the hands before application to the hair. In a different application, such as a body cleanser or shower gel, a lower viscosity is acceptable since the product is often applied to a cloth or sponge.

In Fig. 6, a test solution of 12% SLES and 3% "thickener" was measured for viscosity under standard pH and temperature conditions. The thickener in one case was a conventional alkanolamide, cocamide DEA. In the other case, LPG served as the thickening agent. Sodium chloride was then added at various levels and viscosity measured. The viscosity curve as plotted shows comparable behaviors in the LPG to the alkanolamide. Compared to LPG, the viscosity-enhancing characteristics of DPG are less apparent.

## Effects of Alkyl Polyglycoside on Hair

Shampoos today are formulated not only to clean but also to condition, build volume, add curl, moisturize, add tints of color, and the list goes on. The individual effects of surfactants on the hair may add to or detract from its value in the formulation. Objective measurements can determine the surfactant's value.

One such measurement which demonstrates the influence of surfactants on hair is tensile strength. This measurement is the force required to break a hair strand that has been treated with the test-surfactant solution. As hair is damaged or weakened by a particular surfactant, less force is required to break the hair strand. Conversely, a higher force required means the surfactant solution has had little or no effect on the hair's strength.

To demonstrate the effect of different surfactants, SLES, LPG, and DPG were applied individually to already damaged hair (Fig. 7). Each hair bundle was treated with a 12% active solution and dried. The results show that a 15% greater force is required to break the hair bundles treated with DPG than with the SLES.

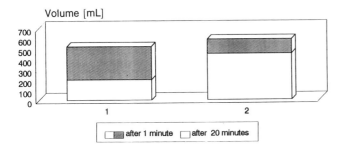

**Fig. 5.** Foam behavior of alkyl polyglycosides. Method, Beating Method (DIN 53902); Water hardness, 0 mmol/L $Ca^{2+}$; Concentration, 1 g AS/L. 1. Disodium laureth sulfosuccinate. 2. Disodium laureth sulfosuccinate/LPG 3:1.

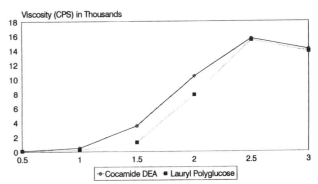

**Fig. 6.** Viscosity potentiation.

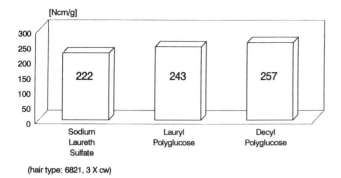

**Fig. 7.** Influence of APG on the hair tensile strength. Method: hair bundle tensile strength/dry hair.

These values indicate that alkyl polyglycosides have a milder, less damaging effect on the hair. It is known that hair treatments, such as permanent waves, have a negative effect on hair. The incorporation of alkyl polyglycosides as a co-surfactant into a shampoo for damaged hair will reduce the harsher effects of traditional surfactants, such as sodium lauryl sulfate (SLS) and SLES. Also the high alkaline stability of alkyl polyglycosides allows the formulator to utilize a milder surfactant with the harsher ingredients necessary in the permanent wave process or in the coloring of hair.

Another measurement of interest is dry combability. This is the force required to comb a hair tress which has been washed and dried with surfactant solutions. To evaluate the influence of alkyl polyglycosides on this parameter, hair tresses were treated with SLES alone and SLES in combination with LPG and with DPG, then rinsed and air dried. The combing force increased in both cases relative to the SLES solution (Fig. 8). This increase in the value implies greater body and holding power that has been given to hair treated with alkyl polyglycosides. This also suggests that alkyl polyglycosides may provide a conditioning effect, in addition to its function as a cleansing agent.

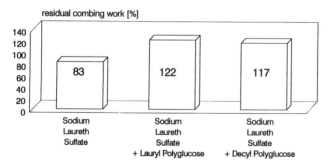

**Fig. 8.** Influence of APG on the hair dry combing force. Method, dry combability of hair strand; concentration, 12% AS (ratio 3:1); pH value, 6.5.

# Evaluation of Alkyl Polyglycoside in Skin Cleansing

Human skin serves as a complex barrier to our external environment. It effectively prevents penetration of foreign organisms, maintains body temperature and, among many other essential functions, prevents loss of moisture.

The primary role of a surfactant is the removal of dirt, debris, and excess body oils. An aggressive surfactant performing this function efficiently can also cause irritation of the skin and mucous membranes. The interactions of surfactants on the skin can be attributed to absorption on the skin surface, penetration to deeper layers, elution of individual lipids, and irritation via cytotoxic effects. To produce safe consumer products, it is necessary to test and assess these effects. These test methods can also be a means to evaluate and determine the relative value of different surfactant types. For example, objective measurement of skin redness, scaling, swelling, and water loss after surfactant exposure allows us to assess mildness and assign relative scores for each surfactant tested.

The Duhring Chamber Test was used to compare DPG and SLES (7). This test is an extended occlusive patch test done on human volunteers. By prolonged exposure to the skin surface, the irritation effects of the test solution are exaggerated. Test solutions of SLES and DPG were applied alone and in mixtures with each other. After the 24-hr period, the test areas were evaluated for negative effects, such as erythema and drying or scaling. In Fig. 9, an improvement in skin tolerance can be seen as DPG is substituted for SLES.

The Arm Flex Wash Test (AFWT) is a skin compatibility test in which substances are tested in a repeated open skin application (8). This method can be characterized as an intensified consumer-use test. It is well correlated to the Duhring Chamber Test. The use of AFWT evaluates a surfactant solution as it would be used in normal consumer use.

The Arm Flex Wash Test was conducted over a 2-week period using 20 volunteers with healthy skin in the inner elbow area. Twice a day, this sensitive skin of the inner elbow was washed with a 2 mL solution containing 10% active substance. Using a circular motion, the volunteers covered a 10-cm dia. skin area, applying the test solution

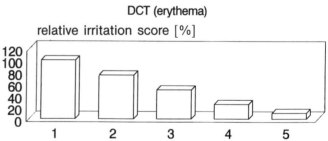

**Fig. 9.** Improvement of skin compatibility of Fatty Alcohol Ether Sulfate by APG. 1 = SLES. 2 = SLES and DPG (3:1). 3 = SLES and DPG (1:1). 4 = SLES and DPG (1:3). 5 = DPG.

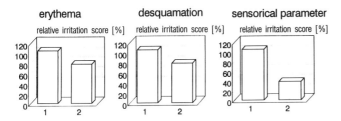

Fig. 10. Effects of fatty alcohol ether sulfate and APG on Skin Arm Flex Wash Test. 1 = SLES. 2 = SLES and DPG (3:1).

for 2 min. The lather was then rinsed off and the washing procedure repeated. After the second application of the test substance, the lather was left on the skin for 2 min. The arm was rinsed and the skin air dried. The visible dermal reactions, such as erythema and desquamation or scaling, were assessed and documented by a dermatologist.

In Fig. 10, the expected reduction in both redness and scaling as a result of substituting 25% of the SLES with DPG can be seen. In addition to these values, a subjective sensory evaluation was made of the panelists asking for their comparative evaluation of burning or stinging with the test solutions. The phenomenon of stinging is distinct from inflammation and does not necessarily lead to inflammatory changes. The results of this sensory survey are even more dramatic than the objective test. The sensory evaluation revealed that the volunteers reported an average reduction of 70% in stinging and burning when 25% of the SLES surfactant solution was substituted by DPG.

Transepidermal water loss is a measure of water lost through the skin. As the skin barrier is damaged, the water loss is greater. In this test, an evaporimeter was used to measure transepidermal water loss. The use of the alkyl polyglycosides solution resulted in a reduction of water loss by more than 20% when compared directly to the SLES solution. Thus drying of the skin is reduced, indicative of a reduction in skin damage.

To determine the comparative effects of AES and LPG on mucous membranes, an in vitro test known as the CAM Assay (9,10), also known as the hen egg test, has been used as a successful predictor of vascular damage and mucous membrane compatibility. In this test, the chorioallantoic membrane of a fertilized and incubated chicken egg is exposed to the surfactant solution. Afterward, a 5 min evaluation is made. In Fig. 11, LPG in a 3% active solution is compared to SLES, and also to a special ether sulfate which has been used in extremely mild products. Alkyl polyglycoside shows substantially less irritation than either of the two products. These data strongly suggest the suitability of alkyl polyglycosides in baby shampoos or extremely mild facial cleansers.

## Cleansing Index

A cleansing product by definition is one which removes soil. Efficient soil removal often means a reduction in mildness. The selection process for a surfactant to be used in a personal cleansing product cannot be based on individual values, such as cleansing ability or mildness alone. As a means of approaching this problem, an indexing method was developed which would consider not only cleansing ability but also irritation potential.

The products tested were conventional surfactants found in most personal care products, and included SLS, SLES, cocoamidapropyl betaine, and LPG. To evaluate the effect of LPG in combination with these surfactants, various mixtures were also made and tested.

After utilizing CAM Assay scores to rank surfactant mildness, a standardized washing procedure was developed utilizing soiled ASTM wool swatches as a substrate. Measurement of reflectance was made to determine soil removal after washing. Using this number, a cleaning factor was calculated as follows:

$$\% \text{ Cleaning} = \frac{R_{clean} - R_{soil}}{R_I - R_{soil}} \times 100$$

**TABLE 1**
**Skin Cleansing Index**

| Ingredient | Cleaning Factor[a] | CAM | Index |
|---|---|---|---|
| SLS | 24 | 25 | 0.96 |
| SLES | 24.9 | 22.5 | 1.11 |
| LPG | 18.1 | 4 | 4.53 |
| CAB | 16.9 | 21 | 0.8 |
| SLS/LPG, 3:1 | 22 | 13.75 | 1.84 |
| SLS/LPG, 1:1 | 22.4 | 12 | 1.87 |
| SLES/LPG, 3:1 | 24.1 | 12 | 2.01 |
| SLES/LPG 1:1 | 23.6 | 10.25 | 2.3 |
| SLS/LPG/CAB | 22.1 | 13.5 | 1.64 |
| SLES/LPG/CAB | 21.5 | 13.25 | 1.62 |

[a]% Cleaning at 1.5% −15.4 (water value)

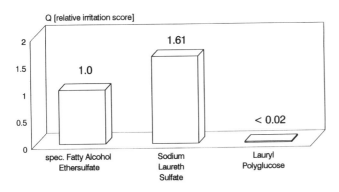

Fig. 11. In vitro local compatibility of APG Hen Egg Test (chorioallantoic membrane). Method, HET-CAM; concentration, 3% AS.

Having this factor, the CAM Assay data, and deducting for the cleansing value of water alone, a skin cleansing index (CI) was calculated as follows:

$$CI = \frac{C_p - C_{H_2O}}{CAM}$$

where  $C_p$ = Cleaning value for product
$C_{H_2O}$ = Cleaning value for water
CAM = CAM Assay score

As expected (Table 1), LPG, because of its low irritation score, has a high index value, when mildness is considered. However, also shown is that by combining SLS or SLES with LPG, equivalent cleaning to the anionic alone could be achieved. This is significant in light of the overall lowered irritation potential for this combination. In ternary systems that could be found in consumer products, good cleaning scores were also found with lowered irritation. It can be said in all cases tested that LPG, when combined with more aggressive surfactants, yielded products in which cleansing ability was not sacrificed and irritation potential was reduced.

By utilizing alkyl polyglycosides as a primary or co-surfactant in shampoo, shower and bath products, or skin cleansers an improvement in mildness is immediately realized with no sacrifice in performance. Its nonionic nature opens wider possibilities of formulation techniques. Having the ability to be based completely on renewable vegetable resources makes this surfactant well suited to the personal care market now and in the future.

## Acknowledgments

The author expresses his thanks to H. Tesmann, H. Hensen, P. Busch, K. Raabe, B. Salka, P. Bator, B. Gesslein, and J. Varvil for their contributions.

## References

1. Richtler, H.J., and J. Knaut, *Seifen-Öle-Fette-Wasche 39* (1991).
2. Fischer, E., *Chem. Ber. 26*:2400 (1893).
3. Fischer, E., *Chem. Ber. 28*:1145 (1895).
4. Siracusa, P., *Happi: 29:10*:100 (1992).
5. Varvil, J., *Alkyl Polyglycoside, A Natural for the Personal Care Industry*, presented at the American Oil Chemists' Society Annual Meeting, June 1988.
6. Kamegai, J., and S. Onitsuka, *Manufact. Chem. 14*:21 (1993).
7. Jackwerth, B., H. Kracter, and W. Matthies, *Parfümerie und Kosmetik 74*:142 (1993).
8. Strube, D.D., S.W. Koontz, R. Murtrata, and R.F. Theiler, *J. Soc. Cosmetic Chem. 40*:297 (1989).
9. Luepke, N.P., *Fed. Chem. Toxic 23*:287 (1985).
10. Sterzel, W., F.G. Bartnik, W. Matthies, W. Kastner, and K. Kunstler, *Toxic. in Vitro 4*:698 (1990).

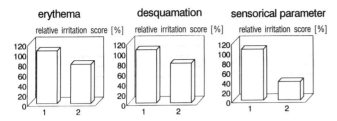

Fig. 10. Effects of fatty alcohol ether sulfate and APG on Skin Arm Flex Wash Test. 1 = SLES. 2 = SLES and DPG (3:1).

for 2 min. The lather was then rinsed off and the washing procedure repeated. After the second application of the test substance, the lather was left on the skin for 2 min. The arm was rinsed and the skin air dried. The visible dermal reactions, such as erythema and desquamation or scaling, were assessed and documented by a dermatologist.

In Fig. 10, the expected reduction in both redness and scaling as a result of substituting 25% of the SLES with DPG can be seen. In addition to these values, a subjective sensory evaluation was made of the panelists asking for their comparative evaluation of burning or stinging with the test solutions. The phenomenon of stinging is distinct from inflammation and does not necessarily lead to inflammatory changes. The results of this sensory survey are even more dramatic than the objective test. The sensory evaluation revealed that the volunteers reported an average reduction of 70% in stinging and burning when 25% of the SLES surfactant solution was substituted by DPG.

Transepidermal water loss is a measure of water lost through the skin. As the skin barrier is damaged, the water loss is greater. In this test, an evaporimeter was used to measure transepidermal water loss. The use of the alkyl polyglycosides solution resulted in a reduction of water loss by more than 20% when compared directly to the SLES solution. Thus drying of the skin is reduced, indicative of a reduction in skin damage.

To determine the comparative effects of AES and LPG on mucous membranes, an in vitro test known as the CAM Assay (9,10), also known as the hen egg test, has been used as a successful predictor of vascular damage and mucous membrane compatibility. In this test, the chorioallantoic membrane of a fertilized and incubated chicken egg is exposed to the surfactant solution. Afterward, a 5 min evaluation is made. In Fig. 11, LPG in a 3% active solution is compared to SLES, and also to a special ether sulfate which has been used in extremely mild products. Alkyl polyglycoside shows substantially less irritation than either of the two products. These data strongly suggest the suitability of alkyl polyglycosides in baby shampoos or extremely mild facial cleansers.

## Cleansing Index

A cleansing product by definition is one which removes soil. Efficient soil removal often means a reduction in mildness. The selection process for a surfactant to be used in a personal cleansing product cannot be based on individual values, such as cleansing ability or mildness alone. As a means of approaching this problem, an indexing method was developed which would consider not only cleansing ability but also irritation potential.

The products tested were conventional surfactants found in most personal care products, and included SLS, SLES, cocoamidapropyl betaine, and LPG. To evaluate the effect of LPG in combination with these surfactants, various mixtures were also made and tested.

After utilizing CAM Assay scores to rank surfactant mildness, a standardized washing procedure was developed utilizing soiled ASTM wool swatches as a substrate. Measurement of reflectance was made to determine soil removal after washing. Using this number, a cleaning factor was calculated as follows:

$$\% \text{ Cleaning} = \frac{R_{clean} - R_{soil}}{R_I - R_{soil}} \times 100$$

**TABLE 1**
**Skin Cleansing Index**

| Ingredient | Cleansing Indices | | |
|---|---|---|---|
| | Cleaning Factor[a] | CAM | Index |
| SLS | 24 | 25 | 0.96 |
| SLES | 24.9 | 22.5 | 1.11 |
| LPG | 18.1 | 4 | 4.53 |
| CAB | 16.9 | 21 | 0.8 |
| SLS/LPG, 3:1 | 22 | 13.75 | 1.84 |
| SLS/LPG, 1:1 | 22.4 | 12 | 1.87 |
| SLES/LPG, 3:1 | 24.1 | 12 | 2.01 |
| SLES/LPG 1:1 | 23.6 | 10.25 | 2.3 |
| SLS/LPG/CAB | 22.1 | 13.5 | 1.64 |
| SLES/LPG/CAB | 21.5 | 13.25 | 1.62 |

[a]% Cleaning at 1.5% −15.4 (water value)

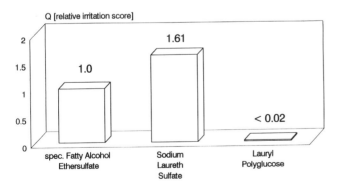

Fig. 11. In vitro local compatibility of APG Hen Egg Test (chorioallantoic membrane). Method, HET-CAM; concentration, 3% AS.

Having this factor, the CAM Assay data, and deducting for the cleansing value of water alone, a skin cleansing index (CI) was calculated as follows:

$$CI = \frac{C_p - C_{H_2O}}{CAM}$$

where  $C_p$ = Cleaning value for product
$C_{H_2O}$ = Cleaning value for water
CAM = CAM Assay score

As expected (Table 1), LPG, because of its low irritation score, has a high index value, when mildness is considered. However, also shown is that by combining SLS or SLES with LPG, equivalent cleaning to the anionic alone could be achieved. This is significant in light of the overall lowered irritation potential for this combination. In ternary systems that could be found in consumer products, good cleaning scores were also found with lowered irritation. It can be said in all cases tested that LPG, when combined with more aggressive surfactants, yielded products in which cleansing ability was not sacrificed and irritation potential was reduced.

By utilizing alkyl polyglycosides as a primary or co-surfactant in shampoo, shower and bath products, or skin cleansers an improvement in mildness is immediately realized with no sacrifice in performance. Its nonionic nature opens wider possibilities of formulation techniques. Having the ability to be based completely on renewable vegetable resources makes this surfactant well suited to the personal care market now and in the future.

## Acknowledgments

The author expresses his thanks to H. Tesmann, H. Hensen, P. Busch, K. Raabe, B. Salka, P. Bator, B. Gesslein, and J. Varvil for their contributions.

## References

1. Richtler, H.J., and J. Knaut, *Seifen-Öle-Fette-Wasche 39* (1991).
2. Fischer, E., *Chem. Ber. 26*:2400 (1893).
3. Fischer, E., *Chem. Ber. 28*:1145 (1895).
4. Siracusa, P., *Happi: 29:10*:100 (1992).
5. Varvil, J., *Alkyl Polyglycoside, A Natural for the Personal Care Industry*, presented at the American Oil Chemists' Society Annual Meeting, June 1988.
6. Kamegai, J., and S. Onitsuka, *Manufact. Chem. 14*:21 (1993).
7. Jackwerth, B., H. Kracter, and W. Matthies, *Parfümerie und Kosmetik 74*:142 (1993).
8. Strube, D.D., S.W. Koontz, R. Murtrata, and R.F. Theiler, *J. Soc. Cosmetic Chem. 40*:297 (1989).
9. Luepke, N.P., *Fed. Chem. Toxic 23*:287 (1985).
10. Sterzel, W., F.G. Bartnik, W. Matthies, W. Kastner, and K. Kunstler, *Toxic. in Vitro 4*:698 (1990).

# Use of Lauric Oil Nitrogen Derivatives in Laundry Products

F.E. Friedli[a], M.M. Watts[a], A. Domsch[b], P. Frank[a], and R.D. Pifer[a]

[a]Witco Corporation, 5777 Frantz Rd., P.O. Box 646, Dublin, OH 43017, USA.

[b]Steinau, Germany.

## Abstract

With the advent of more concentrated laundry-detergent formulations, there is a need for nontraditional surfactants for improved cleaning and formulation properties. A number of coco-based (coconut oil-based) amphoterics and betaines were found to have excellent detergency on a variety of standard stains. Ethoxylated amines are good alternatives to ethoxylated alcohols as nonionic detergents. These materials are very tolerant of hard water and some cleaned grass stains much better than standard anionics and nonionics. Alkanolamides proved to be good at grease removal, while amine oxides were only moderate in detergency. Addition of some betaines offers a synergistic cleaning effect with linear alkyl benzene sulfonates and ethoxylated alcohols.

---

The major use of nitrogen derivatives in laundry formulations is as fabric softeners, where quaternary ammonium salts are used. Unfortunately, lauric based quats do not function well as fabric softeners. However, lauric quats and other analogs are quite useful in the detergent area, which was the focus of this study.

The current trends of household-laundry formulations center on concentrated detergents, both liquid and powder. Packages are smaller, lighter, and more convenient; shipping is less costly, and in the case of powder, fewer inerts enter the environment. For surfactants in general, consumers want milder, safer, and more effective products.

In the laundry area, there is a strong need to identify new surfactants that could offer one or more of the following advantages: Better detergency per gram of surfactant, so concentration could be even higher; good detergency on very difficult stains; and coupling of formulation properties (combined with reasonable detergency) that make a total formulation functional and aesthetically pleasing.

This project started by screening a wide variety of surfactants as prespotters. The theory was that cost is less important for a prewash formulation, since the goal is to remove the spot and save the garment. Currently, most prespotters are dilute formulations of linear alkylbenzene sulfonates (LAS) or nonyl phenol ethoxylates (NPE), and although they are good detergents they are not exceptional. They work primarily because they are placed directly on the spot and may be rubbed into the stain. The best of the prespotters were then examined as liquid laundry detergents along with a variety of standard ingredients.

Laundry detergents are generally composed of LAS and alcohol ethoxylates (AE) plus enzymes, builders, and possibly a bleach. The major surfactants are well known for their particulate- and oil-removing properties respectively, in addition to low cost and high safety factors.

Ether sulfates (SLES) have long been known to be excellent detergents with good hard water tolerance. Alpha-olefin sulfonates (AOS) are widely used in Japan and less widely used elsewhere. Recent studies have shown $\alpha$-sulfomethyl esters are workable detergents, as are alkyl polyglycosides (1) and alkyl glyconamides.

Primarily due to cost or even tradition, other surfactants generally have not been used in laundry detergents. However, these specialty surfactants, namely coco-based nitrogen derivatives, have been included in a number of other household formulations and products. One purpose of this work was to study and summarize a variety of available but rarely used materials as prespotters and detergents.

## Experimental Procedures

### Materials

Stained cloth was obtained from test fabrics as 3″ × 4″ swatches. Surfactants tested were standard products of the Witco Corporation. Detergencies under U.S. conditions were done on a Terg-O-Tometer, and stain removal was measured by a Hunter Lab Colorimeter.

### Testing Procedure—Prespotter

The American Society of Testing and Materials (ASTM) recommends a 2 mL dose of material for testing prespotters (2), which was the amount of a commercial prewash added to the swatch. The commercial prewash had a solids content of 18% (mostly nonyl phenol) with a density of 1.023 g/mL. The actual quantity of surfactant used in this test was 0.37 g.

The use level of a heavy-duty liquid detergent (HDL) in the home is one capful, which holds 120 mL. A washing machine, on the large setting, holds 18 gallons of water. That equates to 1.76 mL of HDL/L of water or 1.95 g/L. In order to see a cleaning difference, the amount of HDL was cut back to 1.0 g/L.

The pretreating of the fabric was started by placing four swatches of each stain on the cotton/polyester fabric onto a piece of aluminum foil. A plastic template (65 mm dia.)

was placed over the fabric to ensure that the surfactant would cover a specific area. This allows the Colorimeter to read nothing but the clean area. The desired quantity of surfactant was added to the fabric and allowed to stand for 5 min.

The HDL was weighed into a small boat. One liter of 100°F 150 ppm hardness water and HDL were poured into the Terg-O-Tometer tub. The treated swatches (four per tub) were placed in the tubs. The swatches were washed for 5 min. at 75 rpm. The swatches were rinsed in 1 L 100°F 150 ppm water for 3 min. The swatches were then labeled and hung to dry. The next day the swatches were read by the Hunter-Lab Colorimeter for $L$ (reflectance), $a$ (redness/greenness), and $b$ (yellowness/blueness) values. These values were then placed into a spreadsheet for calculating the Stain Removal Index (SRI) and the percent Whiteness (3).

$$SRI = 100 - [(L_c - L_w)^2 + (a_c - a_w) + (b_c - b_w)^2]^{1/2}$$

where:

$c$ = unstained washed fabric
$w$ = stained washed fabric

Whiteness ($W$):

$$W = 0.01 * L(L - 5.7b)$$

### Testing Procedure—Detergency

*U.S. Conditions.* Active surfactant (1.5 g) was weighed into a small boat. One liter of 100°F water at the desired hardness and the test surfactant were poured into the Terg-O-Tometer. The four identical soiled swatches were placed in the tubs. The swatches were washed for 10 min. at 75 rpm. After the wash cycle, the swatches were removed, and the water changed. The rinse cycle was 3 min. long. The swatches were then labeled and hung to air dry. The dried swatches were read by the Hunter Lab Colorimeter for reflectance ($L$), redness/greenness ($a$), and yellowness/blueness ($b$).

*European Conditions.* All washing tests were carried out in a LINITEST apparatus at 60°C for 45 min. with 15 steel balls as ballast using 0.5% active of the test materials. The percentage of standard deviation for the reflectance measurements was a maximum of 2.5%.

*Biodegradation.* Biodegradation was conducted by ABC laboratories by the Sturm test (4).

## Results and Discussion

Various properties were observed and conclusions drawn after screening a number of these specialized lauric based nitrogen-containing surfactants.

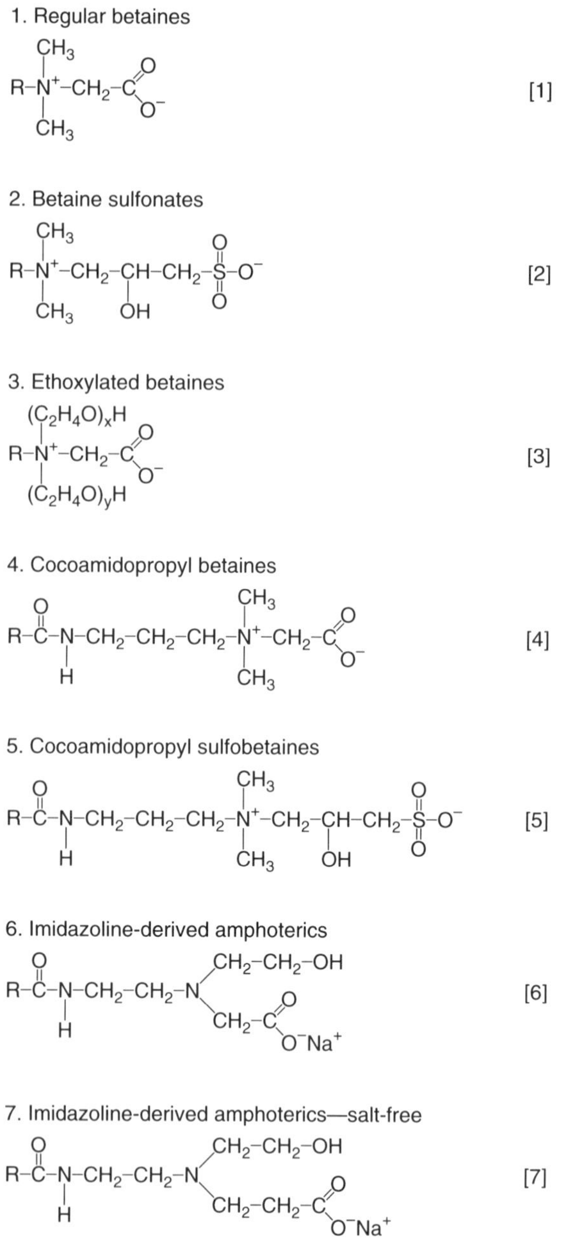

1. Regular betaines

$$R-N^+(CH_3)_2-CH_2-C(=O)O^-$$ [1]

2. Betaine sulfonates

$$R-N^+(CH_3)_2-CH_2-CH(OH)-CH_2-S(=O)_2-O^-$$ [2]

3. Ethoxylated betaines

$$R-N^+((C_2H_4O)_xH)((C_2H_4O)_yH)-CH_2-C(=O)O^-$$ [3]

4. Cocoamidopropyl betaines

$$R-C(=O)-NH-CH_2-CH_2-CH_2-N^+(CH_3)_2-CH_2-C(=O)O^-$$ [4]

5. Cocoamidopropyl sulfobetaines

$$R-C(=O)-NH-CH_2-CH_2-CH_2-N^+(CH_3)_2-CH_2-CH(OH)-CH_2-S(=O)_2-O^-$$ [5]

6. Imidazoline-derived amphoterics

$$R-C(=O)-NH-CH_2-CH_2-N(CH_2CH_2OH)(CH_2-C(=O)O^-Na^+)$$ [6]

7. Imidazoline-derived amphoterics—salt-free

$$R-C(=O)-NH-CH_2-CH_2-N(CH_2CH_2OH)(CH_2CH_2-C(=O)O^-Na^+)$$ [7]

**Fig. 1.** Types of amphoterics and betaines.

### Amphoterics and Betaines

Seven structural types of amphoterics and betaines are commonly known and used.

The general properties of amphoterics and betaines are that they work well as prespotters (Tables 1–3), are compatible with many surfactants, are hard water tolerant, are generally good detergents (Tables 4–6), and have low irritation to skin and eyes (5). A number of the amphoterics and betaines are readily biodegradable (Table 7).

The best laundry detergent tested was cocoamidopropyl betaine [4], while cocoamidopropyl sulfobetaine [5] was also very good (compare [4] to [3], [5], and [6] in Tables

### TABLE 1
Prespotter Test with Dust-Sebum on Rayon

| Surfactant | SRI |
| --- | --- |
| [8] | 93.3 |
| [4] | 93.0 |
| SLES | 92.9 |
| AE | 92.8 |
| Commercial prewash | 91.5 |
| [5] | 91.4 |
| [11] | 86.6 |
| [10b] | 86.2 |
| [3] | 83.2 |
| Commercial HDL | 78.5 |

### TABLE 2
Prespotter Test with Coffee on Cotton Polyester

| Surfactant | SRI |
| --- | --- |
| [4] | 95.5 |
| SLES | 95.4 |
| [5] | 94.6 |
| Commercial prewash | 93.5 |
| [10b] | 93.4 |
| [8] | 90.9 |
| [11] | 89.7 |
| Commercial HDL | 87.3 |
| [3] | 86.6 |

### TABLE 3
Prespotter Test with Ketchup on Cotton

| Surfactant | SRI |
| --- | --- |
| SLES | 91.0 |
| [8] | 90.4 |
| [4] | 90.3 |
| [11] | 90.3 |
| AE | 89.7 |
| [5] | 89.5 |
| [10b] | 89.3 |
| Commercial prewash | 87.6 |
| [3] | 87.1 |
| Commercial HDL | 83.1 |

1–6). Cocoamidopropyl sulfobetaine is also very effective as a hard surface cleaner (6). Our work has also shown [5] and imidazoline-derived amphoterics [6] to be excellent at soap-scum removal. For this particular use, C-8,10 is better than lauryl or coco. Compounds [4] and [5] disperse lime soaps very easily, and a number of these materials also function well as coupling agents in various formulations.

## Ethoxylated Amines [8]

Coco amine with 10 moles ($x + y$) of ethylene oxide behaves as a nonionic detergent. It is an excellent coupling agent and similar amines are used to emulsify oil in water for many agricultural formulations.

$$R-N\begin{pmatrix}(CH_2CH_2O)_xH\\(CH_2CH_2O)_yH\end{pmatrix} \qquad [8]$$

Where:
$$R = coco, x + y = 10$$

The ethoxylated amines exceeded ethoxylated alcohols (AE) in most detergencies (see Tables 1–6) and were the best surfactants that we tested on grass stains (see Table 5).

## Alkanolamides [9]

Alkanolamides are well-known foam stabilizers, and are moderate detergents. Alkanolamides also are readily biodegradable (see Table 7).

### TABLE 4
Average Detergency for 7 Stains[a] on Cotton/Polyester in 150 ppm Hardness Water

| Surfactant | Average SRI |
| --- | --- |
| [8] | 86.0 |
| [4] | 85.2 |
| AE | 84.7 |
| AOS | 83.8 |
| LAS | 83.6 |
| SLES | 83.4 |
| [7] | 83.1 |
| [6] | 81.5 |
| [9] | 80.6 |
| [5] | 77.4 |
| [3] | 76.6 |

[a]Dust-sebum, grape juice, grass, tea, chocolate, coffee, and motor oil.

### TABLE 5
Detergency for Grass on Cotton/Polyester in 150 ppm Hardness Water

| Surfactant | SRI |
| --- | --- |
| [8] | 85.2 |
| [4] | 85.0 |
| AOS | 71.8 |
| AE | 71.7 |
| SLES | 69.7 |
| LAS | 69.4 |
| [7] | 69.4 |
| [5] | 62.5 |
| [3] | 59.9 |
| [6] | 58.5 |
| [9] | 55.3 |

## TABLE 6
### Detergency for Motor Oil on Cotton/Polyester in 150 ppm Hardness Water

| Surfactant | SRI |
|---|---|
| [9] | 58.2 |
| AE | 56.9 |
| [8] | 55.1 |
| [4] | 53.2 |
| LAS | 53.0 |
| AOS | 51.9 |
| SLES | 51.7 |
| [6] | 51.7 |
| [7] | 50.7 |
| [5] | 49.4 |
| [3] | 48.5 |

## TABLE 7
### Biodegradation of Some Surfactants by Sturm Test

| Surfactant | % $CO_2$ Evolved |
|---|---|
| [6] | 67 |
| [4] | 100 |
| [9] | 65 |
| [11] | 91 |

$$R-\overset{O}{\underset{}{C}}-N(CH_2CH_2OH)_2 \qquad [9]$$

Although none of the surfactants that we tested could totally remove these tough stains, they were the most effective against used motor oil (see Table 6).

## Quaternaries [10]

The literature claims that certain quaternaries, namely lauryl trimethyl ammonium chloride, enhance the cleaning power of some anionics (7–9), and some commercial products have contained this substance for a number of years. Most quaternaries are moderate detergents by themselves or with nonionics.

One particular hydrophilic quaternary, cocoamine 15 moles of EO quaternized with methyl chloride [10] is a reasonable detergent (see Tables 1–3).

$$\begin{array}{c}(CH_2CH_2O)_xH\\ \overset{+}{R-N-Me}\\ (CH_2CH_2O)_yH\end{array} \quad Cl^- \qquad [10]$$

Where:

$$x + y = 15$$

## Amine Oxides [11]

In testing cocoamidopropyl amine oxide as a prespotter, mediocre detergency was found (see Tables 1–3). As a result, further testing of this material was not pursued. However, a recent patent shows an amine oxide to be a key surfactant in a HDL (10). In general, this class of surfactants is readily biodegradable (see Table 7).

$$\begin{array}{c}CH_3\\ |\\ R-N\rightarrow O\\ |\\ CH_3\end{array} \qquad [11]$$

Where:

$R$ = cocoamidopropyl

## Overall Detergency

To confirm that the detergency differences seen were valid, a number of duplicate runs were made, and a statistical analysis performed. As can be seen in Table 8, the standard deviations for the SRI and whiteness are very small. Here, [8] easily beats SLES in removing dust-sebum.

## TABLE 8
### Statistical Comparison of Ethoxylated Cocoamine and SLES on Dust-Sebum Stains on Cotton-Polyester in 150 ppm Water

|  | Ethoxylated Cocoamine [8] | SLES |
|---|---|---|
| Average[a] SRI | 88.0 | 83.4 |
| SD | 0.13% | 0.23% |
| Average[a] Whiteness | 57.7% | 45.5% |
| SD | 0.77% | 0.64% |

[a]12 runs.

## TABLE 9
### Effect of Water Hardness on Dust-Sebum Stains on Cotton/Polyester (SRI)

| Surfactant | 50 ppm | 150 ppm | 300 ppm |
|---|---|---|---|
| [8] | 88.9 | 88.6 | 87.6 |
| LAS | 88.1 | 87.6 | 83.5 |
| AE | 87.6 | 86.7 | 86.7 |
| [4] | 84.8 | 84.8 | 84.3 |
| [5] | 83.4 | 81.7 | 79.1 |

### TABLE 10
Effect of Water Hardness on Grass Stains on Cotton/Polyester (SRI)

| Surfactant | 50 ppm | 150 ppm | 300 ppm |
|---|---|---|---|
| [8] | 88.7 | 85.2 | 71.4 |
| [4] | 86.4 | 85.0 | 82.2 |
| AE | 79.1 | 71.7 | 66.2 |
| LAS | 77.7 | 69.4 | 60.0 |
| [5] | 74.0 | 58.5 | 54.1 |

### TABLE 11
Improved Detergency with Amphoterics on Standard European Soil Washed with a European Heavy-Duty Liquid

| Surfactant | SRI |
|---|---|
| HDL | 48 |
| 95% HDL + 5% [4] | 53 |
| 95% HDL + 5% [6] | 58 |
| 95% HDL + 5% [7] | 55 |
| 95% HDL + 5% [5] | 53 |

### TABLE 12
Improved Detergency Using Amphoterics (HDL vs. 95% HDL + 5% [6])

| Stain | SRI | |
|---|---|---|
| Standard | 48 | HDL |
| | 58 | Blend |
| Blood | 70 | HDL |
| | 73 | Blend |
| Cacao | 38 | HDL |
| | 50 | Blend |
| Red wine | 68 | HDL |
| | 83 | Blend |

To better examine the effect of water hardness on the best surfactants, tests were run at 50, 150, and 300 ppm calcium-ion water hardness compared to LAS and AE. Sample [4] performed very well, followed by [8], but AE with [5] and LAS were affected by increased hardness (Tables 9 and 10).

Specialty surfactants can offer some exceptional cleaning properties, particularly on certain stains. They can also work synergistically with standard materials to enhance detergency. Table 11 shows several amphoterics tested under European washing conditions with a commercial European heavy-duty liquid. The imidazoline-derived amphoteric [6] offered the best synergy under these conditions. This was explored further on a wide variety of stains with excellent results (Table 12).

As stated earlier, coco or palm kernel quaternaries are not useful as premium fabric softeners. However, some of the materials can be used in detergent-softeners or "softergents," where compatibility is important and less softening is required. Type [10], where $x + y = 2\text{--}15$, is compatible with most anionic/nonionic detergent systems. Even Types [4] and [8] can impart some softness and static reduction while functioning as detergents. Thus, when reformulating a laundry detergent or household cleaner for better cleaning, improved mildness, higher concentration, multifunctional performance, and other traits, a wide variety of materials are available for consideration.

## References

1. Balzer, D., *Tenside Surf. Det.* 28:413 (1991).
2. ASTM. 1988. D4265-83.
3. ASTM. 1987. E313-73.
4. OECD Method 301B, "Ready Biodegradability: Modified Sturm Test."
5. Palicka, J., *J. Chem. Tech. Biotechnol.* 50:331 (1991).
6. Michael, D.W., U.S. Patent 5,108,660 (1992).
7. Rubingh, D.N., and T. Jones, *Ind. Eng. Chem. Prod. Res. Dev.* 21:176 (1982).
8. Smith, R.J.M. and A.C. McRitchie, U.S. Patent 4,333,862 (1982).
9. Hughes, L.J., U.S. Patent 4,507,219 (1985).
10. Hepworth, P., *PCT Int. Appl.* WO 92 20,772 CA *118*:194054u (1992).

# The Increasing Importance of Methyl Ester Sulfonates as Surfactants

**Ned M. Rockwell and Y.K. Rao**

Stepan Company, 22 W. Frontage Rd., Northfield, IL USA.

## Abstract

From these studies it can be concluded that sulfonated methyl esters (SME) are very efficient surfactants and provide excellent adsorption and micellar characteristics. They perform well in hard water and at low concentrations. They also improve the micellar and detergency behavior of conventional linear alkylbenzene sulfonate (LAS)- and primary alkyl sulfate (PAS)-based surfactants. Although more work needs to be done from a physico-chemical point of view, their highly desirable surfactant properties definitely put them in the category of primary surfactants.

If a single word were required to describe the prevalent trend of market driven global economies, it would most likely be "efficiency". Efficiency drives the choices we make as business people and human beings as we work toward optimization of our company's resources, the responsible use of the world's resources, and the impact today's technologies have on the environment in which we share together. More specifically, with regard to the evolution of the soaps and detergents industry, the concept of efficiency as the primary influence of development is very true in terms of the bulk of new product and technology introduction over approximately the last decade. The principle driving force of today's ingredient and product manufacturers is to provide desired products and services in such a way as to achieve the greatest utilization of sustainable resources while minimizing energy consumption, refuse, and virtually all unfavorable impact to our environment. This readily translates to continually striving for greater efficiency in every action and process we undertake; ranging from the manufacture of ingredients and design of formulations, to the packaging and transportation of products.

It is the responsibility of a raw material or ingredient supplier to deliver materials to product designers and formulators which meet today's and tomorrow's standards for efficiency. Today's standards for efficiency are fairly well known (although perhaps not completely agreed upon.) These criteria include the availability of raw materials, the capability to consistently manufacture high quality products, the fate of the product and its decomposition products in the environment, and the ultimate toxicity of the product to the earth's flora and fauna.

A more specific standard of efficiency is considered with regard to the use of surfactants. Surfactants or surface-active substances provide efficiency in applications by minimizing the amount of work required to maintain or generate surfaces or interfaces. For example with regard to detergency, the reduction of surface or interfacial tension through the adsorption and solubilization capacity of surfactants is generally considered to be one of the basic performance mechanisms of surfactants in the cleaning process.

Although there are a wide variety of surfactants used in a wide variety of applications today, only a few are really considered "workhorse" or primary surfactants. For example, in the relatively mature technologies of detergents and hard surface cleaners, the most commonly used surfactants are anionics and nonionics. Of the anionics, linear alkylbenzene sulfonates (LAS), primary alkyl sulfates (PAS), and alkyl ether sulfates (AES) are generally considered to be the surfactants which provide the primary influence on detergency performance for most of today's detergent formulations. Schirber (1) provides data that is consistent with this statement by estimating that the combined consumption of LAS, AES, and PAS for household detergents in the United States, Western Europe, and Japan was approximately 71% of the total surfactants used in 1990. With alcohol ethoxylates (AE; i.e., nonionics) accounting for about 21% of the remaining 29% of surfactants used in these households, it is clear that the term primary surfactant refers to only these four classes of surfactants.

Since the 1960s, researchers have been working on a different class of anionic surfactants. A class of surfactants which includes surfactant candidates which have strong potential for eventually becoming classified as primary surfactants. The class of surfactants referred to are fatty acid ester sulfonates (FAES). Figure 1 shows the structure of FAES, where the sulfonate group is alpha ($\alpha$) to the carbonyl of the ester.

This paper will focus on the efficiency provided to the marketplace by specifically the $\alpha$-sulfo methyl ester (SME) versions of FAES. Considerable research effort has been invested by many people toward achieving a better understanding of the capabilities of SME in surfactant applica-

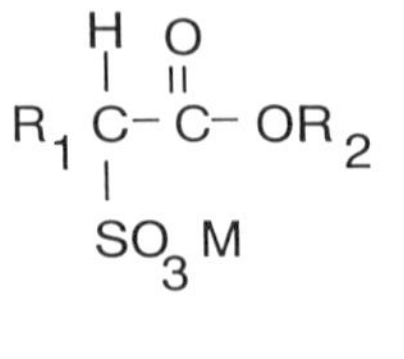

$$R_1-\underset{\underset{SO_3M}{|}}{\overset{\overset{H}{|}}{C}}-\overset{\overset{O}{\|}}{C}-OR_2$$

$R_1 = C_nH_{2n+1}$, $n = 6$ to $16$

$R_2 = C_mH_{2m+1}$, $m = 1$ to $10$

$M = Na, K, Mg, Ca, NH_4$

**Fig. 1.** The structure of fatty acid ester sulfonates (FAES).

tions, as evidenced by the many presentations to industry groups and journal and proceedings publications. It is the goal of this work to shed greater light on the utility and the as yet relatively untapped benefits of SME by linking their use in surfactant applications to the concept of efficiency and providing new data and interpretation to support this relationship.

## Chemistry

As with any major class of chemicals, a variety of compounds exist for FAES surfactants according to chain length and substituent groups. FAES surfactants consist of fatty carboxylates sulfonated α to the carbonyl, the alkyl group of the alkyl ester portion of the molecule, and the inorganic or organic counter-ion to the sulfonate. Many people including ourselves have explored the surfactancy of the molecular permutations available as the fatty group and the alkyl ester are varied (2–4). Additionally, the di-salts of α-sulfo fatty acids (SFA) are commonly included in the discussions of FAES as they are often observed as minor co-products of FAES and have been investigated for surfactancy by their own right (5,6).

Of particular importance are the α-sulfo methyl esters (SMEs), where the fatty alkyl chain is derived from lauric oils and/or tallow. Typically, SMEs are prepared in two basic steps as shown in Fig. 2. In Step 1, a triglyceride oil such as coconut, palm kernel, palm, or tallow undergoes methanolysis to form methyl esters and glycerin, usually with the aid of a catalyst such as sodium methylate. Glycerin is physically separated from the esters and is refined for use in other applications. The methyl esters may be further refined by various means, the most common of which is to separate the different alkyl chain lengths by distillation. Step 2 involves the thin-film sulfonation and subsequent neutralization of the methyl esters to form the α-sulfo methyl ester mono-salt and relatively small quantities of the α-sulfo fatty acid di-salt. It should be pointed out that methyl esters derived from glyceride oils containing olefinic unsaturation require hydrogenation prior to sulfonation, in order to saturate the olefinic compounds and avoid color body formation.

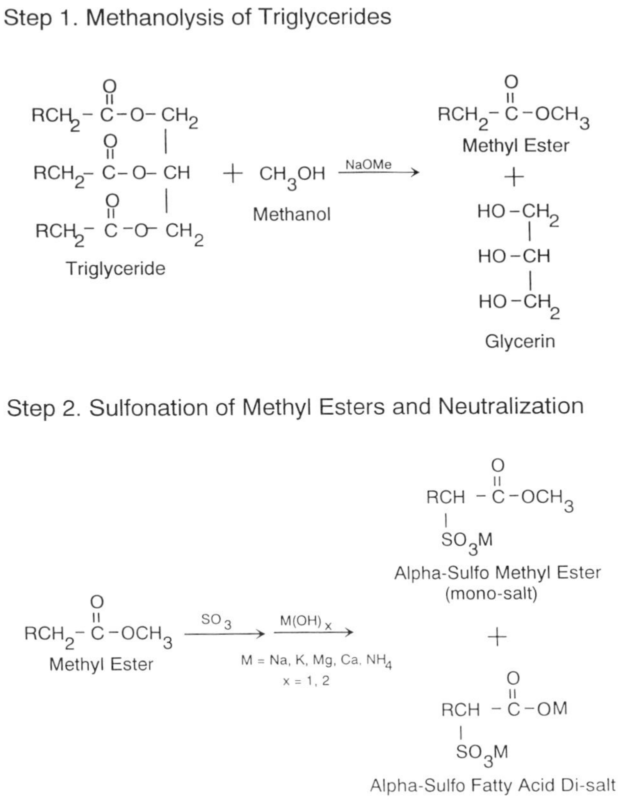

## Raw Material Efficiency

The four existing classes of primary surfactants are derived from both oleo-chemical and petro-chemical resources. Of the four, three are dependent on triglyceride oils or ethylene (Fig. 3). Smith (7) provided in 1989 an insightful discussion regarding the relative efficiency advantage of SME to the supply of ethylene and its derivatives, at a time when there was a particular amount of uncertainty regarding the stability of the ethylene market. Satsuki (8) concluded in 1992 that the rising production of palm and palm kernel oils has a market stabilizing influence which more than offsets the lack of growth of tallow supply and the financial risk associated with the historical cyclicality of coconut oil production. If one were to derive an equation describing the historical cyclicality of coconut oil price as a continuous function of time, it would undoubtedly express a cycling four to five year period of fluctuation, with a downward trend. The coconut oil (CNO) price data displayed in Fig. 4 were taken from monthly highs and lows printed in the Wall Street Journal, in constant dollar terms. This graph shows the reduction in amplitude of price peaks and an overall decrease in price volatility, especially since 1986. Admittedly, in spite of tempered volatility, coconut oil pricing remains difficult to predict. Nevertheless, reduced price volatility of CNO implies improved raw material supply efficiency of SME

Fig. 2. The preparation of α-sulfo methyl esters (SMEs).

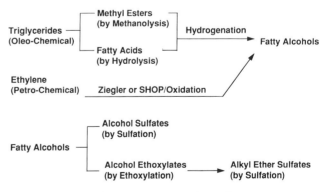

Fig. 3. Primary surfactant preparation from oleo- or petro-chemical resources.

**TABLE 1**
Five-Year Comparison of Estimated Lauryl Alcohol Capacities Worldwide

| Country | Producer | 1988 Capacity (MM lbs) | 1993 Capacity (MM lbs) | New Plant Start-Up (Year) |
|---|---|---|---|---|
| North America/South America | | | | |
| United States | P&G | 190 | 250 | |
|  | Henkel | 0 | 88 | 1992 |
|  | Regional Total | 190 | 338 | |
| Europe | | | | |
| Germany | Henkel | 221 | 385 | |
|  | Condea | 66 | 66 | |
|  | Huls | 22 | 88 | |
| France | Henkel | 66 | 75 | |
| United Kingdom | Albright-Wilson | 44 | 44 | |
| Belgium | Oleofina | 11 | 66 | |
|  | Regional Total | 430 | 724 | |
| Asia | | | | |
| Japan | Kao | 73 | 117 | |
|  | New Japan | 26 | 26 | |
| Indonesia | Salim | 0 | 66 | 1990 |
| Malaysia | P&G | 0 | 88 | 1993 |
|  | Henkel | 0 | 66 | 1991 |
|  | Kao | 0 | 66 | 1990 |
| Philippines | Cocochem | 66 | 66 | |
|  | Kao | 55 | 55 | |
| India | | 0 | 66 | |
| Others | | 26 | 60 | |
|  | Regional Total | 247 | 677 | |
|  | Worldwide Total | 867 | 1739 | (Change = 872) |

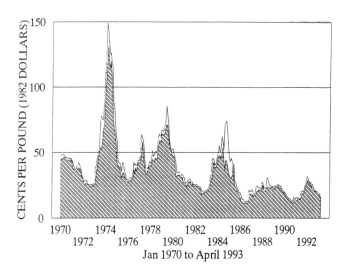

**Fig. 4.** The coconut oil price data (taken from *Wall Street Journal*).

surfactants by lowering the financial futures market risks of CNO purchasing.

Perhaps the most striking development in recent years affecting the supply of oleo-chemicals is the growth of oleo-chemically derived fatty alcohols. As shown in Fig. 3, fatty alcohols are a precursor to PAS and AE. The production of fatty alcohols involves either methyl ester or fatty acid intermediates which are hydrogenated to alcohols. The alcohols are subsequently sulfated, ethoxylated, or ethoxylated and sulfated into surfactants. Our assessment of worldwide lauryl alcohol capacity shows an annual growth rate of approximately 15% from 1988 to 1993. This equates to the world's capacity having more than doubled in the past five years as shown in Table 1. We estimate the distribution of this capacity to be 57% methyl ester and 43% fatty acid based. Jan Vogel (9) estimated in his 1992 presentation to the CESIO conference in London that replacement of the current 2.6 million tons of petroleum based surfactants by oleo-chemical based surfactants would require approximately 85 new manufacturing facilities. Although the cost for all of these new facilities is most likely an unrealistic burden for the ultimate consumers of alcohol derived surfactants to bear, it appears that market efficiency forces are strong enough to be capable of shifting investment toward oleo-chemically derived alcohol, at least in the near term. It also seems reasonable to suggest that with all of this alcohol capacity coming onstream, one should expect the relative volatility of today's oleo-chemical based alcohol supply to be reduced. Perturbations of supply and demand should be more readily absorbable by a greater number of players in the market.

As methyl esters are an intermediate source to the production of PAS, AES, and AE, it is readily observable that an entire process step is avoided by producing surfactants directly from methyl esters. Process efficiency coupled with rampant growth of oleo-chemical derived alcohols, signals favorable overall raw material supply efficiencies for SME surfactants.

## Manufacturing Efficiency

Although batch or continuous oleum sulfonation technology allowed the first synthetic detergents to be made in the 1940s, oleum technology is incapable of manufacturing LAS and PAS to today's high concentration and quality standards. For example, the composition of today's ultra concentrated detergent powders is less tolerant of fillers such as the sodium sulfate by-product of oleum sulfonates and sulfates.

Since its invention in the early 1960s, the thin film sulfonation process using sulfur trioxide has been the most efficient process for manufacturing anionic surfactants. Considerable effort has been made over the years to optimize thin film process technology. The quality of organic feedstocks to the film sulfonation process has also improved dramatically, as demands for better color and higher conversion and specificity to desired products has increased. For example, in spite of the fact that LAS has been manufactured by thin film sulfonation for over thirty years, today's standard for detergent grade LAS color is considerably more demanding than just five years ago. Five years ago the typical color of a five weight percent aqueous solution of LAS was approximately 350 Klett. Today's color standard for LAS is less than 50 Klett. Achievement of such high quality LAS is the result of many years of work by those involved with the manufacture of alkylates and those dedicated to fine tuning the sulfonation process.

In contrast, the evolution of SME manufacturing technology has been much slower than LAS and PAS. Despite early research in the late fifties and early sixties, the development of an efficient manufacturing process for SMEs has been hindered by relatively unusual chemistry. As with most thin film sulfonation processes, the efficient sulfonation of the organic feedstock requires a delicate balancing act in the reactor to achieve high concentrations of active matter and low concentrations of unsulfonated methyl ester and by-products. In the case of methyl ester sulfonation however, efficient sulfonation requires additional processing downstream of the thin film reactor in specialized but relatively noncapital intensive equipment. Only in the last decade have the requirements for this equipment been thoroughly refined to the extent that manufacturing efficiency has improved to achieve commercial grades of SME. Stepan Company now manufactures two grades of SMEs commercially, a lauryl sulfo methyl ester and a stripped coco sulfo methyl ester.

The lack of understanding of the product quality requirements for SMEs has contributed to their slow evolution as commercially desirable detergent ingredients. For example, the relatively small concentration of by-products such as α-sulfo fatty acid di-salts in typical SME products has created considerable uncertainty as to the benefits or detriments these materials may have in applications. Schambil et al (6) demonstrated the relative lack of antagonism that small concentrations of SFA di-salts have with SMEs in detergent applications. In the following section, it will become more obvious that the quality of currently available commercial grades of SMEs are quite adequate for today's detergent applications. They are comparable to LAS and PAS in terms of surfactant efficiency and improve the properties of these surfactants by exhibiting synergism.

## Surfactant Efficiency

The literature is full of information regarding the surfactant properties and applications of SME and SFA. Table 2 provides a brief summary of applications which have been disclosed by many interested parties. It is evident that many industries benefit from the unique properties of these materials, and it would be redundant and tedious to review each application. In this section, studies on the efficiency

**TABLE 2**
α-Sulfo Fatty Acid Derivatives: Use and Application Listing

| | |
|---|---|
| Adhesives | Heavy-duty detergents |
| Amphoteric surfactants | |
| Antistatic agents | Light-duty detergents |
| | Lubricants |
| Bubble baths | |
| | Metal Cleaning |
| Chemical intermediates | Metal working/milling |
| Cleansers | |
| Corrosion inhibitors | Oil recovery |
| Detergents | Paint |
| Dispersing agents | Paper making |
| Emulsification | Shampoos |
| Emulsion polymerization | Softening/softeners |
| Foaming | Urethanes |
| Frothing agents/flotation of ores | |
| | Wetting agents |
| Germicides | |

of SME surfactants will be discussed, which should make it more obvious why SMEs perform as they do in some previously disclosed detergent applications.

Due to their amphiphilic character, surfactants have a tendency to adsorb at various interfaces, which leads to the reduction of surface or interfacial tension. After saturating the interface, these surface active molecules start aggregating in the solution to form micelles at a specific surfactant concentration known as the critical micelle concentration (CMC). The adsorption characteristics and the micellization behavior of surfactants are responsible for the technologically important properties of surfactants such as foaming, detergency, wetting, solubilization, and emulsification. Thus, the efficiency of any surfactant system depends upon choosing the right type of surface active agents that can lower interfacial tension, form micelles, and display good adsorption characteristics by providing good packing of surface active molecules at the interface in very small concentration.

SMEs possess excellent adsorption and micellization characteristics and have been the focus of several scientific investigations as evidenced from several articles (4–8) published in recent years. Although a great deal of work has been reported on this class of compounds, it seems worthwhile to present new information on some commercial grades of SMEs with regard to surfactant efficiency. They will be compared with two of today's primary anionic surfactants, LAS and PAS. Also, the synergistic behavior of these surfactants with SMEs will be displayed.

## Materials and Methods

All the surfactants used in this study are either commercial products or laboratory materials synthesized to commercial composition standards. Table 3 shows the amount of α-sulfo methyl ester mono-salt and α-sulfo fatty acid di-salt and their ratios in the SMEs used here. The sodium salt of α-sulfo methyl laurate ($C_{12}$ SME) has primarily a twelve carbon alkyl chain. All of the stripped cocoates have an alkyl chain length distribution of 12-18 carbon atoms, primarily containing mixtures of α-sulfo methyl laurate and α-sulfo methyl myristate ($C_{12/14}$ SME). All the SMEs have small but significant amounts of di-salts in them. On the other hand, the α-sulfo stripped coco fatty acid sodium di-salt is relatively pure. All the surfactants in this investigation contain at least 0.5% of inorganic salts and unsulfonated organics.

A fully automated Krüss K-12 tensiometer was used to measure the adsorption and micellar behavior of individual and mixed surfactant systems by the Wilhelmy plate method. All the measurements were done at room temperature, and distilled water was used for CMC measure-

**TABLE 3**
**Compositional Analysis of Surfactants Used in This Study**

| Surfactant | Molecular Weight | Mono-Salt (%) | Di-Salt (%) | Mono/Di Ratio | Inorganic Salts (%) | Chain Length Distribution (%) | | | | | | | |
|---|---|---|---|---|---|---|---|---|---|---|---|---|---|
| | | | | | | C8 | C10 | C11 | C12 | C13 | C14 | C16 | C18 |
| α-Sulfo Methyl Laurate*Sodium Salt | 316 | 31.4 | 6.6 | 5:1 | < 2.0 | < 1.0 | < 1.0 | — | > 96 | — | < 4 | — | — |
| α-Sulfo Methyl Stripped Cocoate* Sodium Salt | 338 | 32.6 | 4.7 | 7:1 | < 1.0 | < 1.0 | < 1.0 | — | 55–58 | — | 20–23 | 8–11 | 9–12 |
| α-Sulfo Methyl Stripped Cocoate Magnesium Salt** | 654 | 33.3 | 3.8 | 9:1 | < 2.0 | < 1.0 | < 1.0 | — | 55–58 | — | 20–23 | 8–11 | 9–12 |
| α-Sulfo Methyl Stripped Cocoate Ammonium Salt** | 333 | 40.3 | 1.5 | 26:1 | < 2.0 | < 1.0 | < 1.0 | — | 55–58 | — | 20–23 | 8–11 | 9–12 |
| α-Sulfo Stripped Coco Fatty Acid Sodium Disalt** | 344 | < 2.0 | 98.0 | 1:49 | < 0.5 | < 1.0 | < 1.0 | — | 55–58 | — | 20–23 | 8–11 | 9–12 |
| Linear Alkyl Benzene Sulfonate*, *** Sodium Salt | 346 | N/A | N/A | N/A | < 2.0 | < 1.0 | 20 | 34 | 35 | 9 | 2 | — | — |
| Sodium Lauryl Sulfate* | 302 | N/A | N/A | N/A | < 1.0 | — | — | — | > 96 | — | < 4 | — | — |

*Commercial Materials.
**Laboratory synthesized materials.
***Linear Alkyl Benzene Sulfonate has 2.25–2.75% sodium xylene sulfonate (SXS) as hydrotrope.

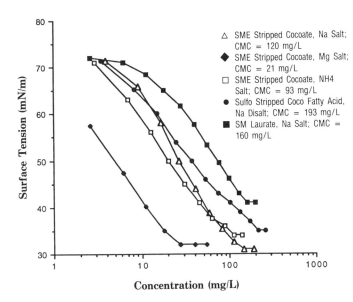

Fig. 5. CMC behavior of α–sulfo fatty acid salts.

ments. Interfacial tension between the 0.3% surfactant solutions and mineral oil was measured using a spinning drop tensiometer as developed by the University of Texas. Detergency measurements were performed using a Terg-O-Tometer.

## Adsorption and Micellar Characteristics

Figure 5 shows the surface tension-concentration plots for all the SMEs used in this study. Surface tension and concentration graphs for the individual and mixed surfactant systems were plotted by monitoring the change in surface tension with increasing surfactant concentration. The critical micelle concentrations (CMCs) for all the surfactant systems was calculated from these graphs. Graphs typical of conventional surface active agents were obtained with the characteristic break in the curve at the CMC, after which the surface tension remains relatively constant. Be aware that it is very difficult to visually determine the CMC associated with each curve. The tensiometer calculates CMC values, and they are tabulated on the figure. With the exception of the sodium salt of α-sulfo methyl laurate, all the SMEs significantly reduce the surface tension of water. The magnesium salt of α-sulfo methyl stripped cocoate shows exceptionally low CMC, which is analogous to the conventional surfactants, as salts of surface active compounds containing divalent ions pack better at the interface. Similarly, the ammonium salt of α-sulfo methyl stripped cocoate showed a lower CMC than sodium salt of α-sulfo methyl stripped cocoate, even though it is most efficient in reducing surface tension. Significantly lower values of CMCs were observed for all the SMEs, which may be due to the presence of small amounts of electrolytes and unsulfonated matter in the materials.

From the knowledge of average molecular weights of these surfactants, surface tension and log concentration graphs in molar concentrations were plotted. Adsorption and micellar characteristics were calculated from the slopes of surface tension-log concentration graphs, to determine the efficiency and effectiveness of these surfactant systems and are given in Table 4. Effectiveness of adsorption (Γ) refers to the amount of surfactant required to saturate the surface. A higher number of molecules over a given area is desired. The area of the surfactant head group at the air/water interface ($a_s$), is an important parameter to study as it indicates the relative ability of surfactants to pack at the interface. The smaller the head group area of the surfactant, the better it will pack. The effectiveness of surface tension reduction of the solvent at the CMC ($\Pi_{cmc}$) was calculated as the difference between the surface tension of the solvent and surface tension at the CMC. Also, the efficiency of surfactant adsorption ($pC_{20}$, i.e., the amount of surfactant required to reduce the surface tension by 20 mN/m) was calculated and expressed as concentration. A lower value of $pC_{20}$ and larger value of $\Pi_{cmc}$ is highly desirable for a surfactant to be efficient and effective at various interfaces. Table 4 also contains the surface activity data for the sodium salt of LAS, sodium lauryl sulfate and their 1:1 mixture with the sodium salt of α-sulfo methyl stripped cocoate. Relatively higher values of area/molecule of adsorbed surfactant at the interface are obtained for SMEs as the chain length decreases. From the table, it is evident that the SME mono-salts of sodium, magnesium, and ammonium demonstrate several desirable surface active properties in the following order:

$$Mg^{++} > NH_4^+ > Na^+$$

They adsorb very effectively at the air/water interface and occupy relatively small area, which helps them pack better and make a more stable surfactant film at the interface. More work is being done in this laboratory to establish the effect of the counter-ion on the surface active properties of SMEs. α-Sulfo stripped coco fatty acid sodium di-salt effectively reduces the surface tension at higher concentrations but does not adsorb and pack as well as the mono-salt at the interface. The presence of two polar groups in the di-salt causes repulsion between the surfactant head groups, resulting in poor packing. This is consistent with the findings of Schambil et al (6) who have systematically compared the physico-chemical properties of pure mono- and di-salts and concluded that mixtures of mono-salts possess more favorable properties than di-salts. They have also observed that impurities of small amounts of di-salts did not produce any undesirable effects in the mono- and di-salt mixtures.

## Synergism

From Table 4, it is clear that all the SMEs except sodium salt of α-sulfo methyl laurate possess far superior surfactant efficiency than the two commonly used anionics, LAS and PAS. But, when these surfactants were mixed with sodium salt of α-sulfo methyl stripped cocoate in a 1:1 weight ratio, greater synergism is observed with PAS, and the surfactancy of the mixed systems dramatically improved. Figure 6 shows the surface tension-concentra-

**TABLE 4**
**Surface Active Properties of Surfactant Systems Used In This Study**

| Surfactant | Critical Micelle Concentration, CMC (mM) (mg/L) | Effectiveness of Adsorption, $\Gamma$ (mol/cm$^2 \times 10^{10}$) | Area/Molecule at Air/Water Interface, $a_S$ (Å$^2$) | Effectiveness of Surface Tension Reduction, $\Pi_{CMC}$ (mN/m) | Efficiency of Adsorption $pC_{20}$, (mM) (mg/L) | Interfacial Tension Between Aqueous Surfactant Solution (0.3%) and Mineral Oil (mN/M) |
|---|---|---|---|---|---|---|
| α-Sulfo Methyl Laurate* Sodium Salt | 0.51 mM 160 mg/L | 2.24 | 74.1 | 31.1 | 0.20 mM 63 mg/L | 0.19 |
| α-Sulfo Methyl Stripped Cocoate* Sodium Salt | 0.36 120 | 2.75 | 60.4 | 41.1 | 0.71 24 | 0.16 |
| α-Sulfo Methyl Stripped Cocoate Magnesium Salt** | 0.032 21 | 2.64 | 62.9 | 39.2 | 0.007 24 | 0.11 |
| α-Sulfo Methyl Stripped Cocoate Ammonium Salt** | 0.28 93 | 2.08 | 79.8 | 38.2 | 0.06 20 | 0.16 |
| α-Sulfo Stripped Coco Fatty Acid Sodium Disalt** | 0.56 193 | 1.95 | 85.5 | 36.8 | 0.08 28 | 0.24 |
| Linear Alkyl Benzene Sulfonate* Sodium Salt | 0.45 156 | 1.21 | 137.2 | 28.1 | 0.11 38 | 0.24 |
| Sodium Lauryl Sulfate* | 0.42 127 | 2.11 | 78.7 | 28.4 | 0.20 60 | 0.24 |
| Linear Alkyl Benzene Sulfonate* Sodium Salt + α-Sulfo Methyl Stripped Cocoate* Sodium Salt (1:1) | 0.40 137 | 1.76 | 94.3 | 36.6 | 0.07 24 | 0.22 |
| Sodium Lauryl Sulfate* + α-Sulfo Methyl Stripped Cocoate* Sodium Salt (1:1) | 0.25 80 | 2.46 | 67.6 | 34.2 | 0.08 26 | 0.22 |

*Commercial products.
**Laboratory synthesized materials.

tion graphs for sodium salt of α-sulfo methyl stripped cocoate, LAS, and their 1:1 mixture. Addition of SME reduced the CMC of sodium LAS from 156 mg/L to 137 mg/L, suggesting a straight forward mixing or an averaging of the two components. More importantly, a greater lowering in surface tension at the CMC was observed as the surface tension dropped by about 8 mN/m. The effect of the sodium salt of α-sulfo methyl stripped cocoate on the surface active properties of PAS, seems to be more significant. As shown in Figure 7, the CMC for PAS dropped from 127 mg/L to 80 mg/L in a 1:1 mixture. This is a very significant drop in the CMC and indicates a very strong synergy between the SME and PAS. This synergy is also observed for all of the other physico-chemical properties investigated.

## Detergency

As physico-chemical data is directionally important with regard to judging surfactant performance, it is still necessary and important to investigate surfactant behavior in end-use applications. Heavy duty laundry was chosen as a key measure for comparing surfactant application performance. As life cycle analyses have demonstrated that heating machine wash water is one of the least efficient processes of cleaning clothes, warm and cold temperature

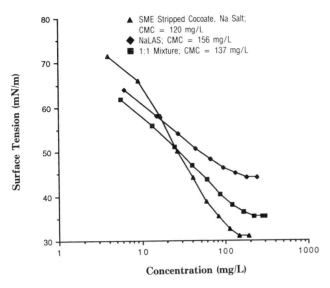

**Fig. 6.** SME stripped cocoate, sodium salt and NaLAS mixed surfactant system.

wash conditions were selected as representing the most relevant conditions for judging future laundry detergent ingredient performance. Fabric and soil selections for detergency comparison were based on typical consumer wash loads. Laundry cleaning performance testing was intended to simulate North American vertical axis machine standards.

All SME surfactants show excellent tolerance towards hard water, as has been reported by several investigators (7,8). As shown in Figures 8 and 9, the addition of SME to LAS or PAS in 1:1 ratio significantly improves the detergency of LAS and PAS for both cotton and poly/cotton fabrics at 60°F and 100°F respectively. The SME/PAS mixture achieves the greatest amount of soil removal. It is

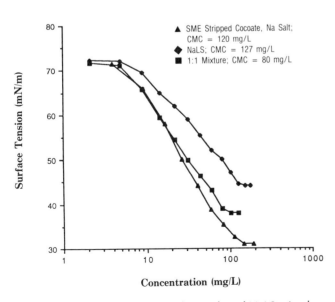

**Fig. 7.** SME stripped cocoate, sodium salt and NaLS mixed surfactant system.

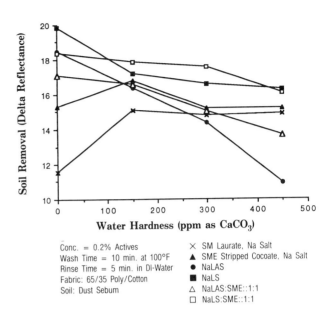

**Fig. 8.** Detergency behavior of single and mixed surfactant systems.

also clear from these figures that α-sulfo methyl stripped cocoate, sodium salt performs consistently across a broad range of low to high water hardness, whereas α-sulfo methyl laurate performs poorer under the same conditions. Satsuki (8) reported similar findings and observed that

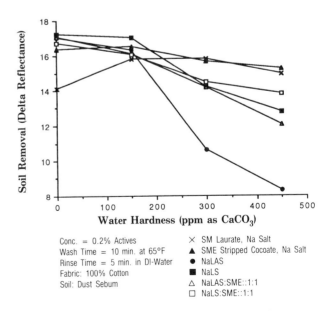

**Fig. 9.** Detergency behavior of single and mixed surfactant systems.

smaller concentrations of SME surfactants are needed to achieve equal or better detergent properties in comparison to LAS and PAS. Also, that SMEs show excellent compatibility with protease enzymes, whereas LAS and PAS reduce enzyme activity to 20% and 35% of original enzyme activity, respectively (8).

### *Interfacial Tension*

The interfacial tension (IFT) data for all the SMEs, LAS, and PAS between the aqueous solutions of surfactants and mineral oil are shown in Table 4. The IFT is less than 1 mN/m for all the surfactants for 0.3% surfactant solutions. The magnesium mono-salt of α-sulfo methyl stripped cocoate has the lowest IFT of 0.11 mN/m. Mono-salts of SMEs reach a constant IFT value in less than two minutes, whereas LAS, PAS, and the di-salt of α-sulfo stripped coco fatty acid reach a stable value of IFT in about 10–15 minutes.

## Environmental Efficiency

Several studies have shown SMEs to biodegrade under aerobic conditions (10–12). SMEs are found to possess poor biodegradability under anaerobic conditions, however they are very unlikely to migrate in ecosystems toward anaerobic environments (13,14). Maurer et al (15) subjected the methyl esters of α-sulfonated fatty acids in the C9 to C18 range to aerobic biodegradation in the presence of sewage micro-organisms. They report that the biodegradability of SMEs increases with increasing chain length. Steber et al (14) demonstrated that both the mono-salt of SME as well as the di-salt of SFA are readily biodegradable. SMEs have also been shown to be resistant against microbial attack (16).

## Toxicological Efficiency

A large volume of literature is available on the toxicological properties of SMEs (17–19). These studies show that these surfactants have very low toxicity towards aquatic organisms (17), exhibit less eye irritation levels at corresponding concentrations to LAS and PAS (2), and their skin irritation index is also very low (2,18,19).

## Conclusions

From these studies it can be concluded that SMEs are very efficient surfactants and provide excellent adsorption and micellar characteristics. They perform well in hard water and at low concentrations. They also improve the micellar and detergency behavior of conventional LAS and PAS based surfactants even in cold water. Although more work needs to be done from a physicochemical point of view, their highly desirable surfactant properties definitely puts them in the category of primary surfactants.

## The Future

The soaps and detergents markets have evolved to a relatively mature state. However, the rate of change of product technologies as well as society's increased concern regarding the maintenance of our environment continues to make the market very dynamic. Tomorrow's standards of efficiency are obviously not completely understood, for it is impossible to have a perfect vision of the future. Hopefully, we are getting better at predicting and controlling the technical shape of the future. The only certainty is that our rapidly changing world will cause efficiency standards to change and subsequently require products and their ingredients to adapt accordingly. It is our sincerest hope that the information provided herein will create a greater understanding among the scientific and business communities towards the development of more efficient surfactant systems.

## References

1. Schirber, C.A., *INFORM* 2:1062 (1991).
2. Knaggs, E.A., J.A. Yeager, L. Varenyi, and E. Fischer, *J. Am. Oil Chem. Soc.* 42:805 (1965).
3. Stirton, A.J., R.G. Bistline, Jr., J.K. Weil, W.C. Ault, and E.W. Maurer, *J. Am. Oil Chem. Soc.* 39:128 (1962).
4. Okano, T., J. Tanabe, M. Fukuda, and M. Tanaka, *J. Am. Oil Chem. Soc.* 69:44 (1992).
5. Lower, E.S., *La Rivista Delle Sostanze Grasse* 63:271 (1986).
6. Schambil, F., and M.J. Schwuger, *Tenside Surf. Det.* 27:380 (1990).
7. Smith, N.R., *Soaps/Cosmetics/Chemical Specialties.* April, p. 48 (1989).
8. Satsuki, T., *INFORM* 3:1099 (1992).
9. Vogel, J. The CESIO conference, London, England (1992).
10. Cordon, T.C., E.W. Maurer, and A.J. Stirton, *J. Am. Oil Chem. Soc.* 47:203 (1970).
11. Maurer, E.W., T.C. Cordon, and A.J. Stirton, *J. Am. Oil Chem. Soc.* 48:163 (1971).
12. Cordon, T.C., E.W. Maurer, and A.J. Stirton, *J. Am. Oil Chem. Soc.* 49:374 (1972).
13. Steber, J., and P. Wierich, *Water Res.* 21:661 (1987).
14. Steber, J., and P. Wierich, *Tenside Surf. Det.* 26:18 (1989).
15. Maurer, E.W., J.K. Weil, and W.M. Winfield, *J. Am. Oil Chem. Soc.* 54:582 (1977).
16. Steber, J., P. Gode, and W. Guhl, *Fat Sci. Tech.* 90:72 (1988).
17. Gode, P., W. Kuhl, and J. Steber, *Fat Sci. Tech.* 89:548 (1974).
18. Stein, W., and H. Baumann, *J. Am. Oil Chem. Soc.* 52:323 (1975).
19. Maurer, E.W., T.C. Cordon, J.K. Weil, and W.M. Linfield, *J. Am. Oil Chem. Soc.* 51:287 (1974).

# A New Generation of Imidazoline-Derived Amphoteric Surfactants[a]

R. Vukov, D. Tracy, M. Dahanayake, P.J. Derian, J.M. Ricca, and F. Marcenac

Rhône-Poulenc Inc., CN 7500, Prospect Plains Road, Cranbury, New Jersey 08512, USA.

## Abstract

Lauric oils have found extensive use as surface-active agents, both as raw materials or as a source of derivatives which may further be used as raw materials. Both oils and their derivatives are widely used for the manufacture of commodity and specialty surfactants. This paper will discuss some specialty surfactants based on lauric oil derivatives, namely amphoteric surfactants. It will address the chemistry of these products, their fundamental surfactant properties, main areas of application, performance in formulated products, some other properties, and recent advances in new product development.

---

Amphoteric surfactants based on carboxymethylation of fatty imidazolines or fatty amidoamines generated by their hydrolysis are well established as extremely mild surfactants (1). They are referenced in the CTFA dictionary as sodium coco (or lauryl) amphoacetate and disodium coco (or lauryl) amphodiacetate. Amphoterics are widely used in mild, tear-free shampoos and sensitive skin cleansers due to their favorable surfactant properties, low irritation profile, and irritation mollifying properties (2). Amphoacetates or diacetates are excellent foaming agents, even in hard water, and they exhibit compatibility with all other types of surfactants.

Although most of the properties and applications of these classes of surfactants are reported widely in the literature, the actual chemistry and composition of commercial products is not well understood. This paper will deal with the most up-to-date knowledge of the chemistry and application properties of imidazoline-derived amphoteric surfactants. The authors' efforts to better understand the chemistry of these products and develop methods for the analysis of amphoacetates and diacetates led to a full understanding of the chemical reactions which occur during the different stages of their manufacture. As a result of this research, new patent-pending technology was developed enabling the introduction of a new generation of high purity amphoterics.

## Historical Background

Synthesis of imidazolinium amphoteric surfactants was reported in a patent awarded to Mannheimer in 1950 (3), in which 1-(2-hydroxyethyl)-2-alkyl-2-imidazoline (hereafter called imidazoline) was reacted with sodium monochloroacetate (SMCA). The first company, the Miranol Company, engaged in large-scale production of these products in 1947, and it was around that time that Miranol's products came to be recognized for their mildness and ability to reduce the irritation of anionic surfactants.

The market launch of the Johnson & Johnson *No More Tears* shampoo containing Miranol's amphoteric surfactant in the early 1960s proved to be the beginning of worldwide use of Miranol® products. Their unique properties and product quality led to worldwide recognition of the Miranol® name, which became synonymous with imidazoline-derived amphoteric surfactants. Over the years through extensive use and testing, a number of desirable properties of these products were recognized. These are low toxicity; biodegradability; good surface activity; hydrotroping ability; compatibility with other types of surfactants; excellent foaming and foam stabilization, particularly in the presence of oil and sebum; free rinsing; hard water compatibility; low irritation; and the ability to reduce the irritation from more aggressively formulated systems.

The Miranol® patents expired in 1973, opening the door for widespread competitive activity in this area. The increasing demands of consumer products for complete identification of chemical structures and impurity profiles, together with the need for products of higher purity and performance led the authors to do research for a comprehensive study of imidazoline-derived amphoteric chemistry.

## Analysis of Imidazoline-Derived Amphoteric Surfactants

Prior to initiating this work, the literature was reviewed for analytical methods suitable to analyze imidazoline-derived amphoteric surfactants. Although numerous wet and chromatographic methods exist, none of these are suitable for direct analysis of the active components. A recent review article (4) provided an overview of published methodology. A number of analytical methods had to be adapted and several new methods had to be developed for this research.

A representative number of commercial samples available in Europe, North America, and Japan was collected and subjected to a battery of analytical tests. These samples included products designated as both amphoacetates and amphodiacetates. The summary of major analytical findings for the two product categories is presented in Table 1.

Analyzing the active surfactant species' structures in these products posed a substantial problem. It was solved,

---

[a]This paper was presented by J. Niu of Rhône-Poulenc Inc.

## TABLE 1
### Representative Samples of Commercial Cocoamphoacetates and Cocoamphodiacetates

|  | Cocoampho-acetate | Cocoampho-diacetate |
|---|---|---|
| Solids % | 45 | 50 |
| Sodium chloride % | 7.5 | 13 |
| Coco fatty acid % | < 0.7 | < 0.7 |
| Glycolates % | < 4 | < 8 |
| Mono/Di ratio | > 9 | > 9 |
| SMCA (ppm) | Variable | 100–2000 |
| Color (Gardner) | < 5 | < 5 |
| Viscosity CPS | Variable | 2–100K |
| Actives as % solids | Variable | 62% |

however, by using nuclear magnetic resonance (NMR) analyses, including proton and $^{13}C$ spectroscopy and two-dimensional experiments. In cocoamphoacetates, for most products the only carboxymethylated structure observable by NMR was structure h, resulting from the carboxymethylation of linear amidoamine b2 (Fig. 1).

Surprisingly, in-depth NMR analyses of representative cocoamphodiacetates identified the same carboxymethylated linear amide as the predominant surfactant species. Thus, both conventional cocoamphoacetates and cocoamphodiacetates have the same active species as a major product. It consists of a hydrophobically modified amino acid and contains a fully functional amino acid element which may account for the mildness of these products and their ability to reduce irritation from more aggressive systems.

Nuclear magnetic resonance studies indicated the presence of dicarboxylated species in variable but small percentages. This compound, after concentration by solvent extraction and chromatographic separation, was identified by NMR spectroscopy and mass spectrometry as g in Fig. 1. It was the result of carboxymethylation of the branched tertiary amide b1, the kinetically controlled product of the hydrolysis of imidazoline. The diacetate typically represents less than 10% of the active species.

A brief study was conducted to test whether the enrichment of the product with cocoamphodiacetate would lead to beneficial surfactant properties. In distilled water the increase in amphodiacetate content did not change foaming behavior. On the other hand, in hard water (300 ppm calcium) the increase in amphodiacetate content closely paralleled the reduction of foam levels (Fig. 2). Thus, the presence of higher levels of cocoamphodiacetate species had an adverse effect on the performance of the amphoteric surfactant.

As seen from Table 1, both types of products contain some coco fatty acid with higher quality products typically containing less than 0.6%. Both types of products also contained some glycolic acid derivatives with products designated as diacetates containing up to 8% of the sodium salt. This indicates that approximately 50% of the SMCA charge may be undergoing hydrolysis in the reaction medium. Cocoamphoacetates were found to contain 1–4% sodium glycolate. Based on the known SMCA charge, it became apparent that some of these products may contain substantial amounts of unreacted amidoamines. Indeed, some commercial samples were analyzed and found to contain as much as 50 mol% unreacted amidoamines.

The sodium chloride content can be determined easily by using wet chemical methods, potentiometry, or by ion chromatography (IC). Ion chromatographic methods were modified to allow for direct measurement of fatty acids, residual SMCA, and sodium dichloroacetate (SDCA) content. Sodium dichloroacetate is an impurity of SMCA that reacts slowly with amidoamines and undergoes some hydrolysis during carboxymethylation. The limit of detection for these impurities is less than 10 ppm.

The level of SMCA was found to vary substantially in commercial products, with some samples containing in

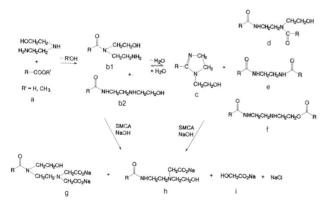

Fig. 1. Major reaction pathways in the formation of imidazoline and its carboxymethylation.

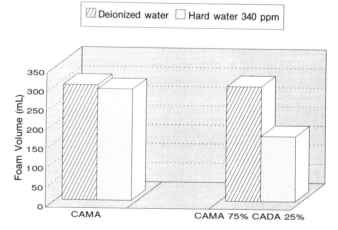

Fig. 2. Impact of addition of cocoamphodiacetate (CADA) (Fig. 1g) on the foam volume of cocoamphoacetate (CAMA) in deionized and hard water (modified Ross-Miles test NF 73-404).

excess of 1000 ppm. Higher quality products typically contained less than 100 ppm.

Traditionally the active content of imidazoline-derived amphoteric surfactants was determined by the expression:

%Active = %Solid − %NaCl

Therefore, traditional amphodiacetates having a 50% solid content and a 12% sodium chloride content were assumed to be approximately 38% active. The expression assumes that other major by-products, such as sodium glycolate, are also active components. In fact, these by-products are not surface active and do not contribute to the surfactant properties of amphoterics. Based on this consideration, a more appropriate expression for the activity is:

%Active = %Solid − %NaCl

− %Sodium Glycolate/Diglycolate

Thus, commercial amphodiacetates having a 50% solid, 12% sodium chloride, and approximately 7% glycolate content have actual activities of 31–32%.

The expression for the determination of active components was described previously, but it does not appear to have met with wide acceptance. At the present time, it is the most appropriate means of obtaining the surfactant content in the absence of rapid and accurate methods for the direct determination of the active surfactant content.

## Chemistry of Imidazoline-Derived Amphoteric Surfactants

The initial literature reports for these products did not recognize that the imidazoline ring opens during the carboxymethylation step (3). These incorrect imidazoline-containing structures are still found in many catalogs and reports. Recently, some papers (5) have identified the actual structures of these products through analytical work based on degradation of commercial products followed by derivatization and various chromatographic techniques.

The synthesis of amphoacetates and amphodiacetates as described in the early Mannheimer patents consisted of two distinct steps, synthesis of the hydroxyethyl-imidazoline and carboxymethylation of the imidazoline with SMCA. In the first step, the fatty acid or the corresponding ester was condensed with aminoethylethanolamine (AEEA) at elevated temperature and reduced pressure (Fig. 1a–c). The reaction proceeded stepwise through the amide stage, Fig. 1 b1 and b2, followed by ring closure. The main component detected after this reaction was 1-(2-hydroxyethyl)-2-alkyl-2-imidazoline, Fig. 1c, with traces of noncyclic amidoamines.

Some by-products may be formed (Fig. 1d–f), depending on the reaction conditions. The diamides exhibiting high melting points and low water solubility (Fig. 1d and e) need to be minimized to get a final product of high clarity and good storage stability. The symmetrical diamide, Fig. 1e, is derived from ethylenediamine which may be present in AEEA and can be eliminated by using high quality raw materials. On the other hand, the asymmetrical amide, Fig. 1d, is formed by the reaction of the primary raw materials and may be minimized by using suitable reaction conditions and maintaining process control. Formation of the amidoesters, Fig. 1f, is only observed during the intermediate stages of synthesis and are not typically present in the finished imidazoline.

During the second step of amphoacetate or amphodiacetate preparation, the imidazoline in Fig. 1c is reacted with SMCA in an aqueous medium. This reaction is carried out under alkaline conditions at moderate temperatures. According to the literature, the cocoamphoacetate is produced by reaction of equimolar quantities of SMCA and imidazoline. Cocoamphodiacetate also can be produced by using the reactants in a 2:1 ratio. Before discussing these reactions, it is worth noting that in water and under alkaline conditions imidazoline undergoes hydrolysis and generates a mixture of amidoamines, Fig. 1 (b1 and b2). Although the formation of two different amidoamines is possible, it was established that the linear amidoamine b2 in Fig. 1 is formed under highly alkaline conditions (pH > 10). The tertiary amidoamine b2 in Fig. 1 may be formed in variable amounts, depending on reaction conditions, but it is typically a minor component. The hydrolysis of the imidazoline ring, to produce high yields of the linear amidoamine b2 in Fig. 1 was observed in the authors' work by NMR analysis. A recent paper (6) supports this conclusion.

It is also important to note that under typical carboxymethylation conditions, SMCA may hydrolyze to form sodium glycolate (see Fig. 1i) which may react further with SMCA to form the corresponding diglycolate. During the carboxymethylation step at least three main reactions occur concurrently: the opening of the imidazoline ring leading predominantly to the linear amidoamine in Fig. 1b; the carboxymethylation of the linear amidoamine by reaction with SMCA which predominantly produces the compound in Fig. 1h; the hydrolysis of SMCA to form sodium glycolate (Fig. 1i); and a possible further reaction of glycolate with this reagent to produce diglycolate, which is typically a minor component. At present, there is no evidence for the direct reaction of imidazoline with SMCA followed by ring opening to produce the final product.

When the carboxymethylation reaction was studied in some detail, it was established that there was no noticeable change in the amount of the monocarboxymethylated species even when the SMCA to amidoamine ratio was substantially increased (from 3:1 to 5:1). Thus, it appears that the reactivity of the monocarboxylated linear amidoamine (Fig. 1h) is diminished due to the electronic effect and increased steric hindrance around the reaction site (6), preventing the quaternization reaction from taking place. These results indicate that the dicarboxylated species can be derived only from a structure such as b1 in Fig. 1 and cannot be formed by further carboxymethylation of the monocarboxymethylated species in Fig. 1h under these reaction conditions.

By combining the information resulting from extensive analytical work with a consideration of various possible chemical reactions, a clear picture relative to cocoamphoacetates and cocoamphodiacetates emerges. Most of the cocoamphoacetates derived from near equimolar quantities of imidazoline and SMCA contain unreacted amidoamine (Fig. 1b2) in addition to the active species (Fig. 1h) and by-products. This is due to concurrent hydrolysis of SMCA during the course of the carboxymethylation reaction.

Cocoamphodiacetates, made by using a larger excess of SMCA, typically contained cocoamphoacetate as the main surfactant component with a diminished amount of the amidoamine in Fig. 1b2 and an increased content of glycolate. Thus, the increased amount of SMCA is used predominantly for driving the carboxymethylation reaction to completion with little effect on the structure of the major surfactant species.

## New Generation of Amphoterics

Based on the analytical results and a full understanding of the underlying chemistry, it became apparent that substantial improvements in the manufacturing technology of imidazoline-derived amphoterics were possible. Therefore, the authors initiated a project with two goals: to develop technology for the production of high-purity monoacetate amphoteric surfactants with consistently low levels of by-products, and to develop a process to ensure that identical products are shipped from all of the company's amphoteric-production facilities. These goals have been achieved and have led to the development of patent-pending technology and commercialization of a new generation of imidazoline-derived surfactants.

The analytical profile of these products is in Table 2. The active species content the new product is the same as

**TABLE 2**
Comparison of Analyses of New Generation with Traditional Amphodiacetates

| | Traditional Amphodiacetate | Miranol® Ultra C-32 | Mirapon® Excel 825 |
|---|---|---|---|
| % Solids | 50 | 40 | 31 |
| % Active | 31 | 31 | 25 |
| Active as | | | |
| % solids | 62 | 82 | 82 |
| % NaCl | 13 | 7 | 6 |
| % Glycolate | < 8 | < 1.2 | < 1 |
| SMCA (ppm) | Wide Variations | < 20 | < 20 |
| Color (Gardner) | 4 max. | 2 max. | 2 max. |
| Viscosity | 2–100K cps | < 500 cps | < 500 cps |
| pH | 9 | 8.5 | 8.5 |

in the previous generation of cocoamphodiacetates, but all other characteristics have been improved. Not only is there a lower salt content, but glycolates have been significantly reduced. While traditional amphodiacetates may contain variable amounts of SMCA and SDCA, the new product contains consistently low levels of these materials (< 20 ppm). The enhanced purity of the sodium cocoamphoacetate (CAMA) may be demonstrated by comparing the active content as a percentage of total solids. In the new amphoteric actives are 82% of the total solids, while in traditional amphoterics actives represent only 62% of the solids content.

The viscosity of the new product is consistently lower than 500 cps, resulting in a product that is easier to handle

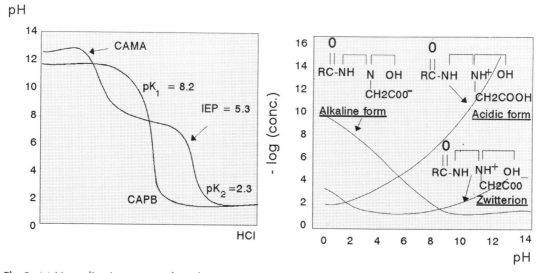

**Fig. 3.** (a) Neutralization curves of amphoterics (CAMA) vs. betaines (CAPB). Both surfactants have been adjusted to alkaline conditions with NaOH prior to neutralization with HCl. (b) Predominant forms of cocoamphomonocarboxylate vs. pH. Note that the zwitterion is the major species between pH 2 and pH 8.

and incorporate into formulations than the traditional products having viscosities that vary from 2000–100,000 cps. In separate experiments, it was established that the viscosity of these amphoterics is not an inherent property of the active species but is due to interaction between the active species and by-products. Therefore it is not surprising that the new generation product, having substantially higher purity, exhibits lower viscosity. The product, nevertheless, responds to increased ionic strength and exhibits a characteristic salt curve.

The new generation product also has a lower color (Gardner 2 Max.) than traditional amphodiacetates (Gardner 5 Max.). This improvement in color has been achieved through process modification without additional treatment. Challenge testings using Gram (+) and Gram (−) bacteria, yeasts, and molds have shown that the new product does not require added preservatives.

## Physico-Chemical and Application Properties

*Isoelectric Point.* True amphoteric surfactants are characterized by their ability to vary their net charge, according to pH conditions. Compared to betaines, which have a permanent positive charge on the quaternized nitrogen atom (Fig. 3), true amphoterics have both a carboxyl group and an amine group which can be protonated. These amphoterics derived from imidazoline can exist in anionic (alkaline conditions), cationic (acidic conditions), or zwitterionic form (around the isoelectric point at pH 5.3). Thus, the molecules can easily adapt their net charge to the skin and eye conditions and could explain their extreme mildness in sensitive areas.

*Surfactancy.* Amphoteric surfactants have the ability to reduce the surface tension of water, as do all other surfactants. The effective critical micelle concentration is 0.58 g/L for the new CAMA. The reduction of surface tension depends on the pH of the solution. At pH 6, close to the isoelectric point, the net charge of the molecule is zero, which allows for closer packing of surfactant molecules at the interface and produces a greater reduction of surface tension. This is clearly illustrated by the actual figures: at pH 10, the surface tension of a 0.1% active solution is 33.1 mN/m; whereas at pH 6 the surface tension is reduced to 27 mN/m.

The new CAMA exhibits decreased wetting time when compared to traditional amphoterics. This is clearly shown by the Draves wetting test (Fig. 4) or by the piece of cotton sinking test (NF T73–406); the wetting speed of the new product is twice the wetting speed of traditional amphoterics.

*Foaming.* Amphoterics are excellent foamers and foam stabilizers under a variety of conditions. Foamability is essentially insensitive to water hardness. The new products exhibit improved foaming in the presence of sebum compared to traditional diacetates (Fig. 5). The quality of the foam generated by amphoterics and especially by the new CAMA is creamy, smooth, and abundant. This can be demonstrated by either consumer-perception studies or physical evaluation of the generated foam. To illustrate this, the authors generated foam in a 1% active surfactant solution (distilled water, pH = 6.5) by exposing the solu-

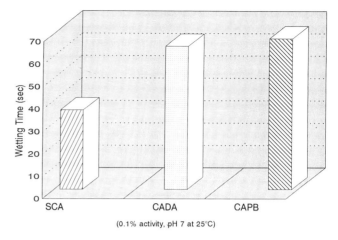

**Fig. 4.** Wetting speed comparison of a new CAMA with traditional cocoamphodiacetate (CADA) and cocoamidopropylbetaine (CAPB), as measured by Draves test.

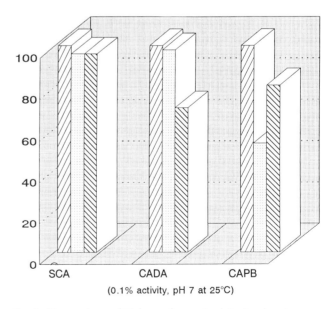

**Fig. 5.** Comparison of % foam change in deionized water, hard water, and in the presence of sebum (Ross-Miles test), for Miranol® Ultra C-32 (CAMA), traditional cocoamphodiacetate (CADA), and cocoamidopropylbetaine (CAPB).

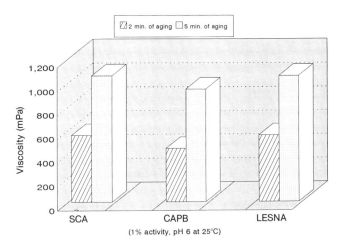

**Fig. 6.** Viscosities of foams generated by agitating for 30 sec, in 1% active solutions (Brookfield RVT, spindle 3, 10 rpm) of Miranol® Ultra C-32 (CAMA), cocoamidopropylbetaine (CAPB), and sodium lauryl ether sulfate (SLES).

tion to agitation for 30 sec in a high-shear mixer. This method reproducibly generated a homogeneous foam which mimicked the quality and appearance of the foam generated by shampooing hair. The viscosity of the foam, related to its firmness and creaminess, was measured with a Brookfield RVT viscometer. The new CAMA was compared to sodium lauryl ether sulfate (SLES), having 2 moles of ethylene oxide (EO), and cocoamidopropylbetaine (CAPB). The results are shown in Fig. 6. The viscosity of the foam from the cocoamphoacetate was 22% higher than the viscosity of the foam from CAPB and comparable to the foam of an SLES known for its good foaming properties.

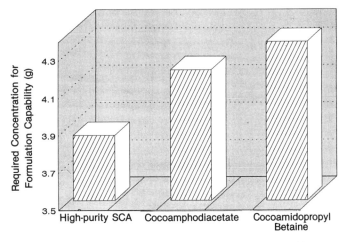

**Fig. 7.** Comparison of hydrotroping ability as measured by the quantity of substance required to homogenize a given solution of 10% NaOH and nonylphenol ethoxylate (9 mole, 5%) for Mirapon® Excel 825 (CAMA), cocoamphodiacetate (CADA), and cocoamidopropylbetaine (CAPB).

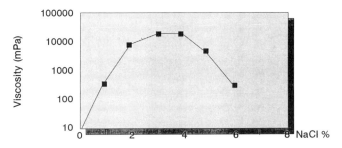

**Fig. 8.** Viscosity response of a mixture of 35% SLES and 8% of a high purity CAMA on the addition of NaCl.

*Hydrotroping Ability.* Although amphoterics are known as excellent hydrotropes, the improved purity of new amphoterics significantly enhances this property. As seen in Fig. 7, the new generation amphoteric significantly outperformed both traditional amphoterics and the representative betaine. This very desirable feature allows for formulation of higher solids products of excellent stability and clarity.

*Association with Anionics.* Amphoterics are compatible with all other types of surfactants. They are commonly used with anionics in order to reduce the irritation of these more aggressive products. The salt curve of a typical simplified shampoo formulation is shown in Fig. 8. Other types of viscosity builders can also be used to modify formulation rheology. These include surfactants (low HLB fatty alcohol ethoxylates, PEG1000 distearate, etc.) or polymers (guar derivatives; cellulose derivatives; and polyacrylates, such as Carbopol or xanthan gum).

### Safety, Irritation, and Environmental Impact

New generation products, such as CAMA, biodegrade easily and completely (Organization for Economic Cooperation and Development [OECD] 302 B testing method). Their relatively low ecotoxicity (25 mg/L for the ID 50 for Daphnae) combined with the biodegradation data indicates that no warning concerning their environmental impact is needed for formulations containing these products.

Additional animal testing is not needed for the new generation amphoterics since they have improvements in purity and quality over traditional amphoterics for which extensive irritation data already exist (1). Nevertheless, it was of interest to assess the mildness of the new products using alternative in vitro testing methodologies. Three different alternative methods were utilized to assess irritation: the red blood cell test (RBC [7]), isolated chicken eye test (ICET), and Eyetex protocol. The RBC test is based on hemolysis of erythrocytes and denaturation of hemoglobin. The correlation between the results of this test and the actual Draize rabbit eye test is described in the literature. The ICET test is based on the swelling of the cornea of viable isolated chicken eyes. In addition to swelling,

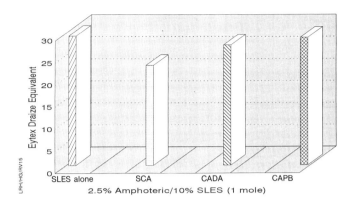

Fig. 9. Comparison of irritation from mixtures of SLES and selected additives by Eyetex Draize equivalent method for Miranol® Ultra C-32 (CAMA), traditional cocoamphodiacetate (CADA), and cocoamidopropylbetaine (CAPB).

changes in the opacity of the cornea induced by products are observed. The Eyetex test is based on the denaturation of the protein matrix and may be related to eye irritation. Although none of these tests could provide us with a definitive answer regarding product's mildness it was considered that this battery of tests will provide a meaningful comparison with established mild systems. The testing was done with 10% active solutions at pH 6, mimicking both surfactant active levels and pH values commonly found in personal care formulations.

When the new CAMA was compared in the three tests with an older CADA known to be a mild surfactant essentially identical results were obtained. They indicated that the new product is at least as mild as the industry standard, with a number of individual data points indicating that the new product is milder. It is difficult to make distinctions between very mild surfactants at the present stage of development of alternative testing technology. The results obtained from the ICET test are listed in Table 3. The lower scores for both swelling and opacity indicate that new CAMA exhibits enhanced mildness when compared with traditional amphodiacetates. Based on the Eyetex test in Fig. 9, in association with anionics, the new CAMA shows a greater capacity to reduce sulfate irritation than cocoamphodiacetate, which is milder than the representative betaine.

## Formulations for Personal Care

Imidazoline-derived amphoterics are widely used in personal care applications due to their multifunctionality and mildness-enhancing properties. The new generation of amphoteric products offers formulators additional benefits over traditional amphoterics. They have enhanced surfactant properties, low color, and consistent low viscosity. Due to a substantially improved purity profile, the new amphoterics have improved hydrotroping properties making them suitable to formulate highly concentrated products with excellent shelf stability and clarity. Most importantly, their foam has greater density and viscosity and was judged superior to those from traditional amphodiacetates and betaines.

Four formulations specifically designed for performance and mildness are shown in the *Formulations* section. Formulation experience gathered so far with the new CAMA indicates that this product has enhanced properties over traditional amphodiacetates and is a product of choice for improving existing personal care products and for the creation of new and superior personal care products.

## Suggested Formulations

### Clear Conditioning Shower Gel

| Ingredients | CTFA | wt % |
|---|---|---|
| Sodium laureth-2 sulfate | Sodium laureth-2 sulfate | 53.40 |
| New CAMA | Sodium cocoampho-acetate | 14.80 |
| Geropon® TC42 | Sodium methyl cocoyl taurate | 7.80 |
| Mirapol® A15 | Polyquaternium 2 | 0.80 |
| Disodium EDTA | Disodium EDTA | 0.03 |
| Hexylene glycol | Hexylene glycol | 1.00 |
| Citric acid, preservative, fragrance | | qs |
| Water | | qs100 |
| pH | 6.5 | |
| Viscosity at 25°C | 500 cps (Brookfield spindle No. 4; 20 rpm) | |

### Clear 3 in 1 Shampoo

| Ingredients | CTFA | wt % |
|---|---|---|
| Jaguar® C-162 | Hydroxypropyl guar Hydroxypropyltrimonium chloride | 0.5 |
| Sodium laureth-3 sulfate | Sodium laureth-3 sulfate | 20.0 |
| New CAMA | Sodium cocoampho-acetate | 10.0 |
| Cocamide MIPA | Cocamide MIPA | 1.0 |
| Octopirox | Piroctone olamide | 0.5 |
| Deionized water | | to 100 |
| Citric acid, preservative, fragrance | | qs |
| pH | 6 | |
| Viscosity at 25°C | 3500–4000 cps (Brookfield spindle No. 4; 10 rpm) | |

**TABLE 3**
**Results of ICET Test**

| | Swelling | Opacity |
|---|---|---|
| Cocoamphodiacetate | 16.7 | 1.33 |
| CAMA | 13.8 | 0.67 |

**Antibacterial Liquid Soap**

| Ingredients | CTFA | wt % |
|---|---|---|
| Sodium laureth-2 sulfate | Sodium laureth-2 sulfate | 25.6 |
| Sodium laureth-3 sulfate | Sodium laureth-3 sulfate | 24.1 |
| New CAMA | Sodium cocoamphoacetate | 6.4 |
| Chlorhexidine® DI | Chlorhexidine digluconate | 0.7 |
| Disodium EDTA | Disodium EDTA | 0.5 |
| Sodium chloride | Sodium chloride | qs |
| Deionized water | | qsp 100 |
| Citric acid, preservative, fragrance | | qs |
| pH | 6.5 | |
| Viscosity$^a$ | 2100 cps (Brookfield No. 4; 20 rpm) | |

$^a$Adjusted with NaCl.

**Make-Up Remover**

| Ingredients | CTFA | wt % |
|---|---|---|
| Phase A | | |
| Hetester PHA | Propylene glycol and Isoceteth-3 acetate | 15.0 |
| Isopropyl adipate | | 5.0 |
| Silbione® 70045 V5 | Cyclomethicone | 5.0 |
| Cremeol PFO | Passion flower oil | 5.0 |
| Phase B | | |
| Deionized water | | 72.1 |
| Rhodicare® S | Xanthan gum | 0.5 |
| Phase C | | |
| New CAMA | Cocoamphoacetate | 2.0 |
| Fragrance, preservative, citric acid | | qs |
| pH | 5.5 | |
| Viscosity$^a$ | 2100 cps (Brookfield spindle No. 4; 20 rpm) | |

$^a$Adjusted with NaCl.

## Conclusion

In-depth analyses of commercial imidazoline-derived amphoterics, cocoamphoacetates, and cocoamphodiacetates has led to the conclusion that both types of products contain identical surface active species—cocoamphoacetate. The major difference between the two groups of products is their by-product content which causes performance differences. Based on this understanding, the authors' company has introduced a new generation of imidazoline-derived amphoteric surfactants. This represents a breakthrough in amphoteric surfactant technology not only in terms of a highly defined product of higher purity and batch-to-batch consistency but also in terms of the benefits and flexibility it offers to formulators: low color, increased purity, improved surfactant properties, consistent low viscosity and viscosity-building behavior, and enhanced mildness.

Consistent low viscosity of the new generation products makes them suitable for shipment and storage in tank truck quantities. Due to the lower salt content, these products exhibit excellent storage stability. They are also easier to pump in manufacturing facilities.

The enhanced hydrotroping ability makes these products suitable for formulation of highly concentrated products having excellent shelf stability and clarity. Their incorporation into the formulated products may also lead to the formation of foam of exceptional density and viscosity. Their viscosity-building profile ensures greater flexibility during the manufacture of formulated products, preventing excursions into high viscosity situations which can occur with traditional amphoacetates, amphodiacetates, and betaines. The combination of properties and multifunctionality makes the new generation amphoterics a cost-effective building block for formulated products and performance concentrates with applications in a variety of end-use products.

## Acknowledgments

The work described in this paper is the result of a team effort involving numerous people from the authors' company facilities throughout the world. The authors would like to thank their many colleagues for sharing their results and for useful discussions. Special gratitude is due to: J. Carr and P. Lees, Leeds, United Kingdom; C. Willemin and J. Gibert, Aubervilliers, France; B. Hendricks, Lyon, France; D. Esdaile, Sophia, France; T. Schamper, M. Corti, J. Li, B. Yang, S. Zhu, H. Mauermann, M. Finney, Cranbury, United States; J. Niu, Singapore; J.P. DalPont, J.L. Schuppiser, Paris, France. The authors also wish to thank Rhône-Poulenc SA for supporting this work and for allowing its publication.

## References

1. Final Report on the Safety Assessments of Cocoamphoacetate, Cocoamphopropionate, Cocoamphodiacetate, and Cocoamphodipropionate, *J. Am. Coll. Toxicol. 9:*2 (1990).
2. Hunting, A.L.L., *Cosmetics and Toiletries 100:*49 (1985).
3. Mannheimer, H.S., and J.J. McCabe, U.S. Patent 2,528,378 (1950).
4. Lomax, E., in *Analysis of Amphoteric Surfactants in Recent Developments in the Analysis of Surfactants,* edited by M.R. Porter, Elsevier Applied Science, London, 1991.
5. Takano, S., and R.J. Tsuji, *J. Am. Oil Chem. Soc. 60:*1807 (1983).
6. Zougsh, L., and Z. Zhuangyu, *Tenside Surf. Det. 31:*128 (1994).
7. Lewis, R.W., J.C. McCall, and P.A. Botham, *Toxic in Vitro 7:*155 (1993).

# Nondetergent Applications of $C_8$–$C_{14}$ Amines and Derivatives

**H.F.G. Patient**

Lion Akzo Co. Ltd, 2-22,1-Chome, Yokoami, Sumida-ku, Tokyo 130, Japan.

## Abstract

The production, mode of operation, and application of short-chain ($C_8$–$C_{14}$) fatty amine derivatives are discussed, with special reference to nondetergent applications. Such products are produced by approximately 12 companies worldwide, with most companies sourcing from one country. Principal raw materials used are fatty acids, but alcohols and fatty acid methyl esters also are used. Basic "building block" products are the primary, secondary, and tertiary amines plus diamines. A brief description of production and possible product combinations is given.

Almost every industry uses $C_8$–$C_{14}$ fatty amine derivatives as chemical intermediates, essential processing aids, and as functional ingredients in many formulations. Such industries can be listed as follows: agrochemicals, fibers/textiles, industrial cleaning, biocides, plastics, water treatment, mining, oilfield chemicals, corrosion inhibitors, demulsifiers, emulsifiers, and paints/pigments. These products derive their versatility from four main functional properties: surface activity, substantitivity, reactivity and antimicrobial properties. Discussion will be confined to the following: antistatic agents for plastics, biostats/biocides, industrial and institutional cleaning, mineral processing/extraction, agricultural, and textiles and fibers.

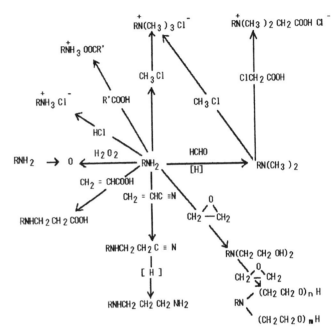

**Fig. 1.** Fatty amine derivatives.

Fatty amines and derivatives have been produced on a commercial scale for over 40 years, originally in the United States, but subsequently throughout the world. Today, there are 12 companies producing such products, with capacities ranging from 5,000 metric tons (t) to 100,000+ t per year, although most of these companies are only sourcing from one country. Over this 40-year period, market applications have expanded to cover practically every industry, from paints to petrochemicals and asphalt to agriculture. At the same time, the original raw material, beef tallow fatty acids, have been augmented by fatty acids derived from vegetable sources, such as coconut, palm, palm kernel, and other oils. In addition, alcohols and fatty acid methyl esters are also used as raw materials in certain instances.

The basic building blocks produced from these raw materials are alkyl nitriles; primary, secondary, and tertiary alkyl amines; and *n*-alkyl propylene diamines. From these "building blocks," via combination of alkoxylation, quaternization, neutralization, and other chemical processes, arise the family of fatty amine derivatives (Fig. 1), of which the lauric oil derivatives are a part. Perhaps the most important production step is the formation of alkyl nitrile by catalytically reacting fatty acids with ammonia at elevated temperatures (280–360°C) in either continuous or batch processes. Catalytic hydrogenation of the nitrile can then yield primary, secondary, or tertiary alkyl amines. Catalytic reductive alkylation of the primary amine with formaldehyde yields the corresponding methyl dialkyl tertiary amine or the dimethyl alkyl tertiary amine, depending upon the reaction conditions. The permutations of amine plus alkoxylation, quaternization, oxidation (to amine oxides), cyanoethylation, neutralization, and other chemical procedures are numerous, giving rise to the possibility of a multitude of tailor-made products. This paper concentrates on the primary use, industrial applications of $C_8$–$C_{14}$ fatty amine derivatives available on the world market.

As previously mentioned, $C_8$–$C_{14}$ fatty amine derivatives are used in almost every industry as chemical intermediates, essential processing aids, and as functional ingredients in many formulations. Such industries and applications can be listed as follows: agriculture, biocides, corrosion inhibition, de-emulsifiers, emulsifiers, fibers/textiles, industrial cleaning, mining, oilfield chemicals, paints/pigments, plastics, and water treatment. This is by no means an exhaustive list.

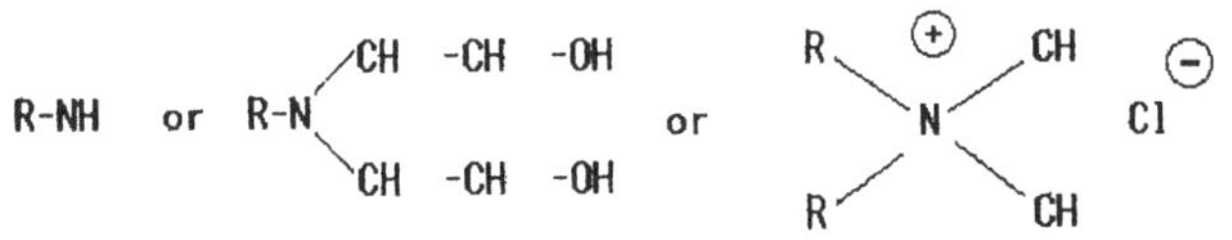

**Fig. 2.** Amino or quaternary ammonium groups with aliphatic alkyl groups from $C_8$–$C_{20}$.

The molecular structure of these fatty amine derivatives is characterized by one or more $C_8$–$C_{20}$ aliphatic alkyl groups, R, together with one or more amino or quaternary ammonium groups (Fig. 2). Due to the number of carbon atoms in the alkyl group, this group is strongly *hydrophobic*. On the other hand, the nitrogen atom is *hydrophilic*, particularly when it is protonated, alkoxylated, or quaternized. When dissolved or dispersed in water or nonaqueous solvents, one part of the molecule is strongly repelled by the surrounding solvent. Due to this repelling force, the molecules tend to orient themselves at surfaces and interfaces, or form micelles (aggregates of oriented molecules [Fig. 3]). Consequently, $C_8$–$C_{14}$ amine derivatives exhibit surface activity and derive their versatility from four main functional properties, i.e., surface activity, substantivity, reactivity, and antimicrobial effects.

Applications arising from these properties are surface activity, emulsification/de-emulsification, foaming, thickening, and wetting.

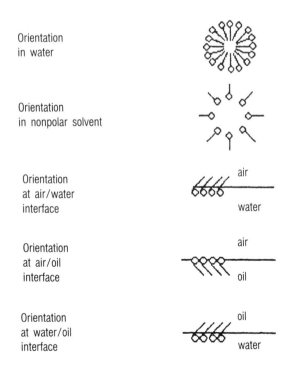

Fig. 3. Orientation of surfactant in several systems.

## Substantivity

This relates to the adsorptive properties, particular to solid surfaces, of the positively charged nitrogen atom to the negative charge characteristic of most surfaces. Resulting in the following functions: adhesion, hydrophobicity, softening, antistatic, lubrication, and corrosion inhibition.

## Reactivity

Reaction properties of cationic surfactants can be identified in several uses, especially when complexed with anionic species on a molecular level. Such complexes often exhibit low water solubility, giving rise to applications in water treatment, sugar refining, and organoclay production. The derived functions are flocculation, ion exchange, and decolorization.

## Antimicrobial Effect

The antimicrobial properties can be recognized by their effect on algae and other microorganisms. The functions may be described as: algicidal, bactericidal, and fungicidal. In addition to the molecular structure, product concentration in use will determine whether the microorganisms are killed or their growth merely inhibited.

## Antistatic Agents for Plastics

Static charges, arising during polymer processing, can give rise to spark formation and dust attraction. The former is clearly a safety concern, and the latter detracts from end-product appearance and can cause problems in later processing stages. Use of cationic antistatic agents (antistat) increases the dissipation of static electricity without seriously altering the mechanical or optical properties of the polymer. Such use may be via internal or external antistatic agents.

An inexpensive and effective method, although somewhat temporary, is to coat the finished articles with an external antistat, such as a quaternary ammonium compound. The articles may be sprayed, dipped, or wiped with a solution of 1–2% antistatic agent in water or other solvents. Care must be taken if the film application has food contact.

A more durable protection against charge accumulation is achieved by using an internal antistatic agent, such as a 2 M ethoxylate of a $C_8$–$C_{14}$ primary amine. This is incorporated into the polymer compound and migrates to the surface of the finished article. The antistatic effect is achieved through interaction with moisture in the surrounding atmosphere, forming a conductive layer that facilitates the dissipation of electrical charges. Because the action of the antistat is based on humidity, antistatic performance improves with increased levels of relative humidity. High purity ethoxylated amines are used widely in PE, PP, ABS and other styrenics at use levels of 0.1–0.15% (PE, PP) and 1.0–4.0% (PS, ABS, HIPS). The time needed to obtain a sufficient level of protection can vary from 0 to 72 hours after processing, depending on type and use level of antistat. If end-products are to be used in food-contact applications, the maximum approved level of addition may be limited due to local legislation.

## Biostats/Biocides

The use of $C_8$–$C_{14}$ amine derivatives for their microbiological performance has been long established (1) and, in fact, was the subject of an earlier paper (2). Perhaps the most common products in commercial use are the alkyl ($C_{12}$–$C_{14}$) dimethylbenzyl ammonium chlorides and the dialkyl ($C_8$–$C_{10}$) dimethyl ammonium chlorides, but also used, depending on application, are n-coco propylene diamine and its salts, tetradecylbenzyldimethylammonium chloride, alkyl (coco) primary amine salts, and other similar compounds. The biocidal activity of such products is highly structure dependent and efficacy against two classes of microorganisms, for example both bacteria and fungi, frequently requires a combination of biocides and biostats to ensure wide spectrum activity. This tends to be the requirement of the food, beverage, and animal husbandry industries. In other cases, control of a specific microorganism is required using the lowest possible concentration. Examples of this requirement are control of fungi in wood preservation, control of *Legionella pneumophila* in air conditioning, control of *Desulfovibrio desulfuricans* in oil recovery, and algae control in greenhouses and gardens.

A recent study (3) has examined the relation between structure and microbiological activity in order to permit better product selection. This study also included biodegradability data relating molecular structure to speed of biodegradation, concluding that "most of the fatty amine derivatives which are effective antimicrobial agents are readily biodegradable."

Table 1 illustrates the biocidal activity according to the DGHM protocol of an alkyl ($C_{12}$–$C_{14}$) benzyldimethyl ammonium chloride against a variety of microorganisms as a function of concentration of active ingredient versus time. The microbiological activity of $C_8$–$C_{14}$ amine derivatives is an extremely important function and is featured in many industrial applications.

**TABLE 1**
**Killing Time in Minutes**

| Microorganisms | Concentration of Active Ingredient (ppm) | | | |
|---|---|---|---|---|
| | 50 | 100 | 200 | 400 |
| *Escherichia coli* | 5–10 | 2 | <2 | <2 |
| *Proteus mirabilis* | 5 | 2 | 2 | <2 |
| *Pseudomonas aeruginosa* | >20 | 20 | 5–10 | <2 |
| *Staphylococcus aureus* | 5–10 | 2–5 | <2 | <2 |
| *Candida albicans* | 15–20 | 5–10 | <2 | <2 |
| *Saccharomyces cerevisiae* | >20 | 2–5 | <2 | <2 |

## Industrial and Institutional Cleaning

This market is characterized by $C_{12}$–$C_{14}$ amine derivatives being used in many different applications for many different functions. The applications include chemical toilets; vehicle and industrial cleaners with functions including sanitizing, emulsification, and thickening. Typical markets are food and beverage companies, general production plants, hotels, hospitals, office buildings, water treatment plants, and metal treatment plants.

Use of tetradecylbenzyl-, cocobenzyl-, and didecyldimethylammonium chlorides is increasing in aircraft, mobile home, and construction-site toilets. The quaternary may be solid or liquid, with the latter being an effective alternative to formaldehyde. Typical quaternary concentrations range from 5–15%, with the balance being water and fragrance. Solid quaternaries have the advantage of ease of handling and can be applied as a powder or packed in a water-soluble sachet (e.g., polyvinyl acetate).

Vehicle-cleaning formulations often contain $C_{12}$–$C_{14}$ amine derivatives as an essential ingredient. For example, an effective degreasing formulation may be obtained by incorporating up to 40% of an amine oxide based on *bis*-[2-hydroxyethyl] cocoamine with 20% coconut fatty acid ester, 6% nonionic surfactant, and with the balance being an acidified water/ethanol mix. Similarly, in an alkaline vehicle shampoo/cleaner, *pentadecakis*-[2-hydroxyethyl]-dimethyl ammonium chloride, at up to 8%, confers excellent wetting and solubilizing characteristics to the formulation.

Alkyl[$C_{12}$–$C_{14}$]dimethyl amine oxides are widely used in personal care and liquid dishwashing detergents where they confer excellent foaming, thickening, and detergent properties (4). They also are used to increase viscosity in hard surface cleaners (5), not only for alkyl- and ether-sulfate solutions, but also for thickening sodium hypochlorite bleach solutions that are used in industries where cleaning and disinfection are required. The increased viscosity permits a longer residence time of the formulation and a thicker layer of active material. In the aggressive bleach environment, the amine oxide is highly stable and fulfills two functions, thickening (in conjunction with anionic surfactant/electrolyte) and perfume solubilization. The latter advantage is due to the excellent hydrotropic properties of these amine oxides. In addition, the surfactant properties of the amine oxide reduces the surface tension to improve wetting, add detergency, and disperse soils that are removed from surfaces.

## Mineral Processing/Extraction

The valuable minerals in ore deposits are usually combined with worthless material, and the ore must be processed to liberate and separate the desired mineral from the worthless gangue. Mineral beneficiation uses $C_8$–$C_{14}$ amine derivatives in two totally different ways. The first is in a mineral froth flotation of nonmetallic ores (historically one of the first applications of fatty amines), and the second is as organic solvents/complexing agents for heavy metal salts.

## Mineral Froth Flotation

While other methods of separating the valuable and gangue minerals rely on differences in their physical properties, such as specific gravity, magnetic susceptibility, or particle size, froth flotation depends on differences in surface properties. Froth flotation involves thorough mixing of air bubbles with an aqueous slurry of ground ore. Suitably conditioned mineral particles attach themselves to these bubbles as they rise through the pulp and are concentrated in the froth, which forms on the liquid surface. Only minerals having hydrophobic surfaces will adhere to the air bubbles and these minerals are, therefore, separated from those with hydrophilic surfaces. With few exceptions, mineral surfaces are naturally *hydrophilic*.

The wide application of the froth-flotation process followed the discovery that the adsorption of certain surfactants on the surface of minerals renders those minerals *hydrophobic*. Such chemicals are known as collectors or promoters. Adsorption of collectors on one or more of the minerals present in the ore pulp leads to selective attachment of only these minerals to the air bubbles and, therefore, to the concentration of these components in the froth. As a consequence, a careful collector choice enables valuable and gangue materials to be separated.

The $C_{12}$–$C_{14}$ amine derivatives, especially those containing a coco-alkyl group, are extremely selective collectors (under the correct flotation conditions) and are frequently formulated to give the best combination of selectivity and economic recovery. Such formulations are used in the beneficiation of the following minerals: calcite, feldspar, kaolin, magnesite, quartz, sands, oxidized zinc ores, and potash (kainite). Either the collector is used to float the valuable minerals, or it is used to float the impurities, this latter process being known as *reverse flotation*. A much wider range of minerals can be floated, of course, using other alkyl ($C_{16}$–$C_{18}$) amine derivatives and blends.

## Heavy Metal Extraction

The use of trialkyl ($C_8$–$C_{12}$) amines and their trialkylmethyl ammonium chloride derivatives in the solvent extraction of heavy metals was the subject at a previous American Oil Chemists' Society (AOCS) conference (6). The heavy metals concerned are, for example, uranium, vanadium, molybdenum, copper, and nickel, and the amine derivatives are an essential ingredient of the extraction process. Effectively, in a counter-current mixer/settler, a water-immiscible solvent (e.g., kerosene) containing the organic extractant (amine derivative) comes into contact with an aqueous solution containing the metal salt to be extracted. The dispersion is then passed to the settler, and phase separation occurs. The organic phase, containing the metal salt-amine complex, is passed to the stripping section where the desired metal salt is stripped from the organic phase by aqueous extraction, and the organic portion is returned to the extraction section.

## Agricultural Industry

The term agriculture is in reference to the use of amine derivatives as processing aids in fertilizer production, chemical intermediates in herbicide/fungicide production, adjuvants and dispersants for pesticides, and wood preservation.

### Fertilizer-Processing Aids

Many fertilizers, produced as prills, granules, or beads, exhibit a strong tendency to cake. During storage, which can last several months, the hitherto free-flowing fertilizer can agglomerate into large lumps, causing severe problems in application. Use of formulated amine derivatives can prevent this "caking" phenomenon. Most anticaking formulations are based on the $C_{16}$–$C_{18}$ derivatives, but shorter chains can be used in specific circumstances.

### Chemical Intermediates

Perhaps the best-known example in this category is the use of dodecylamine for the production of dodecyl guanidine (acetate salt). This latter product (7) is an anticryptogamic (fungicide) used in the treatment of apple scab, a fruit disease. $C_{12}$-alkyl-based primary and tertiary amines are also useful starting points for the modification of plant hormone acids, such as 2,4-dichlorophenoxyacetic acid (2,4-D) and 2,4,5-trichlorophenoxyacetic acid (2,4,5-T), to yield water-insoluble, readily emulsifiable herbicides with good foliage adherence characteristics (8,9).

### Adjuvants/Dispersants

Many herbicides are not readily wet by water and a dispersant is necessary to avoid aggregation and cake formation, resulting in incorrect dosing, at the bottom of the container. Such sediment is difficult to redisperse without a dispersion agent. Mono- and dicocodimethyl ammonium chlorides have proved to be effective in a wide variety of products, combining good dispersion and leaf-wetting characteristics.

Alkoxylated cocoamines are also used as adjuvants with hydrophilic pesticides, particularly with glyphosate herbicides. Typically, the adjuvant is incorporated into a ready to use "tank-mix" product of which a specified amount is recommended for certain applications. The characteristics of adjuvants can be summarized as: influencing permeability of foliar cell structures to enhance pesticide uptake, improving substantivity due to cationic character, reducing surface tension of spray mixture, and offering pesticide/crop selectivity by tailoring adjuvant properties. These characteristics offer solutions for various problems in agrochemical applications as they: replace mineral oil adjuvants at much lower dose rates; improve pesticide activity without increasing pesticide rate; improve efficacy on a wider weed/pest spectrum; improve weather tolerance, for example, rain fastness; and, reduce pesticide rates without losing efficacy. The above is not meant to imply that adjuvants will solve all problems, but rather to indicate the benefits possible, which will vary on a case-by-case basis.

### *Wood Preservation*

Due to growing environmental pressure on the use of wood preservatives such as creosote, pentachlorophenol (PCP), and copper chrome arsenate (CCA), the development of alternatives has been gaining momentum in recent years. Often in combination with other chemicals, monoalkyltrimethyl-, dialkyldimethyl-, and alkylbenzyldimethyl-ammonium chlorides, where the alkyl is based on $C_8$–$C_{12}$, have been finding favor in above ground protection and in antisap-stain formulations. This area of application is, like most agrochemical applications, subject to strict regulation and toxicological control.

## Textiles and Fibers

Alkylbenzyldimethyl quaternary ammonium chlorides, where the alkyl is $C_{12}$- or coco-based, are well known as retarders/dye levelers for acrylic-based fibers. The cationic surfactant functions by preferentially adsorbing onto the most active dye sites, thus permitting a more even uptake of the dye onto the fiber. Also used in this application is *bis*-[2-hydroxyethyl]cocomethyl ammonium chloride for the dyeing of acrylics with modified basic dyestuffs. Because of their cationic nature, such quaternaries are also used (at 0.5 wt% fabric) as antistatic agents for various synthetic textile fibers.

In processes such as spinning, weaving, or knitting, the fiber is often treated with spinning oils and/or avivages. Usually these processing aids are applied as an aqueous emulsion, and since most natural fibers, such as wool, possess an inherent negative surface charge, cationic emulsions are very effective. Clearly, if emulsifiers, such as dicocodimethyl ammonium chloride, can also function as antistats, then this is an additional reason to use the emulsifier. The concentration of the emulsifier is typically 0.5–1.0% of the oil.

Besides use in textile chemical formulations, ethoxylated coco amines have long been used (10,11) in the manufacture of viscose rayon. Viscose rayon is produced by reacting carbon disulfide with alkaline cellulose to form a xanthate. This viscose solution is allowed to "ripen" and then spun, via a spinneret, into a spin bath containing a regenerating solution, which typically contains zinc sulfate. The ethoxylated amine can perform many functions.

The ethoxylated amine can act as a modifier in the xanthation step. Here, it acts as an emulsifier for the carbon disulfide, accelerates absorption, prevents gel formation (which would otherwise block the spinnerets), and acts as a fiber lubricant. It is difficult to generalize the concentrations used, since this is process dependent. If no spin aid is used, a concentration of 0.1–0.2 wt% ethoxylated amine in the cellulose will show significant improvements.

As regeneration of the cellulose occurs in the acidic spin bath, the ethoxylated amine is leached from the fiber, passing into the spin bath and aiding the wetting of the fiber by the coagulating solution. The optimal concentration of cationics is available from the viscose in very few cases, and addition of an aqueous solution of ethoxylated amine is usually required.

## References

1. Domagk, *Dtsch. Med. Wschr.* *61*:828 (1935).
2. Schaeufele, P.J., *J. Am. Oil Chem. Soc. 61*:387 (1984).
3. van Brederode, H., *Proc. 3rd CESIO Int. Surfactants Congress, Sec D,* 244 (1992).
4. Roerig, H., and R. Stephan, *La Rivista Italiana delle Sostanze Grasse 68*:317 (1991).
5. Roerig, H., and R. Stephan, *Proc. 2nd CESIO Int. Surfactants Congress, 8: Sec. D,* 534 (1988).
6. House, J., *J. Am. Oil Chem. Soc. 61*:357 (1984).
7. U.S. Pat. 2,425,341 and 2,362,512 (1944), 3,004,065 (1961), 3,169,146 (1965).
8. Armour and Co., U.S. Pat. 2,900,411.
9. Armour and Co., U.S. Pat. 2,843,471.
10. American Enka Corp., U.S. Pat. 2,572,217.
11. Vereinigreglanztoff Fabrik, DAS Pat. 1,052,054.

# Special Surfactants for Personal Care Products

Kohsiro Sotoya, Yukinaga Yokota, and Akira Fujiu

Wakayama Research Labs, Kao Corp., 1334 Minato, Wakayama C., Wakayama 640, Japan.

## Abstract

Alkyl glycoside (AG) is a surfactant that has a sugar skeleton in its hydrophilic group. Natural, renewable materials derived from plants can be used for both the hydrophobic and hydrophilic groups in AG. An ordinary surfactant's foamability decreases sharply in the presence of an edible oil; however, AG lathers well even in the presence of a small amount of oil. Therefore, AG shows good performance as a surfactant for dish and hand soaps. In personal care applications, attention has to be given not only to detergency but also to the surfactant's effect on hair and skin. AG has a low protein denaturation effect and causes less skin irritation.

Monoalkylphosphates (MAP) are phosphate-based surfactants. They are mixtures of monoalkyl and dialkyl compounds, or so-called sesquiphosphates. When the mixtures are used as base surfactants, the detergents produced do not offer advantages in solubility, foamability, or surface activity. Improving the purity of monoalkyl compounds, however, has opened up new areas of application for personal care products. Monoalkylphosphates are dibasic acids, and have three dissociation states with changing pH. Monoalkylphosphates can be used as detergents at a wide range of pH, from weakly acidic to neutral, and show best performance at neutral pH.

Ether carboxylates (EC) are surfactants that have a polyoxyethylene chain between the alkyl group and the carboxyl group. The solubility in water and the resistance against hard water of EC are superior to those of soap. Ether carboxylates are synthesized by the carboxymethylation of AE. Ether carboxylates can be anionic or nonionic by changing the magnitude of neutralization and the alkyl chain length.

Lifestyle changes in Japan and Western Europe have increased the consumption of body shampoos as soap substitute. Consumption of shampoos has been expanding because of the recent trend toward more frequent hair washing. Surfactants such as AG, MAP, and EC should be important and suitable for personal care products in view of detergency, ecotoxicity, and other factors.

---

Recently the demand for natural materials has increased in the field of personal care. Since personal care products are generally used in direct contact with the human body, the safety of the product is crucial. With this in mind, various products based on "natural ingredients" have been launched on the market (1). These products appeal to consumers because they are gentle to the skin and environmentally safe. As a result, the demand for natural materials is increasing. Additionally, due to changes in lifestyle and in consideration of conserving resources and protecting the environment, consumers are also ready to accept products made from natural materials (2). In the future, we will need to find new properties of natural materials and develop products displaying new functions and features in the personal care field. In this paper, the trend in surfactants using natural materials and the expectations and prospects for these products are discussed.

## Surfactant Materials

Source materials for the surfactants used in the personal care field are oleochemicals and petrochemicals. The term "natural material" refers to fatty acid methyl esters produced from natural fats and the fatty oils and fatty alcohols derived from them. Petrochemical materials are α-olefins produced from ethylene and the higher alcohols produced from those chemicals by oxosynthesis or the Ziegler method.

In the personal care field, surfactants are used mainly in detergents. The type of material itself (oleochemical or petrochemical) does not affect the cleansing function. The technology has been developed on the basis of cost and performance. As the technology has matured, both the number and amount of products available have increased. At the present time, more importance is being placed on the effective use of renewable natural resources (2,3). Review of the material sources and product design from an environmental protection viewpoint has received increased attention.

## Natural Fats and Fatty Oils

Recently, an extensive increase in the production of palm oil and palm kernel oil has been planned in Southeast Asian countries such as Malaysia and Indonesia. Figure 1 shows variations in the production of palm kernel and coconut oil. The production of palm kernel oil has demonstrated a dramatic increase since around 1980. This trend is linked with the production increases planned for higher alcohols and fatty acids used as source materials for surfactants (3,4). In the future, palm kernel oil is expected to be an important starting material for surfactants.

When the price of the natural fats and fatty oils is compared with that of petrochemicals, the latter shows a sixfold increase over the past 20 years while the former shows a stable and predictable pricing pattern, although it is more susceptible to the price of coconut oil, weather,

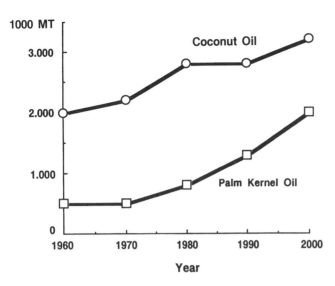

Fig. 1. World production of fats and oils.
Source: INFORM 1:1035 (1990).

and other production-related variables (5). Although petrochemical materials have an advantage in view of their cost and stable supply, natural fats and fatty oils are expected to catch them up in the long run.

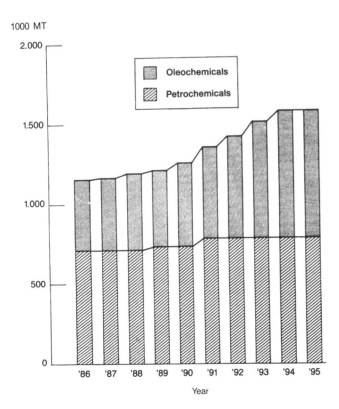

Fig. 2. World surfactant alcohol capacity.
Source: INFORM 2:1064 (1991).

## Natural Higher Alcohols

As shown in Fig. 2, the production of higher alcohols with an oleochemical base was 520,000 MT, and growth to 794,000 MT by 1995 is expected. Table 1 shows production increases planned for natural alcohols released by major world-wide manufacturers. The production of natural alcohols is predicted to increase faster than that of petrochemical alcohols. The increased production of oleochemicals will lead to an abundant supply of materials and will make material procurement easier, both of which are welcome trends in the production of surfactants for the personal care field.

## Surfactants

The technology to develop highly functional surfactants from fatty alcohols has attracted attention in the surfactant industry (6), and various surfactants have been developed (Fig. 3). Surfactants are used in a wide variety of products including cosmetics, pharmaceutical products, foods, fiber and metal processing, quality improvement of plastics, and agricultural chemicals. However, a large amount is used in the household and personal care fields as well as in industrial detergents.

**TABLE 1**
**Production of Natural Fatty Alcohols (1,000 MT)**

| Company | Market | 1990 | 1995 |
|---|---|---|---|
| Kao | Japan | 17 | 17 |
| | Philippines | 24 | 54 |
| | Malaysia | 30 | 85 |
| | Total | 71 | 139 |
| Henkel | Germany | 150 | 150 |
| | France | 60 | 60 |
| | North America | — | 40 |
| | Malaysia | — | 30 |
| | Total | 210 | 280 |
| P&G | North America | 116 | 154 |
| | Malaysia | — | 40 |
| | Germany | — | 36 |
| | Total | 116 | 230 |
| Salim | Indonesia | 30 | 90 |
| | Germany | 20 | 20 |
| | Total | 50 | 110 |
| RWE-DEA (CONDEA-Vista) | Germany | 30 | 30 |
| | Total | 30 | 30 |
| Miscellaneous companies | | 157 | 217 |
| Grand total | | 614 | 1023 |

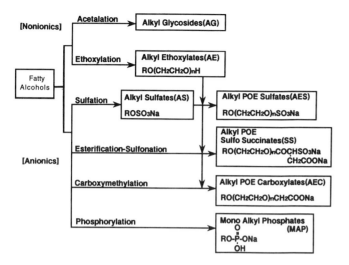

Fig. 3. Surfactants derived from fatty alcohols.

Surfactants are widely used in the personal care field, and in addition to their basic activities; which include emulsification, solubilization, diffusion, moistening, foaming and antifoaming actions; their lubricating, sterilizing, and antistatic actions are also finding applications in many areas. Surfactants demonstrating such a wide range of activities are classified into ionic (anionic, cationic, and amphoteric) and nonionic types, each of which is widely used according to its distinctive feature (7).

The primary alkyl compositions of detergents in the personal care field are in the $C_{12}$–$C_{14}$ range, and properties such as cleansing and foaming features are considered extremely important. The detergent products in the personal care field include shampoo, soap, body shampoo, face wash, and tooth paste.

Soap was the primary detergent for Japanese personal care until around 1955. Although soap is still dominant, products to meet specific needs have been developed for various parts of the body (i.e., shampoo for hair, soap for washing hands, face wash, and body shampoo, etc.). In developing these products, not only the concepts, but also social backgrounds and consumer needs were carefully studied. At the present time, products which can be classified by concept; price; and target, such as type of skin or hair, extent of damage to be repaired, after-wash effect, and finish, have been launched on the market (8).

Alkyl sulfate (AS) and alkyl ethoxy sulfate (AES) have been major materials of shampoos since the 1960s. Although it is still used as the major ingredient, the amount of AES used in compounding is decreasing because of potential skin irritation and solution features. As consumers demanded milder cleansing agents, less irritating surfactants (Alkylglycoside [AG], imidazolin or alkyl amide-type amphoteric surfactants, amino acid-type surfactants, sulfosuccinate-type surfactants, and monoalkyl phosphate [MAP]) have been developed (Fig. 4 [9–11]).

Although soap is still mainly used for body washing, neutral solid soap (for example, AGS) also was marketed (12). However, MAP is more popular in Japan and isethionate (SCI) in the United States as less irritating surfactants. These products are gentle to the skin, giving a refreshed after-wash feeling (Fig. 5).

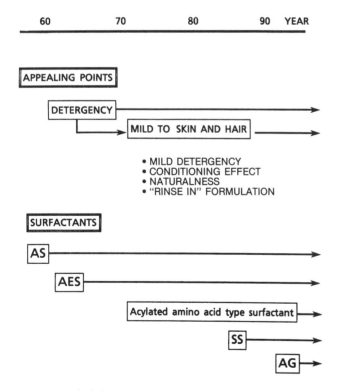

Fig. 4. Trend of shampoos in Japan.

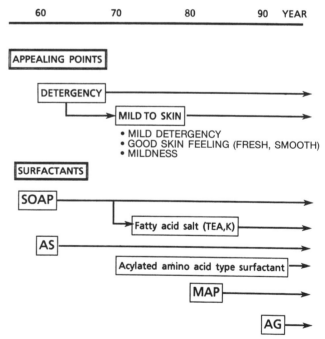

Fig. 5. Trend of soap and body shampoos in Japan.

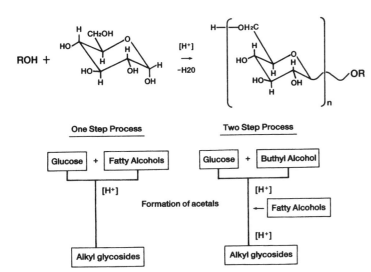

Fig. 6. Manufacturing process of alkyl glycosides.

## Alkylglycoside

Alkylglycoside is a nonionic surfactant having a sugar structure in the hydrophilic part (Fig. 6). This type of surfactant may be manufactured by using natural materials in both the hydrophilic and hydrophobic parts. Alkylglycoside is synthesized by the dehydration of higher alcohols and glucose (Fischer method [14]). This is a method of synthesis by heating and dehydrating alcohols and sugars in the presence of an acid catalyst to produce a mixture of α- and β-anomers. Industrial production of high-quality AG has now become possible by hemiacetalization and various purification processes.

Alkylethoxylate (AE), a representative nonionic surfactant, has a polyoxyethylene chain in the hydrophilic part and is hydrated by hydrogen bonding. On the other hand, no clouding point is observed in the case of AG since it is strongly hydrated. Furthermore, the packing to the interface is more favorable, and it has marked foaming properties due to its compact hydrophilic section.

Alkylglycoside demonstrates excellent foaming characteristics at higher detergent concentration levels, but it rapidly loses these characteristics as the concentration is decreased. In other words, it has a high foaming character as a detergent, requiring less rinsing. As shown in Fig. 7, foaming in most surfactants markedly decreases in the presence of oil, but AG maintains its excellent foaming properties even in the presence of a small amount of oil. This feature makes it suitable for shampoos and facewashes. In the personal care field, the effect of the surfactant on hair and skin is considered as important as foaming and cleansing features. It is known that AG causes less protein degeneration (13) and skin irritation (Fig. 8).

An additional benefit from its molecular structure is that it is easily biodegradable (18). The combination of cleansing features and environmental profile make it a very safe surfactant. With the development of application techniques and new products, growth of AG is expected in the personal care field.

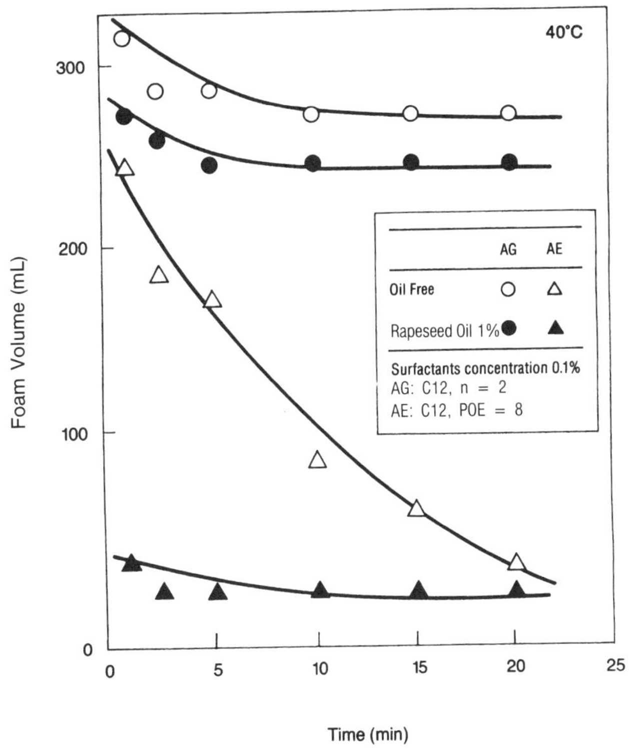

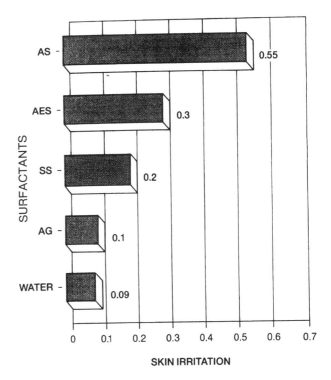

Fig. 7. Foaming ability and stability of AG and AE in the presence of oil.

Fig. 8. Skin irritation scores of surfactants. Surfactant, 7 mM; pH, 7; human arm; 24 h; closed patch test.

## Monoalkyl Phosphate

Alkyl phosphate-type surfactants are manufactured as a mixture of monoalkyl and dialykyl, the so-called sesquiphosphates. The product has found various applications including detergents in many industrial fields. However, as a detergent base, alkyl phosphates are inferior in solubility, foaming, and cleansing characteristics. After it was discovered that its foaming features could be improved by increasing the purity of the monoalkyls (Fig. 9), the compounds have found many applications in the personal care field.

Monoalkyl phosphate is synthesized by reaction of higher alcohols with phosphorylating agents (Fig. 10 [15-17]). Phosphorylating agents include phosphoric anhydride ($P_2O_5$), polyphosphoric acid, and phosphorous oxychloride, as well as complexes with their bases. Phosphoric anhydride is the industry's first choice, but it is now possible to obtain alkyl phosphate with high monoalkyl purity through phosphorylation with polyphosphoric acid.

Because of the solubility and cleansing features required, a weak alkaline soap is preferred. Monoalkyl phosphate is a dibasic acid and demonstrates three types of dissociation depending on the pH. Considering its features as a detergent, application in the slightly acidic and neutral range is possible, but its function is best demonstrated in the neutral range.

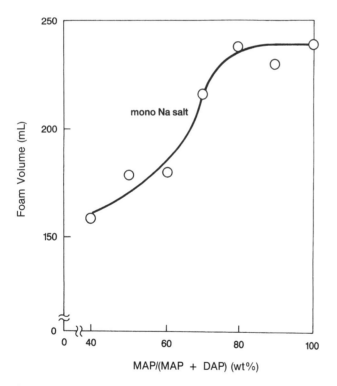

Fig. 9. Foaming ability dependence on purity of $C_{12}$ MAP (clockwise and counterclockwise stirring method, 1,000 rpm for 30 sec at 40°C, 2.0 wt% $C_{12}$ MAP in 4°DH hard water).

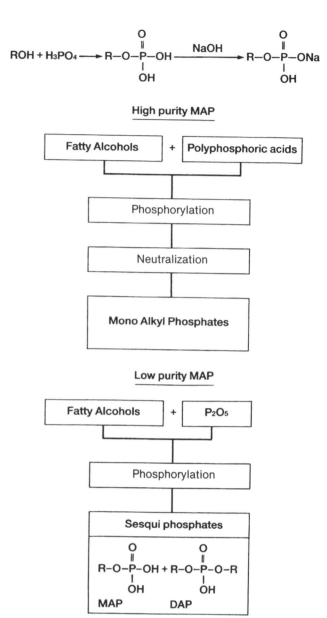

Fig. 10. Manufacturing processes of mono alkyl phosphates.

With ordinary soap, a high pH value is observed on the skin surface after washing. Although the pH drops as time passes, it takes some time to recover. However, since MAP is used in the neutral range, the change in pH is hardly detectable, resulting in an index for alkalinity-induced chapped skin.

Monoalkyl phosphate has excellent foaming characteristics, demonstrating fine, creamy foams like those of soap. It is a mild detergent and removes fewer amino acids and natural moisturizing factor (NMF) from the skin than soap (Fig. 11). It causes little, if any, skin irritation. It is highly biodegradable (18) and has a low toxicity for fish. In other words, it is a very safe anionic surfactant.

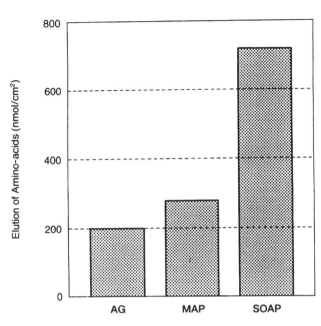

Fig. 11. Effect of surfactants on human skin (0.5% aqueous solution, 15 min., N = 7).

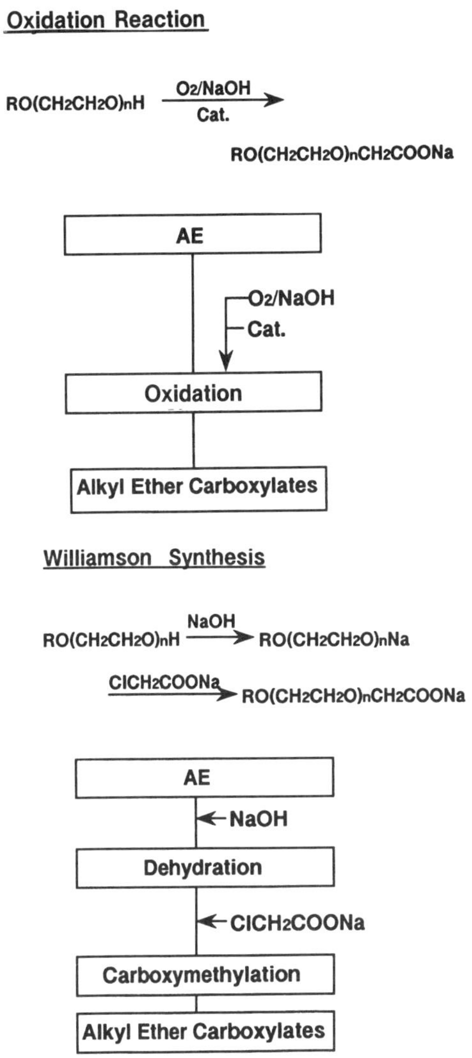

## Alkylether Carboxylate

Since alkylether carboxylate (AEC) has a polyoxyethylene chain and carboxyl group in the hydrophilic section of the molecule, it is a surfactant superior in its water solubility and hard water resistance compared with soap (19,20). Alkylether carboxylate is the generic name of surfactants having an ether linkage and carboxyl group, but in general AEC refers to the terminal carboxymethylated compound of a polyoxyethylene alkylether.

Alkylether carboxylate is obtained by adding ethylene oxide to AE and converting the mixture to carboxymethyl ether. As shown in Fig. 12, the methods of carboxymethylation are classified into two groups, the Williamson method, in which the compound is synthesized through reaction of monochloroacetic acid or sodium monochloroacetate with AE; and the oxidation method, in which direct oxidation with the terminal part of AE is conducted. Presently, industrially produced AEC is manufactured by the sodium monochloroacetate method.

Depending on the length of the oxyethylene chain and the extent of the neutralization of the carboxyl group, AEC demonstrates a wide range of properties from anionic to nonionic. Scum diffusion is known to improve when the oxyethylene chain is longer. Surface tension is higher as the pH and oxyethylene chains increase (Fig. 13).

When mixed with other surfactants, AEC is known to display a characteristic synergistic effect. In the mixture of AEC and AES, it has been confirmed (Fig. 14) that AEC alleviates the skin irritation caused by AES.

As a less irritating surfactant and an agent to alleviate irritation, AEC seems to have features applicable to deter-

Fig. 12. Manufacturing processes of alkyl ether carboxylates.

gents. Its scum diffusion, resistance to hard water (Fig. 15), and cleansing properties are equivalent to those of AES. It is also highly biodegradable and is expected to find many more applications in the personal care field.

The trend for less irritating surfactants has been observed in Japan (21,22), Europe, and the United States in the personal care field (Fig. 16). In Japan, body shampoos using MAP as the main ingredient are popular. Monoalkyl phosphate is less irritating to the skin and therefore gives a refreshing after-wash feel. On the other hand, combination bars and sindet bars are popular in the United States. Less irritating surfactants, such as glyceryl sulfonate (AGES) and SCI, made from coconut oil are used in these products.

Mild, natural, and easy to use shampoos that permit conditioning have been selling well. Demands for less irri-

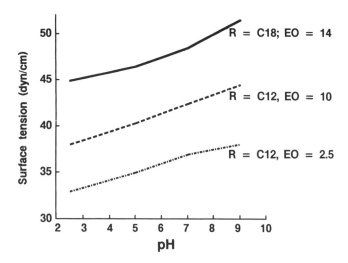

**Fig. 13.** Influence of pH on the surface tension of some AEC. *Source*: Van Paassen (19).

tating detergent bases and for products which keep hair set longer, give gloss and moisture to hair, and prevent split ends are increasing.

## Conclusions

In addition to the use of natural materials, effective cleansing features, and concern for environmental protection, new features, including less skin irritation, are required for products in the personal care field. To satisfy these needs, the role of surfactants, such as MAP, AEC, and AG, derived from higher alcohols is considered important.

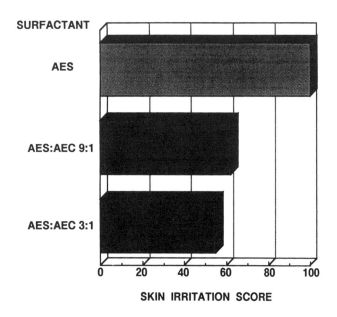

**Fig. 14.** Synergistic effect of surfactant mixture AES/AEC on skin irritation.

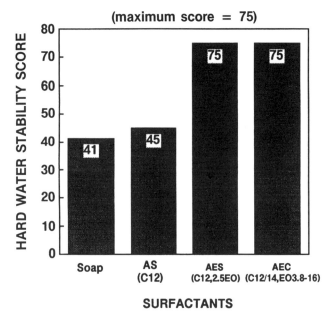

**Fig. 15.** Hard water stability according to DIN 53905. *Source*: E. Stronik (20).

Therefore, the use of oleochemicals is expected to increase. However, not all oleochemicals are environmentally sound, thus, the use of petrochemicals is also necessary.

In the field of detergents, technological developments based on universal concerns are desirable for future products. For example, when new chemicals are launched on the market, it is a corporate responsibility to pay attention to their effects on the human body (irritation, mutagenicity, and other potentially harmful effects) as well as on the environment (i.e., biodegradation, and effects on fish,).

## References

1. Takahashi, K., *Fragrance J. 142*:74 (1993).
2. Tsushima, R., H. Tagata, Y. Yokota, and A. Fujiu, *INFORM 4*:680 (1993).
3. Uehara, H., *Kagaku to Kogyo 66*:152 (1992).
4. Nakamura, K., *Fragrance J. 105*:12 (1989).
5. Baumann, H., *Yushi 43*:56 (1990).
6. Kiyooka, H., *Yushi 44*:42 (1991).
7. Tamura, T., and H. Hirota, *Koushohin Kagaku - riron to jissai*, 1st edn., *Fragrance J.* LTD, Tokyo, 1990, p. 141.
8. Kurokawa, H., *Fragrance J. 106*:82 (1990).
9. Niwase, H., and H. Minamino, *Yushi 44*:58 (1991).
10. Kobayashi, T., and T. Ikeuchi, *Yushi 46*:44 (1993).
11. Yokota, H., *Yushi 42*:39 (1989).
12. Nakamura, K., *Yushi 42*:44 (1989).
13. Kamegai, J., and T. Kurosaki, *Chem. Expresses 21* (1987).
14. Sakakibara, T., *J. Jpn. Oil Chem. Soc. 39*:1 (1990).
15. Kurosaki, T., J. Wakatsuki, T. Imamura, A. Matsunaga, H. Furugaki, and Y. Sassa, *Proceedings of XIX Jornadas del CED/AID*, Spanish Committee of Surface Active Agents, 1988, p. 191.

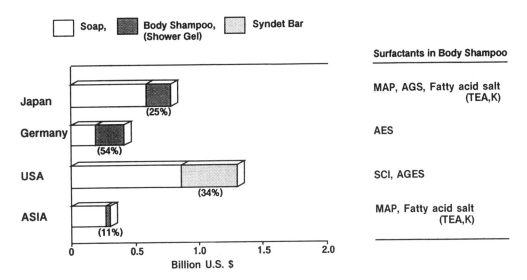

**Fig. 16.** Consumption of body shampoo and soap in the 1990 body cleanser market.

16. Imokawa, G., H. Tsutsumi, and T. Kurosaki, *J. Am. Oil Chem. Soc. 55*:839 (1978).
17. Imokawa, G., *J. Soc. Cosmet. Chem. 31*:45 (1980).
18. Yoshimura, K., Y. Toshima, and N. Nishiyama, *Fragrance J. 115*:59 (1990).
19. Van Paassen, N.A.I., *Sifen-Ole.-Fette-Wachse 109*:353 (1983).
20. Stronik, E., *R. Soc. Chem. 79*:62 (1990).
21. Nozaki, T., *Fragrance J. 124*:15 (1991).
22. *C & T 5*:74 (1992).
23. Drozed, Joseph C., in *Proceedings of World Conference on Oleochemicals: Into the 21st Century,* T.H. Applewhite, Ed., the American Oil Chemists' Society, Champaign, IL, 1990, p. 256.
24. Lorenz, P., in *Proceedings of World Conference on Oleochemicals: Into the 21st Century,* T.H. Applewhite, Ed., the American Oil Chemists' Society, Champaign, IL, 1990, p. 235.

# A Two-Stage Countercurrent Bleaching Process for Edible Oils and Fats

P. Transfeld and M. Schneider

OEHMI Forschung und Ingenieurtechnik GmbH, Berliner Chaussee 66, D-39114 Magdeburg, Germany.

## Abstract

In our countercurrent bleaching process, vegetable oils are bleached in several stages. The bleaching earth and oil are separated after each treatment stage. The separated bleaching earth is fed to the preceding bleaching stage, while the oil flows to the next treatment stage where it is brought into contact with less-spent bleaching earth. Fresh bleaching earth is introduced into the process before the last bleaching stage, and the spent bleaching earth is discharged from the process after the first bleaching stage. Unbleached oil is fed to the first stage, while bleached oil leaves the bleaching plant after passing through the last bleaching stage. Bleaching earth requirements are low (up to 47% lower than for batch bleaching).

---

The principle of our countercurrent bleaching process is shown in Fig. 1. In the countercurrent bleaching process, vegetable oils are bleached in two stages: prebleaching of the unbleached oil by partially spent bleaching earth from the second stage; and final bleaching of the prebleached oil by fresh bleaching earth. The bleaching earth and the oil are separated after each treatment stage. The bleaching earth separated from the second stage is fed to the first stage. After the first bleaching stage the twice-used, spent bleaching earth leaves the process.

The two-stage bleaching process features low bleaching earth consumption, because the bleaching earth is used in two stages as it passes countercurrently through the vegetable oil. Test operations have shown that bleaching earth savings of 47% compared to conventional batch bleaching of palm oil can be obtained. The countercurrent bleaching process performs especially well where strongly colored oils, such as palm oil, are to be treated. The countercurrent bleaching process offers the plant operator several benefits: bleaching earth requirements are low (up to 47% lower than for batch bleaching); only relatively small amounts of oil-containing bleaching earth have to be disposed of; oil losses are limited, due to the small amounts of oil-containing bleaching earth; and the filtration capacity required to separate oil and bleaching earth is comparatively small.

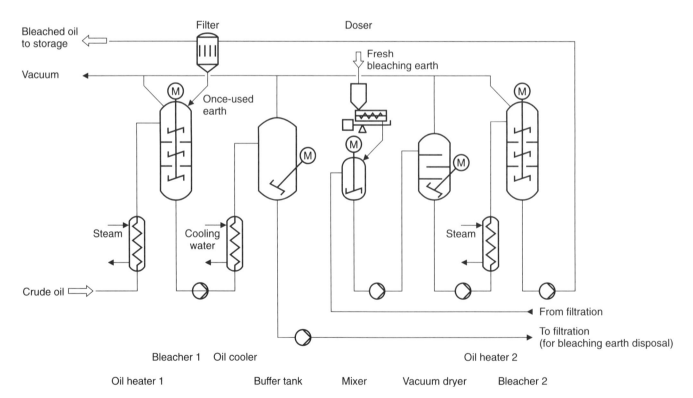

**Fig. 1.** Flowchart for a two-stage, countercurrent bleaching system (OEHMI Forschung und Ingenieurtechnik GmbH).

**TABLE 1**
**Bleaching Test Results**

| Test No. | Process | Bleaching Earth Quantity (%) | Lovibond Tintometer Values after Bleaching | | Carotene Content (mg/kg) |
|---|---|---|---|---|---|
| | | | Red (1") | Yellow | |
| 1 | Batch bleaching | 1.50 | 7.0 | 47.5 | 29.8 |
| 2 | Countercurrent bleaching | 1.48 | 3.0 | 37.0 | 8.4 |
| 3 | Batch bleaching | 0.88 | 13.0 | 64.0 | 72.8 |
| 4 | Countercurrent bleaching | 0.88 | 5.9 | 50.0 | 20.1 |

## Integration into Existing Plants

The options for integration of countercurrent bleaching plants into existing continuous refining plants are integration of the existing filtration systems into the countercurrent bleaching process; and/or integration of the existing continuous bleaching plant into the countercurrent bleaching process.

## Description of a Two-Stage Continuous Countercurrent Bleaching System

To minimize product/air contact, most of the bleaching process takes place under vacuum. The unbleached oil is heated to bleaching temperature in oil heater 1 (Fig. 1). In bleacher 1, the oil is prebleached with once-used bleaching earth from the filter. The suspension is pumped from bleacher 1 through an oil cooler into the buffer tank. The suspension from the buffer tank is pumped into the existing filtration unit. The clear oil from the filtration unit flows into the mixer. In the mixer, the oil is mixed with fresh bleaching earth where the metering unit feeds bleaching earth into a mixer in a continuous operation. The oil/bleaching earth suspension is pumped from the mixer into a vacuum dryer. After drying and deaeration, the mixture is heated to bleaching temperature in oil heater 2. In bleacher 2, the oil is bleached a second time before it is separated from the bleaching earth with a filter. The once-used bleaching earth is fed into bleacher 1, where it serves to prebleach the crude oil. The clear twice-bleached oil leaves the bleaching plant for further processing.

## Test Results

Test results are shown in Table 1, which presents comparisons between batch-type and countercurrent bleaching based on similar bleaching parameters and bleaching earth quantities (1 vs. 2, 3 vs. 4).

# Discussion

There were no questions, so the chairperson commented on the general excellence of the talks, thanked all of the speakers and his co-chairperson.

# Dietary Fat Composition Alters Whole Body Utilization of $C_{16:0}$ and $C_{10:0}$

M. Thomas Clandinin[a,b], Larry C.H. Wang[c], Ray V. Rajotte[d], Margaret A. French[a], Y.K. Goh[a], and Elaine S. Kield[a]

Nutrition and Metabolism Research Group, Departments of Food Science and Nutrition[a], Medicine[b], Zoology[c], and Surgical Medical Research Institute[d], University of Alberta, Edmonton, Canada T6G 2P5.

## Abstract

Six healthy adult males were fed four different diets to determine the effects of the quantity of fat (30 or 40% of energy as fat) and type of fat (polyunsaturated or saturated) on the utilization of fatty acids. Each diet was fed for 15 days. The dietary polyunsaturated to saturated fat (P/S) ratio was formulated at either 0.2 or 1.0 at both levels of fat intake. Subjects provided breath tests to measure background $^{13}C$ and response to [1-$^{13}C$]10:0 and [1-$^{13}C$]16:0 fed with a test meal. Increasing the P/S ratio increased whole-body oxidation of labeled $C_{10:0}$ by 30% after consumption of both low- and high-fat diets. When labeled $C_{16:0}$ was fed, the amount of $^{13}C$ excreted in breath increased by a factor of 2.4 after a low-fat, high-P/S ratio diet compared to the low-P/S ratio diet. The results suggest that the amount and type of fat in the diet affects the utilization of individual fatty acids in normal subjects.

The process of intestinal adaptation has been examined by investigating alteration in villus morphology (1), nutrient transport (2), membrane composition (3), and desaturation of fatty acids (4). Changes in nutrient transport are associated with alterations in brush border membrane (BBM) phospholipid content and composition (5). Previous studies of normal rats have shown that feeding a diet high in polyunsaturated fat increased the transport rate for $C_{16:0}$, $C_{18:0}$, $C_{18:1}$, $C_{18:2}$, and $C_{18:3}$ in the jejunum and ileum (6). However, transport of medium-chain fatty acids was unchanged by dietary manipulation (6).

The morphology of the villi in both rat jejunum and ileum is affected by the polyunsaturated to saturated fat (P/S) ratio in the diet (7,8). Villus surface area is significantly increased with diets high in polyunsaturated fat (7). Since the uptake of palmitic acid is thought to occur in the upper one-third of the villus (9), increasing the P/S ratio in the diet may be associated with the increased uptake rate of palmitic acid.

The current conception of energy utilization assumes that dietary fat oxidation occurs at a rate controlled by mechanisms independent of the individual long-chain fatty acid composition of the diet. Human studies have suggested that different fat substrates are metabolized for energy with different efficiencies (10). These observations suggest that the relative amounts of major dietary fatty acids consumed may affect the net contribution of fat oxidation to total energy production, shifting energy homeostasis and affecting whole-body fat partitioning for adenosine triphosphate (ATP) synthesis or energy storage.

In view of these animal and human studies, an experiment with human subjects was conducted to determine if the total amount of fat and the P/S ratio of the rat consumed affects the amount of $C_{10:0}$ or $C_{16:0}$ absorbed and utilized from a test meal. Dietary medium-chain triglycerides (MCT) are hydrolyzed more efficiently than long-chain triglycerides (LCT) (11), reach the liver more rapidly than long-chain fatty acids (LCFA), and undergo rapid β-oxidation and production of $CO_2$. Thus, one objective of the present study was to determine whether a high or low intake of dietary fat in isoenergetic diets decreases whole-body oxidation of saturated fatty acids that are dependent ($C_{16:0}$) or independent ($C_{10:0}$) on carnitine-mediated transport into the mitochondria. The second objective was to determine if an increased level of polyunsaturated fat in isocaloric diets increases whole-body oxidation of saturated fatty acids, reflecting a shift in the contribution of fat oxidation to total energy production.

## Methods

### Subjects and Diet

Procedures used in this experiment were approved by the Ethics Review Committee for Human Experimentation of the Faculty of Medicine, University of Alberta. Six free-living males (Table 1) between the ages of 21 and 31 years gave informed written consent to participate in this study. Subjects were screened for chronic disease, cigarette smoking, atypical sleeping and activity patterns. Subjects consumed three isoenergetic meals per day consisting of normal foods. The diet provided an energy level equal to each subject's estimated requirements (Table 1). Energy requirements were estimated using the Mayo Clinic Nomogram (12) to determine basal metabolic rate (BMR) and by multiplying this figure by 1.7 for normal activity (13). An additional energy increment was included if the subject participated in extra activity. Subjects were weighed daily, prior to breakfast, and energy intake was

**TABLE 1**
Demographic Data of Subjects Studied

| Subject | Age (yrs) | Height (cm) | Body Weight[a] (kg) | Energy Intake[b] (Kcal) |
|---|---|---|---|---|
| 1 | 21 | 184.0 | 68.88 ± 0.04 | 3400 |
| 2 | 23 | 175.0 | 69.50 ± 0.27 | 3226 |
| 3 | 26 | 170.0 | 63.82 ± 0.14 | 2300[c] |
| 4 | 31 | 179.0 | 67.12 ± 0.16 | 2626[d] |
| 5 | 30 | 167.5 | 51.90 ± 0.29 | 2350 |
| 6 | 26 | 168.5 | 61.49 ± 0.08 | 2650 |

[a]Body weight is expressed as the mean body weight ± SEM over the four experimental periods.

[b]Based on maintenance energy requirements estimated using the Mayo Clinic Nomogram (BMR) × 1.7 + activity factor (15,16).

[c]Started with a 2500 kcal diet but reduced to 2300 after the first 5 days of the study.

[d]Range of 2426–3026 kcal—adjusted throughout study as the subject's weight varied.

adjusted slightly if a sustained weight loss or gain was observed.

The study consisted of four test periods of 15 days each. During each test period, subjects were provided with a different diet treatment. The four diet treatments were high-fat, low-P/S diet; high-fat, high-P/S diet; low-fat, low-P/S diet; and low-fat, high-P/S diet. The diets were formulated based on published food composition tables (14–19) to contain 14.2, 45.4, and 40.4% energy as protein, carbohydrate, and fat, respectively, for the high-fat diets, which are typical values of the North American diet. The low-fat diet contained 14.8, 55.2, and 30% energy as protein, carbohydrate, and fat, respectively, which is the level of fat recommended by health professionals (Table 2). Each fat level was further divided into a low-P/S diet (0.2) or high-P/S diet (1.0), resulting in four diet treatments. Medium-chain triglycerides are found in dairy products with a 10:0/16:0 of about 0.1. In order to provide a similar isotopic dilution in each dietary treatment when labeled 10:0 was administered, the ratio of $C_{10:0}/C_{16:0}$ was formulated at 0.17 for all diets. Medium-chain triglyceride oil was added to the diet to attain this ratio in all diets fed.

Within each dietary treatment, three menus designated as Cycle 1, 2, and 3 were used. The Cycle 1 menu was always consumed 1 day prior to and on all test days. All meals were prepared in the metabolic kitchen and consumed at fixed times. Breakfast and lunch was eaten under supervision and supper was provided on a take-out basis. Subjects were instructed to consume only the foods provided to them. Only decaffeinated coffee, tea, and energy-free beverages were allowed as supplemental foods.

## Fat and Energy Analysis of Diet

On days 9, 10, and 13 of each test period duplicates of the meal consumed by each subject were prepared and homogenized with water for laboratory analysis of nutrient content. Two aliquots were stored at −20°C for subsequent determination of $^{13}C$ content, total fat content, energy content, and fatty acid content of diets fed (Table 2). Thus, the isotopic dilution of the test substrates was constant and known. Fat extraction (20), saponification with KOH and transesterification with boron trifluoride-methanol reagent (21) were carried out. Fatty acid methyl esters were analyzed by gas-liquid chromatography (Vista 6010 GLC and Vista 402 data system; Varian Instruments, Georgetown, ON) as described previously (22). Mean triacylglycerol content was determined based on mean fatty acid chain

**TABLE 2**
Major Fatty Acid Composition of Meals[a] Based on Laboratory Analysis

| | Fatty Acid (wt%) | | | | |
|---|---|---|---|---|---|
| | $C_{10:0}$[b] | $C_{16:0}$[b] | $C_{18:0}$[b] | $C_{18:1}(9)$[b] | $C_{18:2}(6)$[b] |
| High-fat diets[c] | | | | | |
| Low P/S ($n = 45$) | 4.9 ± 0.3 | 27.5 ± 0.3 | 8.9 ± 0.3 | 24.6 ± 0.9 | 8.4 ± 0.2 |
| High P/S ($n = 45$) | 3.9 ± 0.2 | 22.5 ± 0.3 | 6.9 ± 0.2 | 22.3 ± 0.8 | 26.1 ± 0.4 |
| Low-fat diets[c] | | | | | |
| Low P/S ($n = 54$) | 4.9 ± 0.2 | 28.1 ± 0.4 | 9.4 ± 0.3 | 25.0 ± 0.8 | 6.8 ± 0.6 |
| High P/S ($n = 54$) | 4.0 ± 0.2 | 22.0 ± 0.4 | 7.6 ± 0.2 | 23.3 ± 0.7 | 21.5 ± 1.7 |

[a]Meals formulated using food composition tables.

[b]Values represent mean ± SEM.

[c]High-fat diets were formulated to contain the following nutrients/1000 kcal per meal: 35.4 g protein, 113.5 g carbohydrates, and 45 g fat. Low-fat diets were formulated to contain the following nutrients per 1000 kcal per meal: 36.8 g protein, 138 g carbohydrates, and 33.3 g fat. Based on laboratory analysis of meals fed, the high-fat meals provided 42 ± 0.80% of calories as fat, and 1260 ± 20 kcal of energy. The low-fat diet provided 29 ± 0.80% cal as fat and 1135 ± 23 kcal of energy. The high- and low-P/S ratios fed were 0.85 ± 0.02 and 0.2 ± 0.01, respectively. The ratio of $C_{10:0}$ to $C_{16:0}$ in the diet was constant at 0.17.

length. Fatty acid methyl esters were identified by comparison of retention data with that of authentic standards (23) and quantitated by peak area comparison with internal standards (10). Bomb calorimetry (Parr Model 1241, Moline, IL) was performed on aliquots of meal homogenates freeze-dried to constant weight. Atwater factors were applied to obtain metabolizable energy values.

## Experimental Procedures

Each subject participated in all four test periods. The order in which the subjects received the dietary treatments was randomized systematically. The first 9 d of each test period served as a stabilization period, during which time there would be a turnover of at least 2 generations of mucosal cells (24). During days 9, 10, and 13 of each test period, subjects were confined to the metabolic laboratory and provided 6 min breath samples hourly from 07:30 am–4:30 pm for analysis of breath $^{13}CO_2$ enrichment. Analysis of breath $CO_2$ enrichment on day 9 permitted examination of $^{13}C$ contribution from the test diet. On days 10 and 13 of each test period either [1-$^{13}C$] palmitic acid or [1-$^{13}$] decanoic acid was ingested in capsule form at the breakfast meal by the subjects. The order of ingestion of each labeled fatty acid was randomized systematically. The enriched fatty acids were obtained from MSD Isotopes (Merck-Frosst, Canada, Montreal, PQ). The [1-$^{13}C$]16:0 was 99% chemically pure and contained 99.5 atom percent (AP) [1-$^{13}C$]. The labeled decanoic acid was 99% pure and contained 95.2 AP [1-$^{13}C$]. The labels were fed to provide a dose of 0.86 and 0.24 mg of $^{13}C$ per kg of body weight from $C_{16}$ and $C_{10}$, respectively, based on prior analysis of chemical and isotopic purity of the fatty acids at a dose of 15 mg of [1-$^{13}C$]16:0 per kg of body weight and 3 mg [1-$^{13}C$]10:0 per kg of body weight.

### Breath Collection

Hourly breath samples were collected from the reclining subjects for 6 min. between 7:30 am and 4:30 pm on the test days. Each subject breathed through a mouthpiece and Rudolph valve (Roxon Medi-Tech Ltd., Montreal, PQ) into a 100 L latex bag. The bag was then evacuated at 8 L/min. Part of the sample was passed through an infrared detector (Model LB-2 Sensormedics Corporation, Anaheim, CA) at 0.5 L/min to measure the percentage of $CO_2$ in the breath sample. The $CO_2$ analyzer output was monitored on a chart recorder to calculate total volume of air collected. A reference standard was used to calibrate the $CO_2$ analyzer several times daily. The sample outlet from the $CO_2$ analyzer was connected to a 100-cm long spiral trap containing 10 mL of 1 N NaOH. The respiratory sample was bubbled through the NaOH solution for 3 min, trapping all $CO_2$. The resulting NaOH solution was withdrawn into a vacutainer and frozen at −26°C for further analysis. Respiratory $CO_2$ was regenerated under vacuum by mixing the NaOH solution with phosphoric acid (85% w/v) in a Rittenburg tube. The tube was maintained at 25°C until the reaction was complete (approximately 20 min), frozen in liquid $N_2$, immersed in an acetone/liquid $N_2$ slush bath and then attached to the mass spectrometer.

### Mass Spectrometry

Abundance analysis of $^{13}CO_2$ was performed using a dual inlet Isotope Ratio Mass Spectrometer (Finnigan MAT 251, Bremen, Germany). The abundance of $^{13}C$ was corrected for $^{17}O$ and pressure drift (25). Multiple $^{13}C$ abundance analyses ($n = 6$) of $CO_2$ liberated from an NBS #20 limestone standard demonstrated the analytical precision (CV) of this instrument at 0.0036%. The CV of replicate determinations ($n = 12$) was 0.0069% for purified $CO_2$ from combusted, unlabeled palmitic acid. Duplicate analyses were performed for each breath sample collected. Results were expressed $\delta^{13}C$% vs. Pee Dee Belemnite (PDB) standard and were converted to AP$^{13}C$. Background $^{13}C$ values were then subtracted from each sample time on a day when a substrate was fed to yield atom percent excess (APE) values due to oxidation of the substrate (26). Values at the initial time point for each substrate test day were used as the zero reference point. Subsequent points were adjusted accordingly and were used to compare profiles based on similar intakes of $^{13}C$ label and dietary intake of the test fatty acid in the breakfast meal. The percentage of administered dose excreted in the breath per hour was calculated as indicated previously (10) including a correction factor of 1.25 to adjust for uptake of label into the bicarbonate pool (27). The $^{13}C$ enrichment of all four dietary treatments and the different menu cycles within each dietary treatment were measured by combusting a 10 mg aliquot of the freeze-dried meal sample in a semimicro bomb (Parr Instrument Company, Moline, IL) pressurized to 8 atm of oxygen. The resultant $CO_2$ was purified in a vacuum line as described previously (28) and the $^{13}C$ abundance measured.

## Results

Subjects were allowed an interlude of a few days between each dietary treatment which resulted in total compliance in completing the test meals over the 10-week study period. The mean weight variance over the total study was minimal for the six subjects (Table 1). Body-weight fluctuation over the study was 0.22 kg (SEM).

The analyzed fat content of the meals for each dietary treatment was close to the values formulated from food composition tables and was consistent for both P/S ratio diets (Table 2). The analyzed P/S ratio and $C_{10:0}/C_{16:0}$ ratio of the diets were close to that calculated and were consistent within each diet treatment. The $^{13}C$ enrichment of the different menu cycles within the different diet treatments averaged $-24.23 \pm 0.2$% (SEM [$n = 40$]) vs. PDB. The mean $^{13}C$ enrichment of the breakfast meal for cycle 1, which was always fed on test days, was $-23.85 \pm 0.2$%. Therefore, the meals and especially the breakfast meal on test days were consistent in $^{13}C$ content.

After feeding [1-$^{13}C$] palmitic acid, excess respiratory $^{13}C$ was detected in breath $CO_2$ within about 2.5 h (Fig. 1).

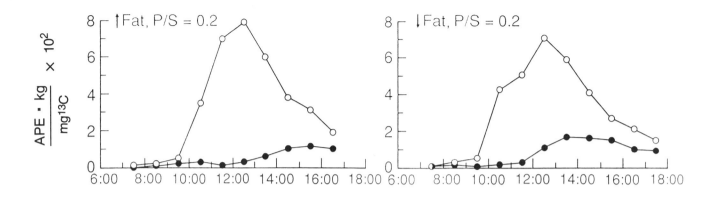

**Fig. 1.** Typical response (Atom Percent Excess $^{13}CO_2$ in breath) of one subject fed four different diets with 30 or 40% total fat as energy and with a P/S of 0.2 or 1.0 versus time. The subject was given either [1-$^{13}$C]10:0 (○) or [1-$^{13}$C]16:0 (●) after a stabilization period on each diet. Values are normalized to correct for different dosages of the two substrates. Error bars from duplicate enrichment measurements fall within the data points.

Excretion of $^{13}$C peaked or reached a plateau during the afternoon. When [1-$^{13}$C] decanoic acid was fed, $^{13}CO_2$ was detected within 1.5 h and peaked within 4 h of feeding. The $^{13}CO_2$ content of breath 24 h after ingesting the labeled substrate indicated that expired $^{13}CO_2$ was within 5% of the background value. Therefore, a minimum of 72 h between the ingestion of labeled substrates was sufficient time for the $^{13}$C content of the breath to return to background levels. No significant difference was found between the order the diet was fed, as well as, the order of label administered.

The APE of $^{13}CO_2$ expired after ingestion of labeled palmitic and decanoic acids over time is illustrated for one subject fed each of the four dietary treatments (Fig. 1). Responses were calculated per mg of labeled fatty acid fed per kg body weight for each fatty acid. Subsequent calculations were normalized for the amount of dietary (including isotope) $C_{10:0}$ or $C_{16:0}$ ingested in the breakfast meal. The response (area under the curve over 9 h) observed for labeled decanoic acid was approximately 50 times that of labeled palmitic acid. The ratio of $C_{10:0}/C_{16:0}$ area was the least for the low-fat, high-P/S diet but did not reach significance compared to the other three dietary treatments.

The amount of $^{13}$C in the breath $CO_2$ was averaged for the six subjects at each sampling time for each dietary period. The response ratio was defined as the normalized average mmol $^{13}$C excreted when fed the high-P/S diet compared to that observed for the low-P/S diet at each fat level (Fig. 2). If the dietary treatment had no effect, the response ratio would be expected to be around 1.0. However, when the substrate fed was labeled $C_{10:0}$, increasing the P/S ratio from 0.2 to 1.0 increased the average response ratio to 1.3 when the normalized, pooled subjects were fed both the low- and high-fat diets, an increase of 30%. Response to the low- and high-fat diets was statistically indistinguishable. When the amount of polyunsaturated fat in the diet was increased after labeled $C_{16:0}$ was fed with a low-fat diet, the pooled response ratio increased to 2.4, an increase in the response ratio of 140%. Therefore, increasing the amount of polyunsaturated fat

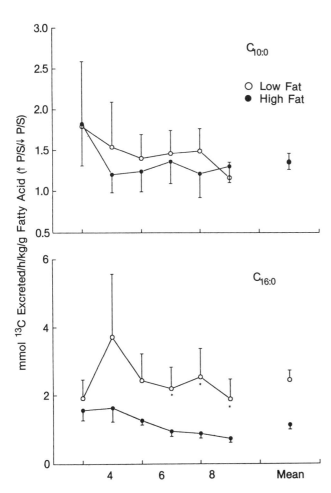

**Fig. 2.** Mean (± SEM) mmol $^{13}$C excreted per h in breath $CO_2$ for six subjects when fed a high-P/S diet relative to a low-P/S diet with 30% fat as energy (○) or 40% fat as energy (●). Whole-body absorption–oxidation of both $C_{10:0}$ and $C_{16:0}$ is increased when fed a high-P/S and low-fat diet (*$P <$ 0.05 between the low- and high-fat diets).

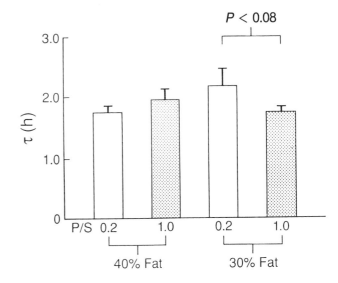

**Fig. 3.** Mean (± SEM) half-life of disappearance of $^{13}$C in respiratory $CO_2$. When consuming the low-fat diet, $^{13}$C disappears more rapidly from respiratory $CO_2$ when subjects are fed a high-P/S diet (*$P <$ 0.08).

while maintaining low levels of fat in the diet increased the whole-body oxidation of medium- and, to a greater extent, long-chain fatty acids.

The half-life of $^{13}$C in breath $CO_2$ was calculated by linear regression of the logarithmic disappearance of $^{13}$C from breath $CO_2$ when subjects were fed labeled $C_{10:0}$. The mean (± SEM) half-life of each fat and P/S level was calculated (Fig. 3). At the high-fat level, no significant difference in half-life was observed for both P/S levels. At low-fat levels, the half-life for elimination of $^{13}CO_2$ from the breath is significantly shorter than when a high-P/S diet is fed. Thus, it is concluded that feeding a low-fat, high-P/S diet results in $^{13}$C disappearing from the breath faster, with the $^{13}$C from $C_{10:0}$ being oxidized faster.

For each dietary treatment, the oxidation rates over 9 h for the labeled $C_{10:0}$ and $C_{16:0}$ feeding days were calculated from the area under the curve, interpolating between data points (Fig. 4). The oxidation of labeled $C_{16:0}$ over 9 h, after feeding a high-P/S, high-fat diet, was 7.2% of the original dose. When subjects consumed the high-P/S diet,

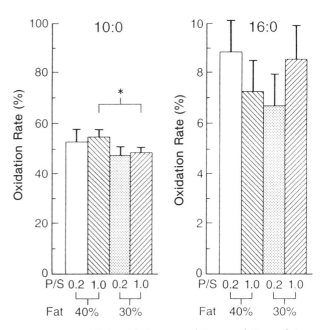

**Fig. 4.** Mean (± SEM) oxidation rate of $C_{10:0}$ and $C_{16:0}$ of six subjects over a 9 h test period. When consuming the high-P/S diet the absorption–oxidation rate observed for $C_{10:0}$ is decreased when the overall fat level of the diet is decreased (*$P <$ 0.05) with a simultaneous increase observed for the absorption–oxidation rate of $C_{16:0}$.

the mean oxidation of $C_{10:0}$ was significantly greater ($p < 0.05$) after feeding the high-fat diet. Concurrently, there is an increase in the oxidation of $C_{16:0}$ when the subjects were consuming the high-P/S, high-fat diet compared to a low-fat diet.

## Discussion

In the present study, normal subjects were stabilized to four diets with different levels of fat and P/S ratios for 15 d. No significant weight gain or loss was observed over the entire diet period of nearly 3 months, such as that observed in a study with fish oil supplementation by Tulleken et al. (29), where an average of 4 kg was lost on low-fat diets fed for several weeks. The variation in weight observed in this study was comparable to other human-feeding studies using similar feeding regimens (28,30). Using the technique of daily weighing and correcting energy intake for any trend in weight change ensured that subjects stayed in energy balance so that any change in whole-body oxidation with different diets was due to a change in energy utilization.

Relative whole-body oxidation of LCFA increased significantly when lower levels of dietary fat were fed in conjunction with a diet high in polyunsaturated fat compared to saturated fat. A Study by Jones et al. (35) found that after feeding high-fat diets, total fat oxidation increased after ingestion of a high-P/S diet compared to a low-P/S diet, but no breakdown of individual fatty acid oxidation was reported.

The half-life of $^{13}CO_2$ in the breath, when labeled $C_{10:0}$ was administered, was found to be significantly shorter when a low-fat, high-P/S diet was fed compared to a low-P/S diet. It is difficult to determine at what stage of metabolism the change in kinetics with dietary fat occurred. When subjects were fed the low-P/S diet, increasing the proportion of fat in the meal shortened the half-life of breath $^{13}CO_2$. This may be accounted for by the faster gastric emptying time and shorter mouth to caecum transit time observed by Cunningham et al. (31), who implied that a test meal would be absorbed more efficiently by a subject fed a high-fat diet. When labeled $C_{10:0}$ was fed with low-P/S diets, there was a trend for the oxidation rate over 9 h to be greater with the higher fat intake than with the lower fat intake, which supports the observations by Cunningham et al. (31). It should be stressed that our diets were isocaloric whereas Cunningham's diets were not, and thus change in total calories may have played a significant role in the observations reported by Cunningham et al. (31).

When subjects were fed a high-P/S, high-fat diet, the observed oxidation for labeled $C_{16:0}$ over 9 h was 7.2% of the original dose. This compares favorably to a $C_{16:0}$ oxidation of 6.6% over 6 h observed by Watkins et al. (32) in normal subjects fed a high-fat, high-P/S lipid emulsion.

The oxidation rates observed after consumption of high-P/S diets suggest that when changing from a high-fat level to a low-fat level, whole-body oxidation of $^{13}C$ from LCFA, such as palmitic acid, is increased at the expense of other substrates, in this case decanoic acid. These results indicated that shifting dietary fat consumption from a low- to a high-P/S diet alters the differential utilization of individual dietary fats. This change in the partitioning of fats absorbed from the diet for energy utilization would also be expected to alter conversion of nonfat substrates to fat for storage, thus altering mechanisms for the net conservation of energy.

## Acknowledgments

The authors wish to thank Grace Hubert and Arlene Parrott for their assistance in food preparation. This work was supported by the Natural Sciences and Engineering Research Council of Canada.

## References

1. Keelan, M., K. Walker, and A.B.R. Thomson, *Comp. Biochem. Physiol. 82A*:83 (1985).
2. Thomson, A.B.R., M. Keelan, M.T. Clandinin, and K. Walker, *J. Clin. Invest. 77*:279 (1986).
3. Clandinin, M.T., C.J. Field, K. Hargreaves, L. Morson, and E. Zsigmond, *Can. J. Physiol. Pharmacol. 63*:546 (1985).
4. Clandinin, M.T., S. Cheema, C.J. Field, M.L. Garg, J. Venkatraman, and T.R. Clandinin, *Fed. Am. Soc. Exp. Biol. J. 5*:2761 (1992).
5. Brasitus, T.A., N.O. Davidson, and D. Schachter, *Biochim. Biophys. Acta 812*:460 (1985).
6. Thomson, A.B.R., M. Keelan, M. Garg, and M.T. Clandinin, *Can. J. Physiol. Pharmacol. 65*:2459 (1987).
7. Thomson, A.B.R., M. Keelan, M.T. Clandinin, and K. Walker, *Am. J. Physiol. 252*:G262 (1987).
8. Sagher, F.A., J.A. Dodge, C.F. Johnston, C. Shaw, K.D. Buchanan, and K.E. Carr, *Br. J. Nutr. 65*:21 (1991).
9. Haglund, U., M. Jodal, and O. Lundgren, *Acta Physiol. Scand. 89*:306 (1973).
10. Jones, P.J.H., P.B. Pencharz, and M.T. Clandinin, *Am. J. Clin. Nutr. 42*:769 (1985).
11. Bach, A.C., and V.K. Babayan, *Am. J. Clin. Nutr. 36*:950 (1982).
12. Committee on Dietetics, Mayo Clinic. *Mayo Clinic Diet Manual*, 2nd edn., Philadelphia, W.B. Saunders Co., 1954.
13. Bell, L., P.J.H. Jones, J. Telch, M.T. Clandinin, and P.B. Pencharz, *Nutr. Res. 5*:123 (1985).
14. Watt, B.K., and A.L. Merrill, *Composition of Foods: Raw, Processed, Prepared. Agriculture Handbook No. 8 (Rev.)*, Washington, DC: Agricultural Research Service, U.S. Department of Agriculture, 1963.
15. Posati, L.P. *Composition of Foods, Poultry Products: Raw, Processed, Prepared. Agriculture Handbook No. 8-5*, Washington, DC: Science and Education Administration, U.S. Department of Agriculture, 1979.
16. Richardon, M., L.P. Posati, and B.A. Anderson, *Composition of Foods, Sausages and Luncheon Meats: Raw, Processed, Prepared. Agriculture Handbook No. 8-7*, Washington, DC: Science and Education Administration, U.S. Department of Agriculture, 1980.
17. Posati, L.P., and M.L. Orr, *Composition of Foods, Dairy and Egg Products: Raw, Processed, Prepared. Agriculture Handbook No. 8-1*, Washington, DC: Agriculture Research Service, U.S. Department of Agriculture, 1976.

18. Marsh, A.C. *Composition of Foods, Soups, Sauces and Gravies: Raw, Processed, Prepared. Agriculture Handbook No. 8-6*, Washington, DC: Science and Education Administration, U.S. Department of Agriculture, 1980.
19. Pennington, J.A.T., and H.N. Church, *Food Values of Portions Commonly Used*, 14th edn., New York, Harper and Row, 1985.
20. Folch, J., M. Lees, and G.H. Sloan-Stanley, *J. Biol. Chem. 226*:497 (1957).
21. Bannon, C.D., J.D. Craske, T.H. Ngo, N.L. Harper, and K.L. O'Rourke, *J. Chrom. 247*:63 (1982).
22. Hargreaves, K.M., and M.T. Clandinin, *Biochim. Biophys. Acta 918*:97 (1987).
23. Miwa, T.K., K.L. Mikolajczak, F.R. Earle, and I.A. Wolff, *Anal. Chem. 32*:1739 (1960).
24. Croft, D.N., and P.B. Cotton, *Digestion 8*:144 (1973).
25. Mook, W.G., and P.M. Grootes, *Int J. Mass Spectrom. Ion Phys. 12*:273 (1973).
26. Jones, P.J.H., P.B. Pencharz, L. Bell, and M.T. Clandinin, *Am. J. Clin. Nutr. 41*:1277 (1985).
27. Irving, C.S., W.W. Wong, R.J. Shulman, E. O'Brien-Smith, and P. Klein, *Am. J. Physiol. 245*:R190 (1983).
28. Jones, P.J.H., P.B. Pencharz, and M.T. Clandinin, *J. Lab. Clin. Med. 105*:647 (1985).
29. Tulleken, J.E., P.C. Limburg, F.A.J. Muskiet, K.M. Kazemier, I.K. Boomgaart, and M.H. van Rijswijk, *Eur. J. Clin. Nutr. 45:*383 (1991).
30. Jones, P.J.H., and D.A. Schoeller, *Metabol. 37:*145 (1988).
31. Cunningham, K.M., J. Daly, M. Horowitz, and N.W. Read, *Gut 32*:483 (1991).
32. Watkins, J.B., P.D. Klein, D.A. Schoeller, B.S. Kirschner, R. Park, and J.A. Perman, *Gastroent. 82*:911 (1982).

# Discussion

Chairperson Yniguez noted that time had passed very quickly and that we had a good week with excellent papers and presentations. He relayed to all that President Ramos was particularly pleased by the outcome of the conference, the delegates' visit to the Palace, and the frank exchange with the smaller group. Yniguez felt that knowledge about lauric oils had been enhanced and updated and he gave credit to the General Chairman, E.C. Leonard, supported by the AOCS staff and the local committees. In closing he thanked all of the visitors in their native tongues.

General Chairperson Leonard note that this was another successful conference in a long succession of such meetings dating back to the first in 1976. He then reviewed a series of economic comments and questions related to this area of the world. For example, how fast are the markets for Asia and Asean growing, both per household and per country? It appears that demand would dictate the construction of a 40,000 MT alcohol plant every year. Is this real or not? Will the nontraditional sources of lauric oil become a reality, and if they do what will be the effect on the Philippine economy? Also, what will be the impact of the continued growth of Malaysian palm kernel oil production? The use of edible oils is declining in the United States, and the tropical oil question appears to be disarmed. Does this mean that the edible use of lauric oils will recover? In the household-surfactant area, petrochemicals are being replaced, and lauric oils have a good price advantage over petrochemicals.

In closing, Leonard noted the great kindness and hospitality of the Philippines. He was particularly honored to have talked to President Ramos firsthand. He especially thanked his co-chairperson, A.D. Yniguez and commented on the excellence of the session chairperson and the speakers. He was most pleased that the conference had covered technical matters that affect economics, particularly agronomics, and that all of this was a reflection of our doing God's work in bettering the lives of the people now and in the future.